D0078492

Post-tensioned Concrete Floors

Harbour Exchange, London

Post-tensioned Concrete Floors

Sami Khan
Director, Bunyan Meyer and Partners

Martin Williams
Lecturer, University of Oxford, Department of Engineering Science
and Fellow of New College, Oxford

Butterworth-Heinemann Ltd
Linacre House, Jordan Hill, Oxford OX2 8DP

A member of the Reed Elsevier plc group

OXFORD LONDON BOSTON
MUNICH NEW DELHI SINGAPORE SYDNEY
TOKYO TORONTO WELLINGTON

First published 1995

British Library Cataloguing in Publication Data

Kahn, Sami
Post-tensioned Concrete Floors
I. Title II. Williams, Martin
693.542

ISBN 0 7506 1681 4

Library of Congress Cataloguing in Publication Data

Kahn, Sami.
Post-tensioned concrete floors / Sami Kahn, Martin Williams.
p. cm.
Includes bibliographical references and index.
ISBN 0 7506 1681 4 (pbk.)
1. Floors, Concrete. 2. Post-tensioned prestressed concrete.
I. Williams, Martin. II. Title.
TH2529.C6K48 1995
690'.16 – dc20 94-36854
CIP

Typeset by Vision Typesetting, Manchester
Printed and bound in Great Britain by
Hartnolls Limited, Bodmin, Cornwall

CONTENTS

INTRODUCTION

This book deals with the design of concrete building structures incorporating post-tensioned floors. Post-tensioning is the most versatile form of prestressing, a technique which enables engineers to make the most effective use of the material properties of concrete, and so to design structural elements which are strong, slender and efficient. Design in post-tensioned concrete is not difficult and, if done properly, can contribute significantly to the economy and the aesthetic qualities of a building. As a result, post-tensioned floors have found widespread use in office buildings and car park structures, and are also frequently employed in warehouses and public buildings. However, in spite of this, most prestressed concrete texts devote comparatively little attention to floors, concentrating instead on beam elements. This book therefore aims to answer the need for a comprehensive treatment of post-tensioned floor design.

The first four chapters of the book give a detailed, non-mathematical account of the principles of prestressing, the materials and equipment used, and the planning of buildings incorporating post-tensioned floors. The following chapters outline the detailed design process, including numerous worked examples, and the book concludes with chapters describing site procedures for construction, demolition and alteration. While the reader is assumed to have a grasp of the basics of reinforced concrete design, no prior knowledge of prestressing is required. The book is thus suitable for use by architects, contract managers and quantity surveyors who may wish to gain an understanding of the principles without going into the mathematical aspects of the design process, as well as structural engineers requiring detailed design guidance. It is also intended for use as an educational text by students following civil engineering, architecture and building courses.

The title of the book reflects the fact that its emphasis is on the behaviour and design of the floors themselves. Thus, while the effect of post-tensioned floors on other structural elements such as columns and walls is considered, detailed guidance on the design of these elements is not given; such information can be obtained from any one of the many excellent reinforced concrete design texts already available. Neither does this book deal with the prestressing of building elements other than floors, such as foundations, moment-resisting columns or vertical hangers. These elements are comparatively rare, or are not usually prestressed. If guidance on design of such elements is required, reference should be made to specialist literature.

In any book on post-tensioning comparisons with reinforced concrete are inevitable. Post-tensioning offers numerous advantages, but it would be foolish to suggest that it is the best design option in all cases. Therefore, detailed guidance is given on the relative merits of post-tensioned and reinforced concrete floors, and the reasons for choosing one or the other in a particular situation are discussed.

Although the post-tensioning technique is now quite well established, research and development activities continue to offer possibilities for future improvements, a recent example being the development of polymer prestressing tendons. Many of these advances require considerable further research before they are ready for practical use. In this book such developments are discussed briefly, but the emphasis is placed firmly on the current practice.

While the general principles and methods of prestressing are universal, detailed design procedures usually follow national code recommendations, which vary from country to country. For a text to be useful as a design guide, it must necessarily make reference to national codes. In this book, full design procedures compatible with both the British Standard BS 8110: 1985 and the American code ACI 318-1989 are described. However, the design methods are always introduced in a way which emphasizes the principles on which they are based rather than simply reiterating the code guidelines. While every effort has been made to describe as fully as possible the provisions of BS 8110 and ACI 318, the book should certainly not be regarded as a replacement of either code. The relevant standard or code of practice should always be consulted to check specific requirements.

As far as possible, the equations and data presented are given in SI units followed by imperial equivalents. Extracts from BS 8110 are presented in metric form only, since it is unlikely they will be employed in countries using imperial units. ACI 318 formulae and data are given in both SI and imperial units.

The help of many individuals and organizations in the preparation of this book is gratefully acknowledged. The finished diagrams were produced with the generous assistance of the British Cement Association, Crowthorne. Most of the photographs and some sketches were contributed by VSL International of Berne, Switzerland. Extremely helpful comments on the manuscript were made by Pal Chana of the Concrete Research and Innovation Centre, Imperial College; B. K. Bardhan-Roy of Jan Babrowski and Partners; Peter Matthew of Swift Structures Ltd; and David Ramsay of DHV Burrow-Crocker Consulting.

Extracts from BS 8110 have been reproduced by permission of the British Standards Institution, 2 Park Street, London, W1A 2BS. Extracts from ACI 318 have been reproduced by permission of the American Concrete Institute, Box 19150, 22400 West Seven Mile Road, Detroit, Michigan 48219, USA. Copies of the codes may be purchased from these organizations at the addresses given. The authors are also grateful to the British Cement Association and the Concrete Society for their permission to include extracts from their publications; to Bridon Wire for supplying data on their products; and to Bunyan, Meyer & Partners for their support.

NOTATIONS

The following symbols are common to all chapters. Further symbols, used locally, are given in the relevant chapters.

A_c	Area of concrete section
A_p	Area of tendon steel
A_s	Area of bonded steel
$A_{s'}$	Area of compression steel
A_{sv}	Area of links per unit length of member
b	Width of concrete in compression
b_r	Width of concrete on tension face, rib width
b_v	width of section effective in shear, rib width for T-, I- or L-sections
C_c	Creep coefficient
D	Overall depth of section
d	Depth of tension steel from compression face
d_c	Depth from extreme compression fibre to centre of compression
d_x	Depth of rectangular stress block
d_n	Depth of neutral axis
d_p	Depth of tendon centroid
d_r	Depth of bonded rod reinforcement in tension
e_p	Eccentricity of tendon
E_c	Modulus of elasticity of concrete at 28 days
E_{ci}	Modulus of elasticity of concrete at stressing
E_s	Modulus of elasticity of steel
f_{cb}	Equivalent stress, assuming a rectangular stress block
f'_{ci}	Initial concrete cylinder strength
f'_c	28-day cylinder strength
f_{ci}	Initial concrete cube strength
f_{cu}	28-day cube strength
f_{pb}	Stress in tendon in the ultimate state
f_{pi}	Initial stress in tendon, after immediate losses
f_{pe}	Stress in tendon after all losses
f_{pu}	Ultimate strength of tendon per unit area
f_{py}	Tendon yield stress
f_{sb}	Stress in bonded reinforcement in ultimate state
f_t	Tensile strength of concrete
f_y	Strength of rod reinforcement

f_{yv}	Strength of steel used in links
I_c	Moment of inertia of concrete section
k	Subgrade modulus
K	Wobble friction coefficient, per unit length
L	Span length
L_t	Tendon length between anchorages
M	Moment
M_{cr}	Cracking moment
M_o	Moment due to self weight of concrete section
M_r	Moment with factored load
M_u	Ultimate flexural strength
P_i	Initial tendon force, after immediate losses
P_f	Final tendon force, after all losses
P_j	Jacking force in tendon
P_o	Tendon force at anchorage
P_x	Tendon force at distance x
p_{av}	Average stress in concrete due to prestress
p_{cu}	Stress in concrete compression block
s_v	Spacing of links
V	Ultimate shear force
V_c	Ultimate shear resistance
V_{co}	Ultimate shear resistance of section uncracked in flexure
V_{cr}	Ultimate shear resistance of section cracked in flexure
V_p	Vertical shear due to prestress
V_r	Ultimate shear resistance of a reinforced, non-prestressed, section
W	Concentrated load or total distributed load
w	Load intensity
w_c	Concrete density
w_e	Equivalent load intensity
Z	Section modulus
Z_t	Section modulus for top fibre
Z_b	Section modulus for bottom fibre
β_1	Ratio d_c/d_n
μ	Coefficient of friction between tendon and sheath
δ	Deflection or displacement
γ_m	Partial safety factor
ν	Poisson's ratio

Sign conventions

Positive signs apply to:

y-axis going upwards
Load acting downwards on concrete member

Support reactions acting upwards on member
Sagging moment
Compressive stress

Definitions and conversions

Rod reinforcement: bonded non-prestressed steel
Assumed relationship between cylinder and cube strength: $f_{c'} = 0.8f_{cu}$

The following approximate conversions are used:

1 metre	(m)	= 39.37	inches	(in)
1 kilogram	(kg)	= 2.2	pounds	(lb)
1 kilonewton	(kN)	= 220.0	pounds	(lb)
1 newton per sq.mm	(N/mm^2)	= 145.0	pounds per sq.in	(psi)
1 kilonewton per sq.mm	(kN/mm^2)	= 145.0	kilopounds (kip) per sq.in	(ksi)
1 kilonewton per sq.m	(kN/m^2)	= 20.89	pounds per sq.ft	(psf)
1 kilogram per cu.m	(kg/m^3)	= 0.0625	pounds per cu.ft	(pcf)

1 THE BASIC PRINCIPLES

1.1 Introduction

Post-tensioning has been in use in floor construction for several decades now, especially in the United States, Australia, the Far East and, to some extent, in Europe. Its economic and technical advantages are being increasingly appreciated, and the proportion of concrete floors being post-tensioned is growing.

In this chapter the basic principles of prestressing are explained, and the various methods of prestressing are briefly discussed. This is followed by comparisons between the alternative forms of using concrete as a structural medium. It answers the questions frequently asked by those interested enough in the subject to wish to know more about it but with no need for a detailed design insight.

Post-tensioning is a technique of pre-loading the concrete in a manner which eliminates, or reduces, the tensile stresses that are induced by the dead and live loads; the principle is further discussed later in this chapter. Figure 1.1 is a diagrammatic representation of the process. High strength steel ropes, called *strands*, are arranged to pass through the concrete floor. When the concrete has hardened, each set of strands is gripped in the jaws of a hydraulic jack and stretched to a pre-determined force. Then the strand is locked in a purpose-made device, called an *anchorage*, which has been cast in the concrete; this induces a compressive stress in the concrete. The strand is thereafter held permanently by the anchorage.

The non-jacking end of the strand may be bonded in concrete, or it may be fitted with a pre-locked anchorage which has also been cast in the concrete. The anchorage at the jacking end is called a *live anchorage* whereas the one at the non-jacking end is termed a *dead anchorage*. To allow the strand to stretch in the hardened concrete under the load applied by the jack, bond between the strand and concrete is prevented by a tube through which the strand passes. The tube, termed a *duct* or *sheathing*, may be a metal or plastic pipe, or it may consist of a plastic extrusion moulded directly on the rope. If extruded, the strand is injected with a rust-inhibiting grease. After stressing, the sheathing, if not of the extruded kind, is grouted with cement mortar using a mechanical pump.

The terms *tendon* and *cable*, are the general and interchangeable names for the high strength steel lengths used in post-tensioning—equivalent to *reinforcement* in reinforced concrete. A tendon may consist of individual wires, solid rods or ropes. It may contain one or more ropes or wires housed in a common sheathing.

Except in ground slabs, tendons do not run in straight lines. They are normally draped between supports with a shallow sag, just as a rope hangs when lightly stretched between two supports. The geometric shape of the tendon in elevation is called its *profile*; it is usually, but not necessarily, parabolic. At any point along its length, the vertical distance between the centroid of the concrete section and the centre of the tendon is called its *eccentricity*; by convention it is said to be positive when the tendon is below the section centroid.

The requirements for concrete, rod reinforcement and formwork for post-tensioning are similar to those for reinforced concrete, except for minor differences. Early strength of concrete is an advantage in post-tensioning; the quantity of rod reinforcement is much smaller; and the shuttering needs a hole in the vertical edge shutter at each jacking end of the tendon, and the anchorages need to be attached to the edge shutters. The differences in the materials, though minor, are discussed in detail in Chapter 2. Post-tensioning also needs stressing tendons to be made of high tensile strength; these and the associated hardware are briefly discussed in this chapter and in detail in Chapter 2.

The basic form of a post-tensioned floor is similar to that of a reinforced concrete floor. Slabs can be solid, ribbed or waffle; beams can be downstand, upstand or strips within the slab thickness.

1.2 Prestressing in principle

In reinforced concrete construction, the lack of strength of concrete in tension is compensated for by providing bonded steel reinforcement near the tension faces of the concrete section. The steel, being strong in tension, bears the tensile forces and the concrete takes the compressive forces. Under no-load condition the steel is unstressed; as a reinforced concrete member is loaded it deforms, inducing compressive and tensile stresses. The stresses in concrete and steel, therefore, vary with the load.

In prestressing, a permanent external axial force, of predetermined magnitude, is applied to the concrete member, which induces a compressive stress in the concrete section. When the service load is applied, the generated tensile stress has to overcome the compressive *prestress* before the concrete is driven into any tension. The tensile strength of concrete is, therefore, effectively enhanced. The prestressing force does not significantly change with the load within the serviceability limit. The principle is illustrated in Figure 1.2(a).

Consider a simple beam, required to carry a downward acting imposed load; at this stage, assume that the self-weight of the beam is negligible. An axial prestressing force is applied at the centroid of the section, which induces a uniform compressive stress across the section, Figure 1.2(a). At the top of the beam, the flexural compression is added to the *prestress* and the concrete on the compression face is subjected to the sum of the prestress and the flexural stress, i.e. the concrete has a higher compressive stress than it would have without the prestress. At the bottom of the beam, the flexural tension is in opposition to the

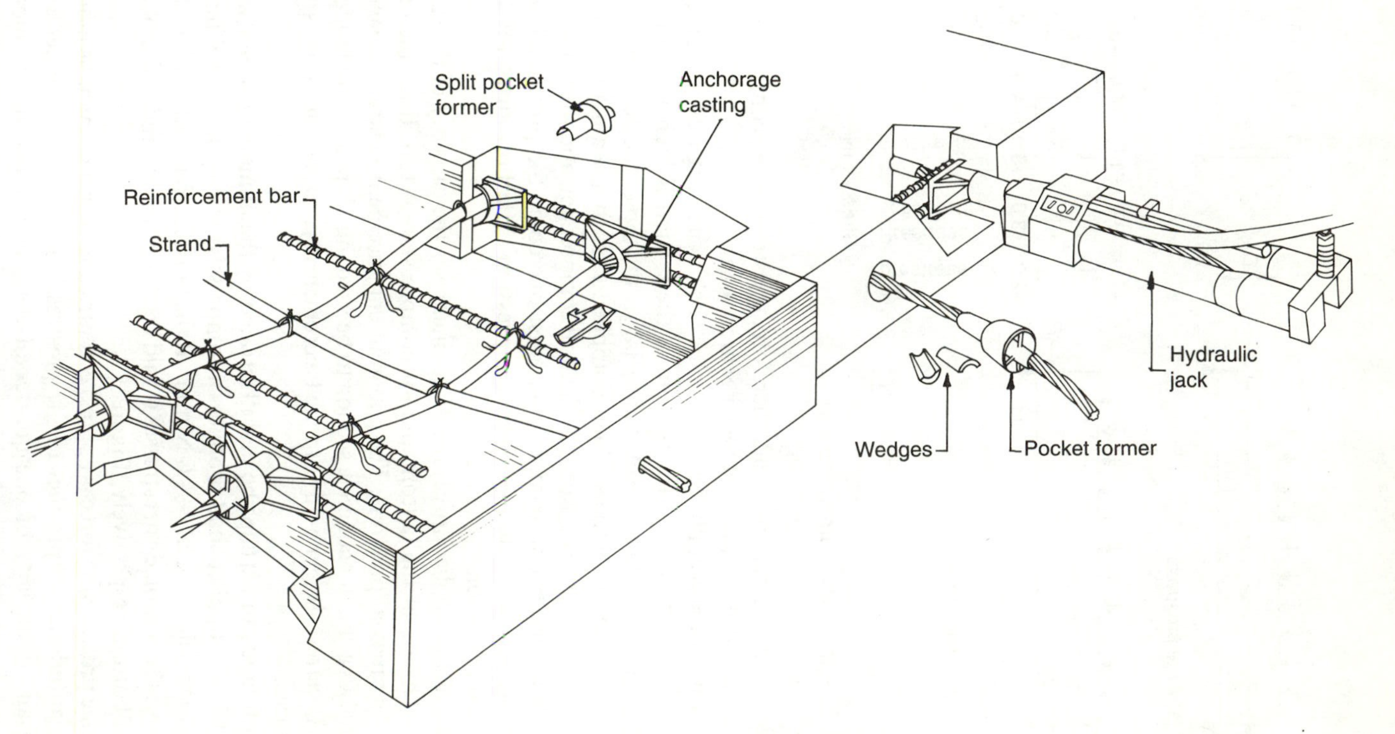

Figure 1.1 *Post-tensioning of a floor*

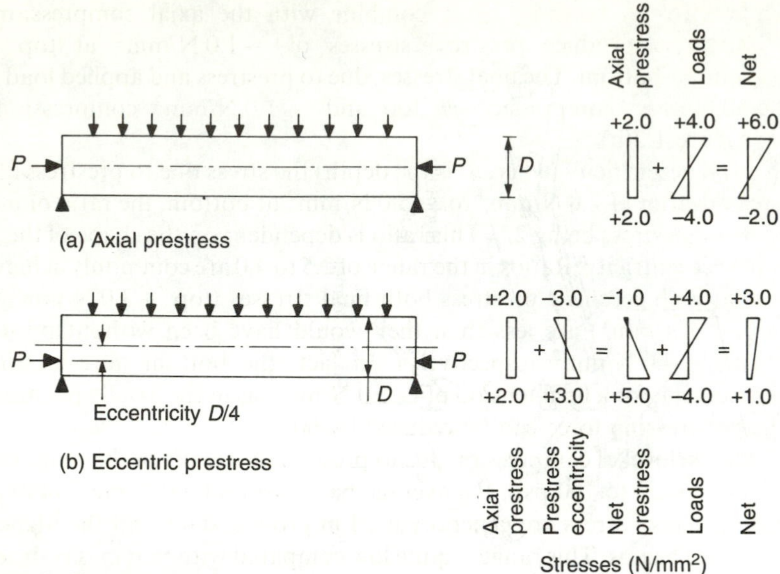

Figure 1.2 *The principle of prestressing*

compression from the prestress and, therefore, the stress in the concrete is lower than the tension it would have under flexure alone.

If the external force is applied eccentrically, as shown in Figure 1.2(b), then the compressive stress induced at the bottom of the section is higher for the same axial force. If the eccentricity is sufficiently large, the top of the beam develops a slight tension. When the imposed load is applied to such a beam, its top fibre is subjected to the difference between the flexural compression and the tension from prestress. Thus the flexural compression must overcome the prestress-induced tension before the concrete goes into compression. The bottom of the beam remains in compression.

An eccentrically applied force, therefore, increases the capacity of the section for flexural tension at the bottom and for compression at top. It is much more efficient than an axial prestress. The apparent enhancement in stress capacity of concrete allows a smaller concrete section to be used than is possible in reinforced concrete. Prestress is generally applied eccentrically, except in very special circumstances.

The advantage can perhaps be best illustrated by the numerical values shown in Figure 1.2. The applied load produces a compressive stress of $+4.0\ \text{N/mm}^2$ at the top and a tensile stress of $-4.0\ \text{N/mm}^2$ at the bottom. With an axial prestress of $2.0\ \text{N/mm}^2$ the combined net stresses would be $+6.0\ \text{N/mm}^2$ and $-2.0\ \text{N/mm}^2$ at top and bottom respectively, Figure 1.2(a).

If the same prestressing force is applied eccentrically then a moment is induced, whose magnitude is the product of the prestressing force and its eccentricity. Assuming an eccentricity of one-quarter of the member depth, the moment produces flexural stresses of $-3.0\ \text{N/mm}^2$ tension at the top and $+3.0\ \text{N/mm}^2$

compression at the bottom. These combine with the axial compression of $+2.0\,N/mm^2$ to produce prestress stresses of $-1.0\,N/mm^2$ at top and $+5.0\,N/mm^2$ at bottom. The final stresses, due to prestress and applied load, are now $+3.0\,N/mm^2$ compression at top and $+1.0\,N/mm^2$ compression at bottom, Figure 1.2(b).

With a $D/4$ eccentricity (where D is the depth) the stress due to prestress alone has increased from $+2.0\,N/mm^2$ to $+5.0\,N/mm^2$ at bottom, the ratio of maximum to average stress being 2.5. This ratio is dependent on the shape of the section and the eccentricity. Ratios in the range of 2.5 to 4.0 are commonly achieved.

Note that with eccentric prestress both final stresses (top $+3.0\,N/mm^2$ and bottom $+1.0\,N/mm^2$) are less than they would have been without prestress ($+4.0$ and $-4.0\,N/mm^2$ respectively). In fact, the bottom fibre is still in compression and, for a final tension of $-2.0\,N/mm^2$ as in the axially prestressed case, the prestressing force can be reduced by 60%.

In floors, the level of compression due to prestress is usually in the range of 1.0 to $5.0\,N/mm^2$ (150 to 700 psi), the average being around $3.0\,N/mm^2$ (450 psi). The lower levels of stress are generally used in ground slabs and the higher in post-tensioned beams. This range is quite low compared with that in, say, bridges where the average stress may be much higher, because of the longer spans and the higher loads.

1.3 Stress reversal

In continuous structures, if the self-weight of the floor is small compared with the applied loads or if there is a wide variation in span lengths, then it is possible for the load-induced stresses in a span to be reversed under a certain load combination; the stress at the bottom in the middle of the span may be compressive and at the top it may be tensile. In such a member in reinforced concrete, sufficient tension steel would be provided on each face to cope with the reversal. Reinforcement at each face would be designed independently of the reverse moment because the two conditions cannot exist simultaneously.

Consider what happens if the applied load is reversed in a prestressed member. Taking the example in Figure 1.2(b), assume that the reverse load produces a tension of $-4.0\,N/mm^2$ at the top and a compression of $+4.0\,N/mm^2$ at the bottom. The final stresses in this case would be as given in Table 1.1.

In the absence of any prestress, the flexural stresses for the reverse load (acting upwards) would be $-4.0\,N/mm^2$ at top and $+4.0\,N/mm^2$ at bottom; with prestress they are -5.0 and $+9.0\,N/mm^2$ respectively. Clearly, this prestressed member is worse off with the reverse loading.

The problem is caused by the high eccentricity, which induces a tension on the top face. This tension gets added to the flexural tension of the reverse load. With a lower eccentricity the stresses under the reversed load are also lower, though the stresses under the normal load would increase. In this example the reverse load is equal in magnitude to the normal load and so the best results would be obtained

Table 1.1 *Effect of load reversal*

	Prestress	+	*Bending*	=	*Final*
Normal load					
Top	−1.0	+	4.0	=	+3.0N/mm^2
Btm	+5.0	−	4.0	=	+1.0
Reversed load					
Top	−1.0	−	4.0	=	−5.0N/mm^2
Btm	+5.0	+	4.0	=	+9.0

with an axially applied prestress. If the prestress produced a uniform compression of +2.0 N/mm^2 over the whole section then the final stresses would be −2.0 and +6.0 N/mm^2 in each case, as in Figure 1.2(a). This is an improvement on the eccentrically applied prestress but not an efficient use of materials.

This clearly illustrates that prestressing is not so effective when reversal of load is involved. Fortunately, load reversal seldom occurs in building floors and when it does, the dead load is usually sufficient either to keep the net load still acting downwards, or to greatly reduce the effect of the reverse loading. In continuous spans, of short length carrying heavy live loads, such as in warehouses, the dead load may not be large enough to avoid stress reversal and, therefore, reversal may be a critical condition. In such cases, a reduced eccentricity of prestress provides the better solution.

1.4 Tendons

In prestressed structures, the external prestressing force is generally applied by stretching steel rods, wires or ropes (*strand*) against the concrete section, which goes into compression. The high strength steel rods, wires, or strands are collectively called *tendons* or *cables*. In post-tensioned floors, however, use of strand is now almost universal.

The term strand can be rather confusing—it applies to the rope consisting of a number of individual wires wound together, it does not mean the individual wire comprising a rope. A typical strand consists of 7 wires wound into a rope; the commonly used sizes are nominal 13 mm and 15 mm (0.5 and 0.6 in) in diameter. The actual sizes and strand characteristics are given in Chapter 2.

In post-tensioning, because the prestress is applied after the concrete has gained sufficient strength, bond cannot be allowed to develop between the concrete and the tendons before stressing, and therefore, the tendons are housed in a bond-breaking *duct* or *sheathing*. More than one wire or strand may be housed in one common duct; the group of one or more is also called a *tendon* or *cable*.

The strand is similar to rod reinforcement with regard to its modulus of elasticity and coefficient of thermal expansion but it is about four times stronger than reinforcement steel. Strand may have an ultimate strength of 1860 N/mm^2

(270 ksi) compared with 460 N/mm^2 (67 ksi) for rod reinforcement. At service loads the strand may have a stress of 1100 N/mm^2 (160 ksi) while rod reinforcement may carry only about 250 N/mm^2 (36 ksi).

In addition to the tendons, some bonded rod reinforcement is also provided in post-tensioned floors—around the anchorages, in the slab as ties, as secondary reinforcement, or if needed to achieve the required ultimate strength.

The weight of steel (strands and rod reinforcement) required in a post-tensioned floor is typically only half, or less, of that needed in a reinforced concrete floor. Those used to working with reinforced concrete are surprised on their first visit to a post-tensioned floor site because there is so little steel. Instead of heavy reinforcement bars spaced at 150 mm centres, they see a couple of 15 mm strands every metre or so.

Strand being so much stronger than rod reinforcement, the question arises as to why it is not used in reinforced concrete. At a working stress of 1100 N/mm^2, the average tensile strain in the strand would be approximately 0.0056. If used as non-prestressed bonded reinforcement, the cracks in concrete resulting from this strain may approach 1.0 mm in width, which would be unacceptable. To limit the cracks to acceptable widths the strand would have to be used at a much reduced and inefficient stress level. In prestressing, the tension in the tendon is used to induce compression in the concrete and, therefore, advantage can be taken of its relatively high strength.

1.5 Prestress losses

In prestressing, the compressive stress in the concrete due to the prestress is maintained by the tension in the tendons, so that any change in the concrete strain is reflected in the tendon forces and vice versa. If the tendons are released or allowed to slip then their tensile force is lost or reduced, and the concrete compression undergoes a corresponding loss. A similar effect occurs if the length of the prestressed member reduces, for example due to shrinkage or creep of concrete; the tendon length, being equal to that of the concrete member, also reduces and some tension in the tendon (and compression in the concrete) is lost.

During the stressing of a tendon, friction between the tendon and the duct causes a loss in the tendon force so that it reduces away from the jacking end. The mechanical action of securing a tendon in the anchorage allows the tendon to draw in, typically by 6–8 mm ($\frac{1}{4}$ to $\frac{3}{8}$ in), which results in a further loss in the tendon force. Also, as each successive tendon in a member is stressed, the length of the concrete member reduces by a small amount, with the consequence that the previously stressed tendons lose some tension. In addition to these immediate losses, post-tensioned concrete undergoes a gradual reduction in length because of shrinkage and creep, which in theory continues for its life, though a significant proportion of it occurs within the first few weeks. Steel tendons also undergo long-term creep elongation (called *relaxation*) but of a smaller magnitude. These actions cause a long-term loss in the tendon force.

The losses can be seen to occur in three distinct stages.

- The prestressing force is at its maximum when a tendon is being stressed and before it is anchored. Friction reduces the force away from the jacking end. The previously stressed tendons, of course, undergo a loss due to elastic shortening of concrete as each successive tendon is stressed.
- When the tendon is anchored then the draw-in causes an immediate loss in prestress.
- Shrinkage and creep of concrete, and relaxation of steel, cause a further loss but this is long term.

The magnitude of losses depends on many factors, such as the concrete properties, the age at which the prestress is applied, the average prestress value, the length of the tendons and the tendon profiles.

1.6 Initial and final stresses

The loss in prestress arising from friction, anchorage draw-in and immediate elastic shortening of the member is termed *initial loss*. The total long-term loss is called *final loss*; it includes the initial loss and the losses due to shrinkage and creep of concrete, and relaxation of strand.

As eccentric prestress is applied to a member, the member tends to deflect upwards so that its self-weight is borne by the prestress. In fact, prestress almost never acts alone; it is accompanied by the self-weight of the member producing stresses opposite to those of the prestress.

In the example shown in Figure 1.2, no allowance was made for losses. Assume a 5% loss due to anchorage draw-in and a further 15% thereafter. Also, assume that the self-weight of the floor induces stresses of $\pm 1.6\ \text{N/mm}^2$ and that the imposed load produces a further $\pm 2.4\ \text{N/mm}^2$. The stresses at various stages are shown in Table 1.2

Note that the bottom of the section is in compression at jacking and initial stages; as load is applied the compression reduces. If the applied load is further increased then the bottom of the member may go into tension. Often, the highest compressive stress exists at the bottom of the section at the jacking or initial

Table 1.2 *Initial and final stresses*

Stage	*Top/btm*	*Prestress*	*Self-wt*	*Imposed*		*Net stress*
Jacking	Top	−1.00	+1.60	+0.00	=	+0.60
	Btm	+5.00	−1.60	+0.00	=	+3.40
Initial	Top	−0.95	+1.60	+0.00	=	+0.65
	Btm	+4.75	−1.60	+0.00	=	+3.15
Final	Top	−0.81	+1.60	+2.40	=	+3.19
	Btm	+4.04	−1.60	−2.40	=	+0.04

stage, and this occurs at an early stage in the life of the concrete when it is not mature. In such cases post-tensioned floors can be considered to have been load-tested during construction.

The situation is reversed at the top of the section; if the self-weight is small then the top fibre may be in tension initially, and would go into compression as further load is imposed. Therefore, the initial and final stresses represent two extremes in the serviceability loading of a post-tensioned floor. Stresses are computed for both stages in design and are generally required to be within prescribed limits. For the initial stage the load consists of only the self-weight of the member; no other imposed load is considered. At the final stage, the various loads are combined, in accordance with the national standard, to arrive at the most adverse condition.

1.7 Pre-tensioning and post-tensioning

The tendons can be stressed either before casting the concrete or after the concrete has been cast and has gained some strength. In *pre-tensioning* the wires or strands are stressed against external anchor points (or sometimes against the mould) and concrete is then cast in direct contact with the tendons, thus allowing bond to develop. When the concrete has gained sufficient strength, the tendons are released from the temporary external anchorages, thereby transferring the force to the concrete, inducing a compressive stress in it. The tension in the tendons and the corresponding compression in the concrete is then solely dependent on bond between concrete and tendon, and no other mechanical device is used.

Pre-tensioned tendons usually run in straight lines. In order to deviate from this, external deflecting devices are needed. With them, a profile consisting of a series of straight lines can be obtained. These devices slow down the manufacturing process and they add to the costs. The requirement for external temporary anchors and the problems in profiling the tendons make pre-tensioning difficult for application *in situ*. The process is almost exclusively confined to precasting, and is not discussed any further.

In *post-tensioning*, concrete is not allowed to come in contact with the tendons. The tendons are placed in ducts, or sheaths, which prevent bond, and concrete is cast so that the duct itself is bonded but the tendon inside remains free to move. When the concrete has gained sufficient strength the tendons are stressed directly against the concrete and they are mechanically locked in *anchorages* cast at each end. After this stage, tension in the tendons, and hence the induced compression in the concrete, is maintained by the anchorages.

In *bonded* post-tensioning the duct is grouted after the tendons have been stressed, so that the stressed tendons become bonded. In *unbonded* post-tensioning, as its name implies, the tendons are never bonded. A detailed comparison between bonded and unbonded systems is given in Section 1.9.

1.8 Reinforced and post-tensioned concrete floors

Reinforced concrete technology is widely available and is well understood. Post-tensioning is an advance on reinforced concrete technology and it is often discussed in the context of reinforced concrete.

1.8.1 General

Post-tensioning offers some very useful technical and economic advantages over reinforced concrete, particularly for long spans, where control of deflection is desirable, or if construction depth must be minimised. It is, however, not the best solution in all circumstances and the various alternative forms of construction should be carefully considered for each structure before making the choice.

For post-tensioning, it is important to consider availability of the hardware and the technical expertise required. Excepting very special design objectives, post-tensioning is unlikely to be economical for short spans. Often a combination of post-tensioning and another form of construction offers a good solution. For example, in a floor consisting of rectangular bays, if the short span is small enough, the best solution may be to span the slab in the long direction in post-tensioned concrete and use reinforced concrete beam strips in the short-span direction.

Economy of construction varies from one site to the next, depending on accessibility and availability of material and labour, and of course on the design loading and constraints that may be imposed by other disciplines, such as a restriction on the depth of the structure. It is, therefore, not possible to give a general cost comparison between the two forms of construction. The authors have found that for span lengths of about 9 m (30 ft) the cost of a floor carrying a superimposed load of 5 kN/m^2 (100 psf) is similar in reinforced and post-tensioned concretes. Post-tensioned concrete is more economical than reinforced concrete above this span length. In cases of restricted floor depth or high loads, the span length for equal costs may be as low as 7.5 m (25 ft). Further savings result from the lighter weight and lesser construction depth of the post-tensioned floor.

For reinforced concrete, only the ultimate strength calculations are normally carried out and deflection in the serviceability state is deemed to be satisfied by confining the span-to-depth ratio within limits prescribed in the national standards. Only in rare cases is it necessary to calculate deflections. Crack control is usually governed by deemed-to-satisfy rules for bar spacing.

In post-tensioned concrete design, serviceability calculations are carried out for the initial and final loading conditions, for deflection and cracking, and the ultimate strength is checked after this. Structural design of prestressed concrete, therefore, requires more effort.

The shallow depth of a post-tensioned floor is a particular advantage in multistorey buildings; in some cases it has been possible to add an extra floor where there was a restriction on building height. Even where there is no such restriction, the reduced building volume generates savings in the cost of services and cladding, and in subsequent running and maintenance costs. The reduction in the weight of the building generates further savings in the cost of foundations; the weight of concrete in floors in a multistorey building may be as much as half of the total weight of the building. The cost per unit area of a post-tensioned floor, considered in isolation, may be higher than that of a reinforced floor of the same span and carrying the same load, but it is quite possible for the building with post-tensioned floors to prove the more cost effective when the other savings are taken into account.

1.8.2 Design and serviceability

There are numerous differences between the behaviour of floors under service loads in the two forms of construction:

- As discussed earlier, post-tensioned concrete is at a slight disadvantage in floors where stresses due to applied load can be reversed. This would be the case for short continuous spans subject to heavy applied live loads compared with the self-weight of the floor, such as in warehouses.
- Prestressed concrete undergoes more shortening of length compared with reinforced concrete, because of the initial axial compression. In a reinforced concrete member, the creep simply affects its deflection, but in prestressed concrete it also affects the length of the member.
- Reinforced concrete floors on average tend to have a span-to-depth ratio between 20 and 25 whereas post-tensioned floors are usually in the range of 30 to 40. It is feasible for the ratio to approach 60 for lightly loaded long spans. Post-tensioning is rarely used in spans under 6 m (20 ft) because the shallow depths do not provide sufficient eccentricity for the efficient use of prestressing.
- The upward force exerted by a curved tendon acts against the applied loads. The deflection of the floor is, therefore, lower because it corresponds to the net difference between the applied downward load and the upward force from the tendons.

  The drape of the tendons and the prestressing force can be tailored to control deflection where so desired. It is possible to design a post-tensioned floor which will have no deflection under a given load, although this is unlikely to result in economical use of prestressing.
- In post-tensioned construction, the concrete section under working load is either in compression or it has a small amount of tension on one face. In either case it is unlikely to have any cracks, and if any do develop they will not penetrate deeply into the section. By comparison, in reinforced concrete construction the concrete must crack before the reinforcement can be stressed to the design level.

  The whole of the post-tensioned concrete section, being uncracked, is effective in flexure, so that a post-tensioned floor will have less deflection than a reinforced concrete floor of the same depth and subject to the same load.
- Post-tensioning keeps the concrete in compression, which controls shrinkage

cracking and reduces the possibility of opening up of construction joints. When tensile stresses do develop in a post-tensioned member, their magnitude is much smaller than in an equivalent reinforced concrete member. A post-tensioned floor, therefore, has better watertightness than a reinforced concrete floor. This is particularly important in car parks where de-icing salts often cause corrosion of reinforcement in a reinforced concrete floor. The hairline cracks over supports in a reinforced concrete continuous floor may allow water to penetrate and freeze, causing spalling of concrete.

- The uncracked concrete of a post-tensioned floor provides a better protection against corrosion of steel than that given by a cracked reinforced concrete section. In unbonded post-tensioning the grease packed plastic extrusion provides excellent protection to the strand.
- A post-tensioned floor, being lighter than a reinforced concrete floor of similar span and carrying the same applied load, imposes smaller loads on the columns and foundations.

  The columns particularly benefit from the post-tensioning of the floor, because the tendon curvature and the higher creep of the floor combine to reduce the column moments, as shown in Chapter 5. The size of a column and its reinforcement are usually governed by the bending moment. A reduction in moment can, therefore, result in significant savings.

  Stiff columns and walls may attract significant magnitudes of lateral forces. These should be checked.
- Draped tendons directly carry some of the shear force—numerically equal to the vertical component of the tendon force near the support. The concrete section, therefore, carries a smaller shear force and so drop panels are less likely to be needed in post-tensioned construction. Of course, this effect can be offset by the fact that post-tensioned floors are shallower than reinforced concrete floors.

  The presence of an axial compressive stress on the concrete section enhances its punching strength.
- At high temperatures, say above 150°C, strand loses its strength faster than rod reinforcement. This is compensated for by specifying a deeper concrete cover to tendons than to rod reinforcement in reinforced concrete.
- In reinforced concrete, micro-cracks must develop before the reinforcement can function at its required level of stress. Post-tensioned concrete, as stated earlier, is expected to remain crack-free in service. In case of an isolated overloading causing cracks in a post-tensioned floor, the cracks are expected to close once the overloading is eliminated.

1.8.3 Materials and equipment

Before discussing the site operations, consider the materials and equipment used in reinforced and post-tensioned concrete.

- Both forms of construction use similar grades of concrete, but early strength is a definite advantage in post-tensioning. Compaction, finishing and curing are identical. Post-tensioned construction requires less concrete; this can lead to significant savings, as the placing of concrete becomes more expensive as the building height increases.

- Formwork is also similar, except that post-tensioning requires holes to be drilled in the vertical edge boards for the tendons to project through, and the anchorage blocks need to be attached to the formwork.
- Post-tensioning requires special materials—anchorages, ducts and strand, which have to be procured and stored on site. Special equipment is also needed, such as stressing jacks and grout pumps, which need to be stored and moved from one position to another.

1.8.4 Site activities

The site activities for reinforced and post-tensioned concretes are now briefly compared.

- A major advantage in favour of reinforced concrete is that knowledge of its construction is readily available. Post-tensioning normally needs the services of trained operatives for installation, which involves an additional trade on site.
- Access to live anchorages at floor edges is needed for stressing tendons, and grouting ducts for bonded tendons. Usually a platform about a metre in width is sufficient. This would not be a problem where scaffolding is erected for other trades. Where no scaffolding is provided, the access platform is a special requirement and purpose-made staging may have to be erected at each floor level for the duration of the stressing and grouting operations, or over-runs for flying forms may be utilized for the purpose.
- Post-tensioned concrete floors require much less work for assembly of steel. The reason is that the quantity of steel in tendons and rod reinforcement for a post-tensioned floor is about half that of the reinforcement in an equivalent reinforced concrete floor. Apart from the smaller quantity of steel to be stored, handled and assembled, there is a further saving in time because the same tendons usually run over a series of spans, unlike reinforced concrete where short lengths of bars, or cages, are hand-assembled in each span and over each support.
- Post-tensioning allows large floor areas to be cast in one operation. This also means fewer construction joints.
- Additional sitework is certainly involved in the operations of laying and stressing of tendons, and subsequent grouting and making good of the anchorage pockets. However, this work is normally carried out within a period of three or four days for single stage stressing or within a fortnight for two stage stressing (see Section 1.10 for discussion of stressing stages). Besides, these operations do not impede construction of columns and the floor above and so they can effectively be programmed out of the critical path.
- Prestress is applied within three or four days of casting concrete and at this stage the floor becomes capable of supporting its own weight. Soffit shutters

can, therefore, be removed earlier for re-use elsewhere. Some of the vertical props are retained for construction loads. Fewer sets of soffit shutters would be needed for a post-tensioned floor.

- Some of the special equipment, particularly for use in bonded post-tensioning, may need crane time for handling, transporting and in use.

1.9 Bonded and unbonded post-tensioning

Post-tensioned floors may use bonded or unbonded tendons, or both types of tendons may be used in a floor. This section introduces the two types and then compares them for their differences and merits.

Overall figures are not readily available for the total area of floors constructed using unbonded and bonded systems but, judging from the sale of hardware, in the UK and in the USA both systems are widely used. Unbonded post-tensioning appears to be more in favour in the UK and the USA. In Australia the trend is much more towards bonded systems in floor construction.

1.9.1 General

In post-tensioning, bond between concrete and tendons is prevented until after the concrete has acquired adequate strength and the tendons have been stressed. To achieve this, the tendons are housed in *sheathing* or *ducts* and concrete is cast around them. The ducts themselves are bonded to the concrete but they preclude contact between concrete and the tendons, which remain free to move. The tendons are stressed at the suitable time and are then locked in position in *anchorage* assemblies.

In *bonded post-tensioning* the ducts are grouted after stressing and so bond is established between the tendons and the concrete section, with the ducts as intervening components. As its name implies, in *unbonded post-tensioning* the duct, usually referred to as a sheathing in this case, is never grouted and the tendon is held in tension solely by the end anchorages.

Ducts for use in bonded post-tensioned floors may be circular in section, 50 to 75 mm (2 to 3 in) in diameter, or oblong, 75 × 20 mm (3 × 3/4 in), and made of corrugated galvanized iron or plastic. The corrugations impart to the duct a longitudinal flexibility, so that it can be easily bent to a desired profile, while retaining rigidity to keep the section circular. With a corrugated duct it is possible for the tendon to be curved to a radius as small as 2.5 m (8 ft) whereas with an uncorrugated metal duct it is difficult to have any profile other than a virtual straight line.

The actual size of the duct varies, depending on the manufacturer and the number of strands to be housed in it. Circular ducts, for use in normal floors, are in most systems available in several diameters; the most common sizes accommodate 2 to 4 strands, 5 to 7 strands and 8 to 12 strands. The oblong or *flat* duct is normally designed to accommodate up to a maximum of four or five strands.

After the tendons have been stressed and the excess length of strand trimmed, the ducts are grouted with a neat cement paste or a sand-cement mortar containing suitable admixtures, which bonds the strands to the duct. Sand is not included in the grout for the small diameter ducts normally used in post-tensioned floors. The duct itself being already bonded, the tendon becomes effectively bonded to the concrete section. Bond is capable of carrying the total prestressing force, should the anchorage become ineffective for some reason.

For unbonded post-tensioning use, as part of its manufacturing process, the strand has a thin polypropylene sleeve extruded onto it and the space between the strand and the sleeve is grease packed. These measures prevent bond with the concrete and inhibit corrosion. This strand is cast directly in the concrete without an intervening duct and it remains ungrouted throughout the life of the structure. As in the bonded system, the tendons are stressed when the concrete has gained sufficient strength but the prestress is maintained solely through the mechanical anchorages at each end.

In both systems, bonded and unbonded, the anchorages are set in pockets at the floor edge, and stressing is carried out using hydraulic equipment. After stressing, the strands are cut close to the anchorage face using a high-speed disc cutter. The anchorage assemblies are then sprayed with a corrosion inhibitant and covered with grease-packed plastic caps. Lastly the pockets are made good with cement sand mortar for protection against corrosion and fire. The corrosion inhibitant, the grease cap and the mortar provide adequate protection to the anchorage assembly in a post-tensioned floor.

Bonded tendons require metal or plastic ducts and the additional site operation of grouting, but several strands are anchored in a common assembly, thereby saving on the number of individual anchorages. The additional cost of the duct and grouting may be offset by the saving in the cost of anchorages; this depends on the relative costs and the tendon lengths. Notwithstanding the cost, bonded tendons are considered to be more secure against accidental damage. They are, therefore, very useful in breaking a floor into smaller areas for confining accidental damage. Bonded tendons are often used in beam strips whereas unbonded tendons may be used in slabs supported on these beam strips.

1.9.2 Design and serviceability

- Bonded tendons usually consist of several strands placed in a common duct. The duct is larger in diameter than a single sleeved strand and when placed in position it gives a smaller eccentricity than an unbonded tendon. Therefore, where maximum eccentricity is desired, bonded tendons are less efficient.
- A given concrete section with bonded tendons would have a higher local ultimate strength than the same section using unbonded tendons. Approaching ultimate load, the strain, and hence the tensile stress, in a bonded tendon must increase with the strain in the adjacent concrete. An unbonded tendon can slip relative to the concrete, so that the increase in its tensile stress is much smaller.

  An unbonded tendon may carry a larger force in the serviceability state because of lower friction.
- Grouting of ducts in the past has not always been carried out with the care it

deserves and there have been cases of tendon corrosion resulting from trapped air and excess water. Freezing of water trapped in a duct may cause bursting of the duct and spalling of the surrounding concrete; this is a possibility where the floor is exposed to icy conditions, such as the upper decks of a car park.

These problems have been discovered during demolition of post-tensioned buildings or when a refurbishment required some tendons to be cut. Awareness of the potential problem has caused the installers to pay much more attention to grouting than they did in earlier days. However, the fact remains that once a duct has been grouted there is no easy way of checking its adequacy.

- The unbonded tendons remain totally dependent on the integrity of their anchorages; if an anchorage suffers damage, accidental or due to corrosion, then all prestress in the particular tendon is lost. A grouted tendon is bonded to the concrete section and if its anchorage gets damaged then the bond should be able to retain the prestressing force beyond the bond length.
- If an unbonded strand is damaged in a continuous floor then the whole series of associated spans suffer from the loss. For a bonded tendon the loss is confined to the particular span where the damage has occurred. A sufficient quantity of rod reinforcement should be provided in unbonded floors to contain such damage.

  A similar situation may arise if tendons are deliberately cut, for instance, to make a hole. The adjacent continuous spans would in this case require propping if the tendons were not bonded.
- In case of damage to a strand in a completed floor, for short lengths it is possible to withdraw the unbonded strand from its extruded sheathing and insert another, possibly a compact strand instead of the normal—the compact strand has a slightly smaller diameter. Replacement, of course, is not possible if the tendon is bonded.
- In case of fire, the bonded tendon has the protection of the additional concrete cover, because the duct is much larger than the strand area and stressing tends to pull the tendon into the mass of the concrete, away from the surface. The duct itself acts as a heat sink to a small extent.

  An unbonded tendon has less concrete cover but the sheathing has a limited insulating value.
- In the case of overloading, the distribution of cracks is similar in reinforced and bonded post-tensioned floors. With unbonded tendons, the cracks are wider and further apart, unless bonded rod reinforcement has been provided to distribute the cracks.

1.9.3 Material, equipment and site activities

- Bonded tendons need extra hardware in the form of metal or plastic ducts with the consequent extra handling and site storage requirement.
- In a bonded system, ducts are usually assembled in position first and then the strands are threaded through. Subsequently, the ducts need grouting. In an unbonded system the tendons, consisting of individual strands, are assembled directly on the formwork.

- Bonded tendons being multistrand, their stressing jacks are too heavy to be handled manually, so crane time is required during the stressing operation.
  The grouting equipment, required for the bonded system, may also need cranage to move it from one position to the next.
- In the bonded system all strands contained in a duct are stressed in one operation, instead of one at a time in the case of unbonded tendons. Therefore, fewer operations are needed to stress bonded tendons.

1.10 Stressing stages

Concrete, though weak in tension, is capable of withstanding a tensile stress of the order of 10% of its compressive strength. In reinforced concrete the tensile strength cannot be utilized, because the high strains on the tension face cause the concrete to crack. Once cracked, the concrete is assumed incapable of carrying any tension. In post-tensioning, the strains on the tension face of concrete are low and the tensile strength of concrete is utilized in calculations for the serviceability state.

It is essential to avoid microcracks in concrete which is to be post-tensioned; they tend to develop during the setting of concrete and within the first few days of casting thereafter, before it has gained sufficient strength to withstand the temperature and shrinkage strains. Prestress, by inducing compression in concrete, reduces the probability of shrinkage cracks, and the normal practice in post-tensioning is to apply some of the prestress at as early an age as practicable. Too early a stressing, before the concrete has attained sufficient strength, can cause cracks through overstressing. The actual timing of the stressing operation is often a compromise between the two conflicting considerations—desirability of stressing early and the need to reduce the possibility of damage from early stressing.

In general, floors are stressed within three or four days of casting. At a cube strength of 25 to 30 N/mm^2 ($f_c' = 3000$ to 3500 psi) it is possible to apply full prestress. Where full prestress cannot be applied in one operation, a proportion, say 50%, is applied within three days. At the first opportunity thereafter, usually within 14 days of casting concrete, the prestress is increased to its full design value.

In the first stage of the two-stage stressing operation, either all tendons may be stressed to half the design force, or half of the tendons may be fully stressed. The latter requires the concrete to be strong enough for the bearing stresses of the anchorages, in which case all anchorages can be stressed, and the only advantage of two-stage stressing would be that of a slight reduction in prestressing losses. Most post-tensioning anchorages are designed for stressing to a 100% force at a concrete cube strength of 25 to 30 N/mm^2 ($f_c' = 3000$ to 3500 psi), although some manufacturers supply a physically larger anchorage range which can be stressed at a concrete strength of about 15 N/mm^2 ($f_c' = 1750$ psi).

Two-stage stressing is used with some reluctance because it involves two visits to each floor, the platform for access to anchorages must be maintained for a longer period and the tendons to be bonded remain ungrouted for a longer period with the consequent danger of water getting into the ducts. Two-stage stressing, however, has a design advantage in that some shrinkage and creep will have already occurred before the second stage and so the tendons will undergo a smaller loss in prestress from these causes.

Two-stage stressing is required where a relatively high level of prestress is needed, for example when a particularly slim concrete section is desired or the applied load is much higher in intensity than the self-weight of the member, such as in transfer beams. Floors containing lightweight concrete are almost always stressed in two stages.

It is desirable to stress all tendons fully in one operation. BS 8110 specifies a minimum concrete strength of 25 N/mm^2 at stressing.

1.11 Construction tolerances

The permissible construction tolerances for deviation in member size and in positioning of the steel are normally specified in the national standards. These tolerances are not reproduced here.

The tolerances specified for post-tensioned floors, with regard to the depth of the member and placing of steel are generally similar to those for reinforced concrete.

Ultimate strength calculations for post-tensioned and reinforced concrete follow identical procedures for shear and for flexure. Therefore, a given dimensional variation can be expected to have a similar effect on the ultimate strength in both systems.

1.12 Fire resistance

Post-tensioned concrete, being largely free of micro-cracks, provides better protection to steel than reinforced concrete. From the durability point of view, therefore, the concrete covers normally specified for reinforced concrete are quite adequate for post-tensioned concrete.

In case of fire, however, at high temperatures post-tensioning tendons lose a much greater proportion of their strength than reinforcement, and, for this reason, most of the national standards specify an increased cover in post-tensioned concrete. The increase in cover depends on the floor configuration—beam or slab, solid or ribbed, simply-supported or continuous. The required additional cover varies from nil for a half-hour resistance to a maximum of 20 mm for a simply supported beam designed to resist two-hours fire.

1.13 Holes through completed floors

Holes, not planned at the design stage, may be required through a completed floor by a tenant for additional services, for installing new lifts, staircases or air conditioning ducts.

Post-tensioning is often said to limit the flexibility of a building by restricting the possibility of making holes through a floor after its completion. For small service holes, a post-tensioned floor, in fact, offers a greater flexibility than a reinforced concrete slab, whether it is in a steel frame or a concrete frame. The tendons are normally spaced much farther apart than the rod reinforcement is in a reinforced concrete floor, and larger holes can, therefore, be made in a post-tensioned floor without cutting any of the steel.

Care, of course, needs to be exercised to ensure that a tendon is not damaged. Concentration of tendons usually occurs along the beam lines which should not be too difficult to identify on site or from the construction drawings. Sometimes, tendon positions are marked on the floor soffit. In any case, a tendon can be easily located on site with a cover meter or a metal detector; in fact this should be done to confirm the tendon position even when it is known.

Large holes are difficult to make in any floor construction, post-tensioned or not. It is almost certain that a large hole will require a new system of beams to support the cut edges, and that the beams will have to be supported directly on the existing columns. If the proposed hole requires one of the existing beams to be cut then the problem becomes even more complex. This requires a careful design check on the existing structure irrespective of whether the floor and the beams are of structural steelwork, reinforced concrete or post-tensioned concrete.

Post-tensioning specialist trained operatives are needed to cut and secure the tendons. Bonded tendons are relatively easy to cut, because bond prevents loss of prestress. Unbonded tendons may need adjacent continuous spans to be propped before the tendons are de-tensioned, cut and re-stressed using new anchorages.

1.14 Post-tensioning in refurbishment

Post-tensioning has been successfully used in the refurbishment of existing reinforced concrete buildings, where the floors needed strengthening or upgrading.

The load capacity of an existing reinforced or post-tensioned floor can be increased by the application of prestress in the manner shown in Figure 1.3(a). Pockets are cut at the end of the slab for anchorages, and sloping holes are drilled, through which the strand is threaded. At each point where a tendon changes direction, the strand is seated on a mandrel to avoid stress concentration which would otherwise occur. The mandrels also provide a simple means of controlling the tendon eccentricity; the tendons are located below the slab and, therefore, they are utilized in a very efficient manner. This method increases the depth of the structural zone on the underside of the floor; the top level remains unchanged, and, therefore, the staircases and lifts are not affected.

A similar arrangement can be used for strengthening the downstand beams in a floor. In this case, the mandrels are located in holes drilled through the downstand web of the beam. The tendons remain within the existing depth of the structure.

For a ribbed floor, an arrangement similar to Figure 1.3(a) can be used, but

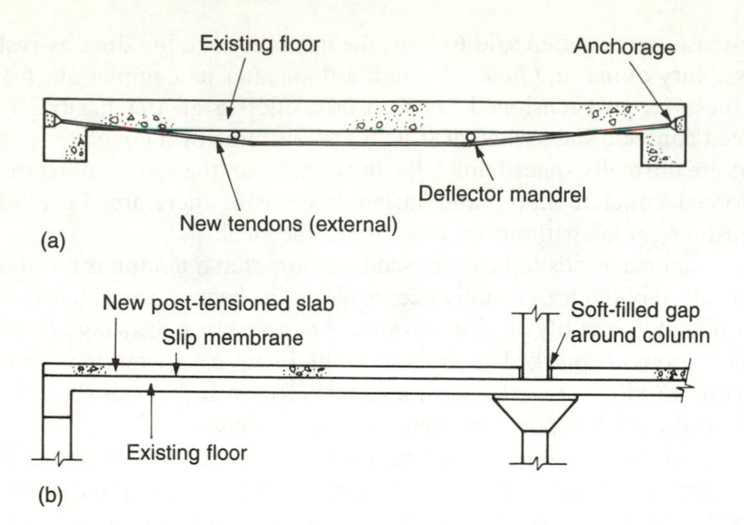

Figure 1.3 *Strengthening of a floor*

with the tendons located between the ribs. In this case there is no increase in the structural depth.

The main advantage of using post-tensioning in the above manner is that most of the work is carried out from the outside, or from below; the floor itself can continue to be used during the strengthening operations. The method also uses a minimum of materials and very little wet trades. It is basically applicable to single spans where access is available to the slab edges. In continuous spans, the tendons may be located in specially prepared grooves over the supports but this would need access to the top.

In the above methods, the tendons being external and unbonded, some protection must be provided to prevent physical or fire damage, and to contain the strand in case of a failure.

Where an existing slab has deteriorated, it is possible to use it as a permanent shuttering and cast a new post-tensioned floor on top, see Figure 1.3(b). The prestress can be so designed that the new slab does not impose any load on the old one in service, except at the supports where the old slab may be required to carry the shear at columns. This arrangement is particularly suitable for upgrading old reinforced concrete flat slabs. Its disadvantage, of course, is that of the increase in the structural thickness above the slab top level, which affects staircases, lifts and ramps.

1.15 Some misconceptions about post-tensioned floors

Prestressing is sometimes conceived as application of a large compressive force to concrete in its green state, subjecting it to the possibilities of sudden failure and buckling. Formwork, it is assumed, must be strong enough to sustain the initial prestressing force. Anchorages are also considered to be a potential source of danger—they may shoot out as high velocity missiles in case of failure of a tendon or an anchorage.

While each of the above possibilities exists in prestressing in general, they do not occur in post-tensioned floors.

Formwork

Moulds are occasionally used to anchor the prestressing tendons in the precast pretensioning process, where the tendons are stressed before the concrete is cast, the force being transferred to concrete by releasing the anchor when the concrete has gained sufficient strength. In this case the moulds must be designed to sustain the prestressing force, with a safety margin.

In an *in situ* post-tensioned floor, the stressing is carried out *after* the concrete has gained sufficient strength to safely sustain the prestressing force. The force is applied directly to the concrete; the formwork is *never* required to carry any of the prestressing force. Apart from the few holes needed for the tendons, the design of formwork is the same as for reinforced concrete.

Buckling under prestress

Concern has, quite incorrectly, been expressed about the possibility of buckling of a post-tensioned floor under the prestressing force.

In fact, buckling is impossible to occur when the tendons are in contact with the concrete, as they are in post-tensioned floors. The possibility of buckling would exist only at the time of stressing where the prestress was applied through straight bonded tendons, because at that stage the tendons would, theoretically, not be in contact with the sheathing. Once bonded, the tendons provide a lateral support to the concrete opposing any tendency to buckle; the arrangement becomes self-correcting.

The situation in practice never arises because:

- For buckling to occur, the slenderness ratio of the slab and the level of prestress need to be much higher than is commonly used.
- Long straight tendons in slender members are used only in ground slabs. However, ground slabs almost always use unbonded tendons, which are in contact with concrete. Even if bonded tendons are used, the stress level is too low, of the order of 1.5 N/mm^2 (about 200 psi), to overcome the self weight of the floor.
- For floors containing unbonded tendons, the possibility of buckling does not exist because the tendons are always in contact with the concrete.

Sudden failure

Sudden and brittle failure of a concrete member occurs when it is subjected to high compressive stresses approaching the limit of its strength. In bridges comprising precast girders, the girders are often shaped and designed to make the maximum use of concrete, which is heavily stressed immediately after, or during, stressing. The possibility of failure occurs at this initial stage, before the girder is put in service; thereafter, service loads usually *reduce* the stresses. Sudden failure can occur during service if a substantial area of the concrete is lost from the compression zone due to damage.

In post-tensioned floors, the level of prestressing is generally too low for the concrete to approach failure in compression. The average stress in a slab is of the order of 2 N/mm^2 (300 psi) compared with 8 N/mm^2 (1200 psi) or more in a bridge girder.

Tendon failure

In case of a local damage to a bonded tendon, the grout should be able to take the full force in the tendon, and thereby contain any tendency of the tendon to move which would cause a loss of prestress away from the damage. The mechanical anchorage assembly of a bonded tendon may in this sense be considered redundant. If a bonded tendon or its anchorage fails, it is unlikely to be noticed at the time. This is evident from the number of defective tendons in road bridges discovered only when they are refurbished, strengthened or demolished. Usually, the first indication of a problem is an increase in deflection.

With unbonded tendons, experimental studies (see Chapter 13) have shown that the extruded plastic sheathing absorbs most of the energy released when a strand is broken. The maximum movement of the strand at an end is of the order of 150 mm (6 inches).

The energy being stored in the strand and not in the anchorage, if an anchorage fails then it will not fly out.

In buildings, the anchorage is covered with a concrete plug in the stressing pocket and the cladding of the building. The possibility of any missiles emerging in case of failure of a tendon or an anchorage is so remote that it can be considered non-existent. This has been confirmed both by full-size tests on actual structures and under laboratory conditions.

Figure 1.4 *Hang Seng Bank – Hong Kong*

2 MATERIALS AND EQUIPMENT

Post-tensioned floors use all the material required in a reinforced concrete floor—formwork, rod reinforcement and concrete—and, additionally, they use high tensile steel strand and the hardware specific to post-tensioning.

As a material, rod reinforcement in post-tensioned floors is exactly the same as that in reinforced concrete in every respect. The normal high tensile steel, as used in rod reinforcement, has a yield stress of 460 N/mm^2 (66 000 psi) and a modulus of elasticity of 200 kN/mm^2 (29 000 ksi). It has a Poisson's ratio of 0.3 and a coefficient of thermal expansion of 12.5×10^{-6} per °C (7×10^{-6} per °F). The strength of high tensile steel is affected by the rise in temperature, dropping from 100% at 300°C to only 5% at 800°C.

The technology for the production, compaction and curing of concrete is well understood and is not discussed here. Only the properties of concrete which are important for post-tensioning are considered.

Normal dense concrete, of 2400 kg/m^3 density (150 pcf), is more common in post-tensioning. Lightweight concrete, however, has certain advantages in the right circumstances. Both are dealt with in separate sections. The properties of the two concretes are quite different and it is not a good practice to use the two side by side; there may be problems from differential movement and the difference in their moduli of elasticity, shrinkage and creep.

2.1 Formwork

In a post-tensioned floor, the vertical edge boards of the formwork need to have holes drilled through for the tendons to pass at live anchorages. During stressing, the concrete undergoes a slight reduction in length due to the axial component of the prestress. Though trapping of formwork between any downstands is not a serious problem, the design of the formwork should recognize the possibility. During stressing, the post-tensioned slab lifts off the formwork, so that there is a re-distribution of its weight. This may impose heavier loads on parts of the formwork than those due to the weight of the wet concrete. In other respects, the formwork for a post-tensioned floor is similar to that for a reinforced concrete floor.

Each live anchorage is set in a recess, or *anchorage pocket*, in the slab edge. The pocket is formed using proprietary plastic formers supplied by the specialist prestressing hardware supplier. The formers are removed when the formwork is stripped. Expanded polystyrene blocks have sometimes been used for this purpose but, in the authors' experience they are often difficult to remove afterwards and pieces of polystyrene are left in the anchorage pocket. Attempts to remove these by burning causes heating of the anchorage and the strand at a critical point and use of chemicals may have a detrimental effect on concrete or steel.

The pockets are normally tapering in shape to allow removal of the former, and are larger than the size of the anchorage in elevation to allow sufficient room for the jack to be coupled. The actual size depends on the anchorage dimensions and the clearance required for the jack; both of these vary between the various prestressing systems. The relevant data can be obtained from the particular manufacturer. The pocket depth is sufficient to accommodate the anchorage, the projecting strand (about 30 mm, 1.25 in) and the grease cap, and to provide adequate cover to the assembly.

Anchorage castings are temporarily attached to the vertical edge board. A single strand casting may be attached by long nails passing through holes at the corners of the anchorage casting, or by proprietary means. The nails get cast in the concrete and cannot be removed when the edge board is stripped. Anchorages for multistrand tendons are heavier and are supported by bolts which pass through sleeves of the same length as the pocket. The sleeve itself gets concreted and cannot be removed but the bolt is removed when the edge shutter is stripped.

A reinforced concrete floor does not undergo any significant longitudinal shortening during the first few days of casting; shrinkage does not occur while the concrete is wet during curing. In contrast, a post-tensioned floor does shorten in length during stressing. The strain depends on the average prestress level. For a slab with an average stress of 2.0 N/mm^2 (300 psi), the strain may be about 0.0001, i.e. a ten-metre bay may shorten by 1 mm. A ten-metre length of beam stressed to an average of 6.0 N/mm^2 (900 psi) would shorten by 3 mm ($\frac{1}{8}$ inch). These strains are not large, but they may just trap the formwork between two vertical faces of a downstand. Removal of such a trapped soffit shutter may need some force, which can damage the arrises of the downstands.

The difficulty can be avoided by a filler strip in the formwork, which can be removed before stressing, or by incorporating a strip of compressible material. In ribbed and waffle floors, removal of the forms is easier if the sides of the ribs are given a generous slope, say not less than 10°.

A post-tensioned floor lifts off its formwork when prestress is applied—usually three or four days after casting. Soffit forms, therefore, become redundant at this stage and are normally removed; some props are retained for construction loads. Early removal of formwork allows a faster turn-around, and it is normally possible to use fewer sets than would be required in a similar project in traditional reinforced concrete. Each formwork set would, therefore, be used many more times and it should be designed accordingly.

2.2 Dense concrete

Concrete for a post-tensioned floor is similar to that in a reinforced concrete structure. In reinforced concrete the important properties are its strength and durability; other properties such as rate of development of strength, elastic modulus, shrinkage and creep are of secondary importance. For post-tensioning, the desirable properties of concrete are high early strength, low shrinkage and creep, and high modulus of elasticity. These properties and their significance in post-tensioning are discussed below.

Air entrained concrete may be a solution for structures subject to freeze-thaw cycles. In each case, the short-term and the long-term properties of the particular concrete should be determined.

It should be appreciated that better quality control needs to be exercised for post-tensioned floors. It is important to avoid the need for repairs arising out of defects such as surface crazing, shrinkage cracks or broken arrises, because, unlike reinforced concrete, the tensile strength of concrete is usually relied upon in serviceability state calculations for Class 2 structures designed in accordance with BS 8110. A crazed or cracked concrete or a repaired patch cannot be relied upon to withstand any tension even if its appearance is acceptable. Measures should therefore be taken to minimize the possibility of the concrete cracking after the initial set.

2.2.1 Thermal properties

In reinforced concrete, the coefficient of thermal expansion is conveniently assumed to equal that for steel, 12.5×10^{-6} per °C (7×10^{-6} per °F). In fact, its value for concrete varies with the type of aggregate, the proportion of cement paste, the moisture content, the age of concrete, and even the temperature of concrete. Because the thermal properties of plain concrete differ from those of steel, the coefficient of expansion of reinforced concrete is also affected by the reinforcement content. With so many variables, it is difficult to give a realistic guidance on the value for the different concretes.

Neville (1981) gives the coefficient of thermal expansion for neat cement as 18.5×10^{-6} per °C (10×10^{-6} per °F), that for cement paste as varying between 11×10^{-6} and 20×10^{-6} per °C (6×10^{-6} and 11×10^{-6} per °F) depending on the sand content, and the values given in Table 2.1 for 1:6 concretes made with different aggregates.

Table 2.2 shows the BS 8110 values for the coefficients of thermal expansion of typical rock groups and concretes made from them.

2.2.2 Concrete strength

Concrete used in post-tensioned floors generally has a 28-day cube strength f_{cu} in the range 25 to 50 N/mm^2 (cylinder strength $f_c' = 3000$ to 6000 psi), the most common being 40 N/mm^2 ($f_c' = 4500$ psi). Higher strengths, well over

Table 2.1 *Coefficient of thermal expansion (from Neville, 1981)*

Aggregate	*Air-cured*		*Wet-cured*		*Air-cured and wetted*	
	$10^{-6}/°C$	$10^{-6}/°F$	$10^{-6}/°C$	$10^{-6}/°F$	$10^{-6}/°C$	$10^{-6}/°F$
Gravel	13.1	7.3	12.2	6.8	11.7	6.5
Granite	9.5	5.3	8.6	4.8	7.7	4.3
Quartzite	12.8	7.1	12.2	6.8	11.7	6.5
Dolerite	9.5	5.3	8.5	4.7	7.9	4.4
Sandstone	11.7	6.5	10.1	5.6	8.6	4.8
Limestone	7.4	4.1	6.1	3.4	5.9	3.3
Portland	7.4	4.1	6.1	3.4	6.5	3.6
Blast-furnace	10.6	5.9	9.2	5.1	8.8	4.9
Foamed slag	12.1	6.7	9.2	5.1	8.5	4.7

Table 2.2 *Coefficient of thermal expansion (from BS 8110)*

	Typical coefficient of expansion $\times 10^{-6}/°C$	
Aggregate type	*Aggregate*	*Concrete*
Flint, quartzite	11	12
Granite, basalt	7	10
Limestone	6	8

100 N/mm^2 ($f_c' \approx 12\,000$ psi) are being produced for specialized use and the trend in post-tensioning is also towards higher grades.

With higher strength concrete, it becomes possible to use a shallower section, which may be more economical even if the cost of a unit volume of concrete is high. However, the modulus of elasticity does not increase in direct proportion with the strength and, therefore, a shallow floor made from high strength concrete deflects more than a deeper one containing a lower grade of concrete. In post-tensioning, deflection can be controlled to a large extent by choosing a suitable combination of prestressing force and eccentricity. High strength concretes may, therefore, prove more economical if post-tensioned than reinforced. The shallower section would have a lower natural vibration frequency, possibly creeping into the undesirable perceptibility range; this is discussed in Chapter 9.

For post-tensioning, concrete is designed to gain high early strength because it permits stressing at an early age, thereby reducing the possibility of shrinkage cracks. It also allows formwork to be released early for use elsewhere. In post-tensioned construction, a minimum strength is specified at which the prestress can be applied, and cubes or cylinders are tested before stressing, usually about three days after casting.

The rate at which concrete gains strength depends on the properties of its constituents and the manner of curing; the most important constituent is the cement. As a general guidance for the rate of development of strength, BS 8110

Table 2.3 *Strength of concrete and age*

Cube strength N/mm² at various ages 7 days	*28 days*	*2 months*	*3 months*	*6 months*	*1 year*
20.0	30	33.0	35.0	36.0	37.0
28.0	40	44.0	45.5	47.5	50.0
36.0	50	54.0	55.5	57.5	60.0

gives the values shown in Table 2.3. The second column represents the nominal grade of concrete. No value is given for an age of three days because the variation in test results is too wide to be reliable.

The 28-day strength of concrete is determined by tests on cubes or cylinders which are crushed by applying a compressive load in one direction only. It is then assumed that the strength applies to stresses on one axis as well as to biaxial and triaxial stresses.

Concrete, commonly assumed to have no tensile strength, in fact is capable of sustaining a significant level of tension. It also has the property that if, under an abnormal loading of short duration, micro-cracks appear on its tension face then on removal of this load the cracks heal over a period, provided that the concrete is maintained in compression. The process is not fully understood but it probably is due to the free lime in concrete.

Traditionally tests are not carried out for measuring the tensile strength of concrete, perhaps because it is not utilized in reinforced concrete design. It is, however, of particular interest in post-tensioning. The tensile strength is a function of the strength of mortar between aggregate pieces and it does not depend on the strength of aggregate to the same extent as the compressive strength of concrete does. In normal weight concretes, the aggregate is generally stronger than the mortar, so that tension cracks develop at the aggregate-mortar interface.

Tests show that concrete is capable of withstanding tensile stresses of the order of 10% of its compressive strength; the relationship, however, is not linear. The rate of gain of tensile strength with age slows down faster than that for compressive strength, with the result that the ratio of tensile-to-compressive strength reduces with age. The tensile strength almost reaches its peak at an age of about one month.

Attempts to measure tensile strength directly have proved unreliable. A more satisfactory method is to apply two line loads to a concrete cylinder along two diametrically opposite sides; analysis indicates that under this loading a state of almost uniform tension exists in the cylinder. Another method is to test a 4in × 4in × 12in (100 × 100 × 300 mm) plain concrete beam to bending fracture; the tensile strength thus determined is termed the *modulus of rupture*. It varies between 1.75 and 2.25 times the direct tensile strength and a value of 2.0 can be assumed for this factor. It should be noted that the direct tensile strength and the modulus of rupture are not interchangeable; the former applies to direct tension situations, such as the principal tensile stress, and the latter to flexure, such as the cracking moment.

ACI 318 specifies a modulus of rupture value of $0.7(f_c')^{0.5}$ for normal weight concrete. BS 8110 does not give any relationship between the modulus of rupture and the cube strength. It allows a principal tensile stress of $0.24(f_{cu})^{0.5}$, which includes a partial safety factor, presumably 1.25.

The tensile strength can be improved by adding glass, polypropylene or steel fibre. Glass fibre is attacked by the alkali in concrete. Special material has been developed which is more resistant than normal glass, but the strength of concrete containing glass fibre tends to fall with time. Short length, fine polypropylene fibre is claimed to significantly reduce early plastic movement of concrete; the strength of mature concrete is not affected. Steel fibre is more suitable for use in post-tensioned floors; a 2% content appears to increase the tensile strength by about 50%, though it may not be economical.

In a real structure, the stresses are almost always biaxial, and concrete strength in one direction is affected by the stresses in the other direction. Tasuji, Slate and Nilson (1978) have given the ratio of biaxial to uniaxial strength shown in Figure 2.1. The effect of transverse stress on tensile strength is at present ignored in the design of post-tensioned concrete.

2.2.3 Elastic properties

Poisson's ratio may be needed in calculations for unusually complex structures only. Its value for concrete varies between approximately 0.11 and 0.21, higher concrete strengths have the lower value. A value of 0.20 is commonly used for all grades in post-tensioning.

The modulus of elasticity, also called *Young's modulus*, of concrete (E_c) is the most important elastic property required in calculations. It is a measure of the short-term strain produced in concrete by an applied stress. It is required for calculating elastic behaviour of a structure, such as deflection, and for estimating losses of prestress. Concrete with a low modulus of elasticity has a larger strain for a given stress; a floor containing such concrete has a larger deflection and more of its prestressing force is lost through elastic shortening, and of course through shrinkage and creep deformations, of concrete.

The value of modulus of elasticity is determined by compression tests and the same value is assumed to apply in tension. Modulus of elasticity depends on a number of factors, such as quality of aggregate and cement, admixtures, manner of curing and the age of concrete. British and European code guidelines have traditionally proposed a cube root relationship between E_c and the 28-day concrete strength (Neville, 1981), but in BS 8110 a simple relationship was adopted for general use, given in Equation (2.1).

$$E_c = K_o + 0.2 f_{cu} \text{ kN/mm}^2 \quad (2.1)$$

where K_o = a constant representing aggregate quality
f_{cu} = the 28-day cube strength in N/mm^2

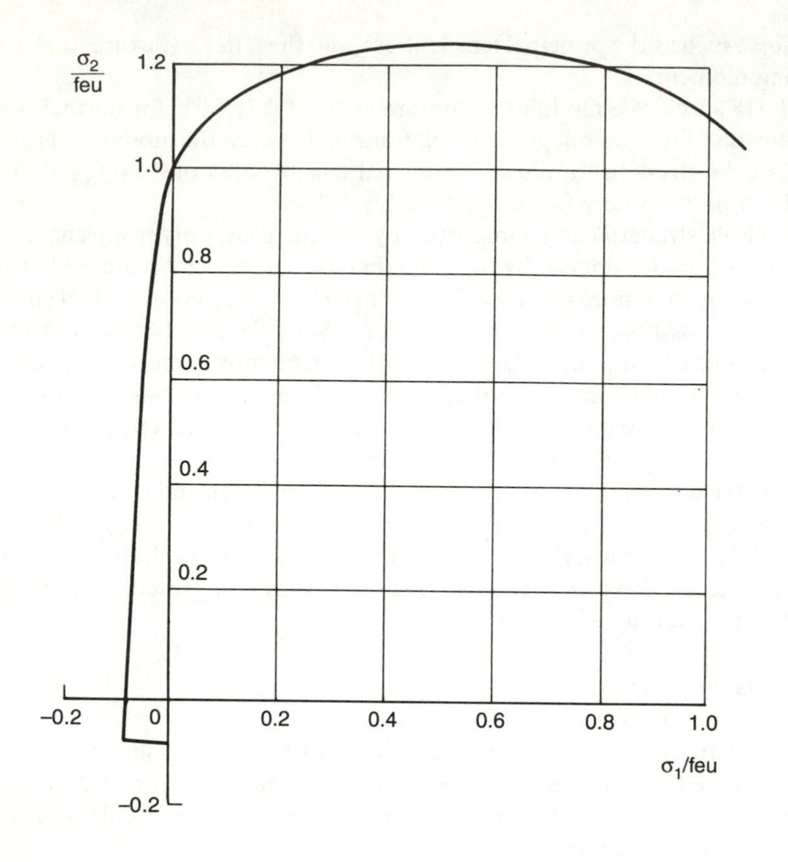

Figure 2.1 *Biaxial strength of concrete*

The value of K_o is closely related to the modulus of elasticity of the aggregate. It varies between 14 and 26 kN/mm^2 for the UK aggregates and is taken as 20 kN/mm^2 for normal weight concrete.

In order to assess prestress loss, the modulus of elasticity value is required at an age other than 28 days. Equation (2.2) gives the BS 8110 relationship.

$$E_{ct} = E_c(0.4 + 0.6f_{ct}/f_{cu})\ \text{kN/mm}^2 \tag{2.2}$$

where E_{ct} = modulus of elasticity at the desired age
f_{ct} = cube strength at that age in N/mm^2

Based on the strengths shown in Table 2.3 and on the above equations, moduli of elasticity are tabulated in Table 2.4 for different concrete grades at various ages with $K_o = 20$ kN/mm^2. It must be remembered that the particular concrete being used may not develop the strengths at the rate assumed in Table 2.3.

It should be noted that BS 8110 does not permit an increase in strength beyond 28 days in satisfying limit state requirements. Where calculations of deflection or

Table 2.4 *Modulus of elasticity of normal weight concrete* (kN/mm^2) (from BS 8110)

Grade	*7 days*	*28 days*	*2 months*	*3 months*	*6 months*	*1 year*
30	20.8	26.0	27.6	28.6	29.1	29.6
40	23.0	28.0	29.7	30.3	31.2	32.2
50	25.0	30.0	31.4	32.0	32.7	33.6

deformation are to be made, strength at higher age may be used. In calculating E_c from Equations (2.1) and (2.2), the actual value of K_o for the particular aggregate should be used. For unknown aggregates a range of values varying from 14 to 26 kN/mm^2 should be used; this is equivalent to a variation of ± 6 kN/mm^2 in the 28-day value of E_c for all grades of concrete.

ACI 318 recommends a square root relationship between E_c and f'_c.

$$\begin{aligned} E_{ct} &= \text{the modulus of elasticity at the desired age} \\ &= 57000(f'_{ct})^{0.5} \text{ in psi units} \\ &= 4700(f'_{ct})^{0.5} \text{ in N/mm}^2 \text{ units} \end{aligned} \qquad (2.3)$$

where f'_{ct} = cylinder strength at the age.

It is generally accepted that the actual value of E_c may vary by $\pm 20\%$ from that given by equations (2.3) depending on the modulus of elasticity of the aggregate.

A comparison between Equations (2.1) and (2.3), assuming a ratio between cylinder and cube strengths of 0.8, is shown in Figure 2.2.

2.2.4 Shrinkage

Water is needed in concrete for the chemical reactions to take place for the initial setting of the mortar and development of strength. With only sufficient water for this purpose, the concrete is too dry for proper placing and compaction; therefore, the actual quantity of water in a mix is much more than that required solely for the chemical reaction. The excess water gradually migrates through the pores and evaporates from the surface. This loss of water in turn causes *shrinkage*—a reduction in the volume of the hardened concrete. Conversely, if the moisture content of hardened concrete is increased then it expands. Incorporation of some plasticizers and polypropylene fibres is claimed to reduce the early shrinkage of concrete.

Shrinkage is generally expressed as a strain. The total extent of shrinkage and its rate depend on the amount of water initially present in the mix, on the ambient humidity and on the section depth. Certain aggregates are also affected by humidity and concretes containing aggregates with high shrinkage properties have a correspondingly higher shrinkage.

If the concrete surface is sealed at an early age, such as during curing or by application of chemicals to give the surface any desired property or by a screed,

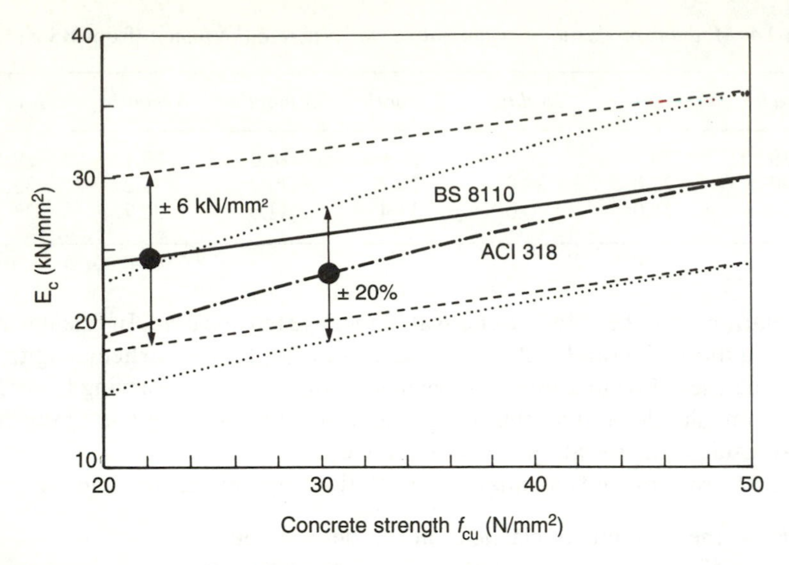

Figure 2.2 *Variation of E_c with 28-day strength*

then the rate of shrinkage is reduced. An arbitrary period of 30 years is, however, normally assumed to elapse before concrete reaches a stable state.

Seasonal variation of humidity has a corresponding effect on concrete if it is exposed to the elements and it shows the seasonal effect superimposed on the normal long-term shrinkage. In the UK climate, the seasonal shrinkage variation in particularly exposed locations can be as much as 40% of the long-term shrinkage strain.

For normal indoor exposure of plain concrete, an ultimate shrinkage strain of 300×10^{-6} is an accepted figure for mixes containing about 190 litres of water per cubic metre of concrete (19% by volume). For outdoor exposure a value of 100×10^{-6} is generally taken. Where concrete is known to have a different water content, shrinkage may be regarded as proportional to water content within the range 150 to 230 litres/m^3 (15 to 23% by volume).

For a more accurate assessment, Figure 2.3 shows the BS 8110 relationship between 30-year shrinkage, humidity and the effective thickness of concrete. The figure relates to plain concrete of normal workability made without water reducing admixtures, with a water content of 190 litres per cubic metre (19% by volume). In the UK, relative humidities of 45% and 85% are considered suitable for indoor and outdoor exposure respectively. The example, shown dotted, indicates that at 60% humidity a 300 mm thickness of plain concrete is expected to have a total shrinkage strain of 325×10^{-6}. In other countries, or where high shrinkage aggregates are to be used, shrinkage characteristics should be obtained from the concrete supplier if known at the design stage.

For non-rectangular sections, the effective thickness may be taken as twice the ratio of volume to exposed surface area and where only one surface is exposed for

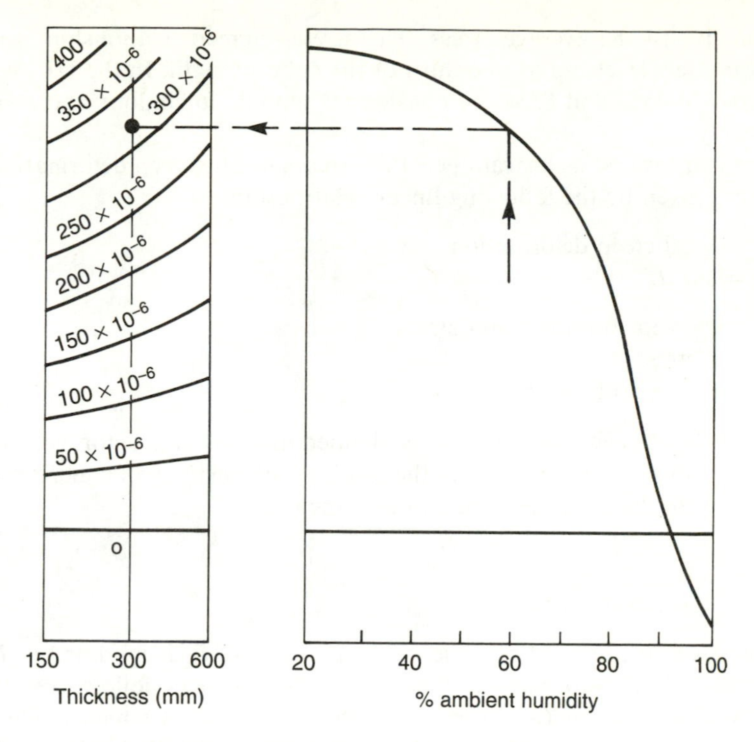

Figure 2.3 *Ultimate shrinkage of plain concrete*

moisture evaporation, such as in ground slabs laid on waterproof membranes, the effective thickness should be doubled.

Shrinkage strain is also affected by the reinforcement content of concrete. This is discussed in the chapter dealing with prestressing losses where it is of more immediate use.

ACI 318 combines the shrinkage strains with creep and its provisions are discussed in the next section.

2.2.5 Creep of concrete

Creep is the gradual change in length of concrete under a sustained load. It is a long term phenomenon, like shrinkage; it is quite rapid if the concrete is loaded at an early age and its rate is negligible after about 30 years. It is partly recoverable; if the loading is removed then after one year the recovery may be of the order of the strain corresponding to a stress of 30% of the actual stress reduction. For the UK conditions, for the concretes normally used in floors, it can be assumed that 40%, 60% and 80% of the final creep develops during the first month, 6 months and 30 months. For higher grades of concrete, say above 80 N/mm^2 (f'_c = 9000 psi), creep occurs at an earlier age.

The final creep is dependent on the same factors that affect shrinkage and, additionally, on the average stress. The stress-to-creep relationship is nearly linear for stress levels up to one-third of the cube strength. In the UK, relative humidities of 45% and 85% are considered suitable for indoor and exposure respectively.

Creep is measured as a strain per unit stress and the creep deformation in a member is given by the following linear relationship.

$$\delta_c = \text{total creep deformation}$$
$$= \varepsilon_c.\sigma.L \qquad (2.4)$$

where ε_c = creep strain per unit stress
σ = stress
L = length of member

Alternatively, the effect of creep can be defined in terms of a factor, called *creep coefficient*, which is used to modify the modulus of elasticity of concrete. Using this approach, the creep deformation in a member is:

$$\delta_c = C_c.(\sigma/E_c).L \qquad (2.5)$$

where C_c = creep coefficient

This approach is preferred because, once the value of E_c is modified to E_c/C_c, calculation of all deformations including deflection can follow the normal methods. The value of C_c varies as shown in Figure 2.4 which is based on recommendations given in BS 8110. The example shown dotted indicates that at 60% humidity for concrete loaded at an age of 28 days and of 300 mm thickness the creep coefficient is 2.0.

It is important to note that the deformation given by Equations (2.4) and (2.5)

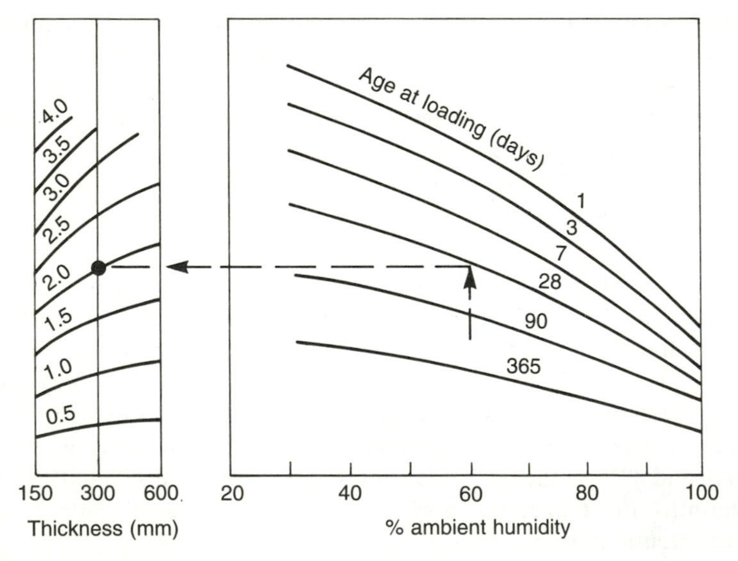

Figure 2.4 *Creep coefficient for plain concrete*

Table 2.5 *Creep coefficient* C_c (from ACI 318)

Load duration	≤ 3 months	6 months	12 months	≥ 5 years
Coefficient C_c	1.0	1.2	1.4	2.0

is that due to creep only and it does not include the immediate elastic deformation. If total deformation including the immediate elastic deformation is desired then C_c in Equation (2.5) should be replaced with $(1+C_c)$.

In ACI 318, for reinforced concrete members the multiplier is $C_c/(1 + 50.p')$, where p' is the compression reinforcement ratio. Here C_c depends only on the duration of loading as shown in Table 2.5. For post-tensioning the same value of the multiplier may be taken as for reinforced concrete.

2.3 Lightweight concrete

The term lightweight concrete applies to concrete containing lightweight aggregates, to aerated concrete and to no-fines concrete. Only the first mentioned concrete is generally used in post-tensioned floors.

Dense concrete, of 2400 kg/m^3 density (150 pcf), at present tends to be the first choice for a post-tensioned floor, because this is the traditional material, it is readily available and it is cheaper. Lightweight concrete, of 1800 to 2000 kg/m^3 density (110 to 125 pcf), is chosen in preference where it is important to save on the self-weight of a floor, such as on a site with poor ground conditions, or on very long spans where the self-weight would otherwise be the dominant load.

2.3.1 General

A lightweight reinforced concrete floor has a larger deflection than a normal weight concrete floor of the same section, because of its lower modulus of elasticity. In long spans where deflection is critical, the lightweight reinforced concrete floor may, therefore, have to be deeper. This limitation does not apply in post-tensioning, because prestress can be designed to cope with the otherwise marginally high deflection.

Using lightweight concrete, a post-tensioned solid floor can be shallower, and hence lighter than a normal weight concrete floor on two accounts—lesser depth and lesser concrete density. In a waffle floor the overall depth of the structure is determined by the beams; if the slab panel is made equal in depth to the beam, then it is deeper than it need be from the structural point of view. Using lightweight concrete, the saving in weight can be significant and, therefore, lightweight concrete can be a very efficient material in such an application.

Another reason for choosing lightweight concrete is environmental. Natural aggregates for use in dense concrete are being depleted in some areas whereas lightweight concrete often makes use of aggregates manufactured from waste products. Therefore, it is seen as a more environment-friendly alternative.

Lightweight aggregates are more expensive than natural gravel, and lightweight concrete needs more cement paste for the same strength than the normal weight concrete. These factors contribute to the higher unit cost of lightweight concrete.

Lightweight concretes have many alternative aggregates, with differing characteristics and, consequently, there is a larger range of variation in their properties. However, the properties for any particular type of lightweight aggregate can be easily and accurately established by tests. It is, therefore, important that the values given here are taken for general guidance only and that the characteristics are determined by tests on the particular concrete to be used. Alternatively, published data for the particular concrete may be used.

Characteristics of lightweight concrete are often defined in terms of those of the dense concrete of the same strength. Compared with dense concrete, it has a slower rate of development of strength, a higher shrinkage strain and a higher creep coefficient. These characteristics differ from the properties described earlier as desirable for post-tensioning. Nevertheless, measures can be taken to reduce the undesirable elements of their influence and lightweight concrete can be, and is, successfully used in post-tensioned floors.

The basic characteristic of all lightweight aggregates, natural or man-made, is their high porosity, which translates into high permeability. Lightweight concrete, compared with a normal weight concrete of the same mortar paste content, would be prone to a higher rate of carbonation and would be more permeable to chemicals, such as chlorides. These apparent drawbacks are, however, partly negated by the higher cement paste content in structural lightweight concretes designed for a given strength. Also, the elastic properties of mortar are often nearer to those of lightweight aggregate than natural aggregate; the mortar in a lightweight concrete, therefore, remains in a closer contact with the aggregate. It is much less prone to the development of micro-cracks at the aggregate-paste interface than normal dense concrete. The normal practice, however, is to increase the concrete cover to steel by about 10 mm (0.375 in). Car park floors, where salt may be used in winter, should be given some protection, as indeed those in dense concrete should be. With these precautions, there is no reason why lightweight concrete should be less durable, particularly in the internal environment of an enclosed building. In fact, lightweight concrete has been used in oil platforms where the environment is much more severe than in a normal building.

Lightweight aggregate has a lower resistance to abrasion because of the porous nature of its surface. It is advisable to use some form of surface finish in high traffic areas.

Lightweight concrete has a better fire resistance than concrete consisting of normal aggregates, because of its lower heat conductivity, and its lower rate of strength loss with rise in temperature.

For coefficient of thermal expansion, see Section 2.2.1.

2.3.2 Production and handling

Lightweight aggregates, being porous, have a larger capacity to hold moisture than normal aggregates; consequently, the workability of lightweight concrete is quite sensitive to its water content. If the moisture content of the aggregate is too low at the time it is placed in the mixer, then it rapidly absorbs water and the concrete mix is too harsh. On the other hand, if the moisture content of the aggregate is too high then the concrete is too wet and prone to segregation. Determination of the optimum quantity of water to be used in a concrete is made rather difficult by the differences in absorption capacities and rates of absorption of the various aggregates used in producing lightweight concrete. It is important to test the characteristics of the particular aggregate to be used, and to accurately measure its moisture content when the concrete is produced.

Care is also needed in handling lightweight concrete on site. It needs more vibration than normal concrete, because its lower density releases a correspondingly smaller amount of potential energy during compaction and because the lightweight aggregate tends to have a rough surface. On the other hand, the specific gravity of lightweight aggregate is lower than that of mortar paste and, therefore, a wet mix or excessive vibration causes segregation through the lighter aggregate floating upwards.

The workability of a dry, harsh mix is often improved by air entrainment, but this reduces the concrete strength and must be compensated for by more cement. Workability and compactibility can be improved with suitable plasticizers.

Lightweight concrete has a lower heat conductivity and, therefore, it has a steeper temperature gradient immediately after casting. The situation is aggravated by its high cement content because more heat has to be dissipated for a given volume of concrete.

Lightweight concrete is more vulnerable to cracking at an early age than dense concrete. Its rate of gain of strength in the first few days after casting is slower and it has a higher shrinkage strain. These factors can set up large tensile strains in the body of the lightweight concrete before it has gained sufficient strength. The problem is alleviated by care in handling, compaction and curing, and by application of some prestress at an early age. Two-stage stressing is commonly used for post-tensioned floors, usually at 15 and 25 N/mm^2 cube strengths ($f_c' = 1750$ and 3000 psi) for the commonly employed Grade 40 concrete ($f_c' = 4500$ psi).

2.3.3 Strength

Lightweight concrete used in post-tensioned floors generally has the same 28-day strength as that of dense concrete, i.e. in the range 30 to 50 N/mm^2 ($f_c' = 3500$ to 6000 psi); the most common being 40 N/mm^2 (4500 psi). Higher strengths are possible but are not in common use at present in post-tensioned floors.

The ratio of its tensile to compressive strength is of the same order as that for normal dense concrete. The elastic properties of lightweight aggregate are of the same order as those of cement mortar. Therefore, a better aggregate-mortar contact exists in lightweight than in normal concrete. Lightweight aggregates are weaker in strength and so tension cracks tend to go through the aggregate particles rather than around them as happens in normal concrete.

Aeration improves the tensile strength. The discussion on the rate of gain of tensile strength for normal weight concrete also applies to lightweight concrete, see Section 2.2.2.

For lightweight concrete, the modulus of rupture is taken as 1.8 times the cylinder splitting strength, if specified; if the splitting strength is not specified then the modulus of rupture is taken as 80% that of the dense concrete of equal strength.

The rate of gain of strength of lightweight concrete is similar to that for normal concrete for the same constituents in mortar paste. In lightweight concrete, Portland cement is sometimes partially replaced by other cements, and sand in fine aggregate may be replaced by other material, such as ground coarse aggregate. Such composite cements, and to a lesser extent the fine aggregate, affect the rate of development of strength.

2.3.4 *Elastic properties*

Poisson's ratio for lightweight concrete is assumed to be 0.2, the same as for dense concrete.

For modulus of elasticity of lightweight concrete, BS 8110 recommends the relationship given in equation (2.6).

$$\begin{aligned} E_c &= \text{modulus of elasticity for lightweight concrete} \\ &= E_{cn} \times (w_c/2400)^2 \text{ kN/mm}^2 \qquad (2.6) \\ &= 0.63 \times E_{cn} \text{ for } w_c \text{ of } 1900 \text{ kg/m}^3 \end{aligned}$$

where E_{cn} = modulus of elasticity for normal dense concrete having the same strength

w_c = density of lightweight concrete in kg/m^3

The Institution of Structural Engineers and the Concrete Society (1987) give a simple expression, Equation (2.7), relating the modulus of elasticity to the concrete density and cube strength.

$$E_c = w_c^{\,2} \times (f_{cu})^{0.5} \times 10^{-6} \qquad (2.7)$$

ACI 318 recommends a more direct relationship for the modulus of elasticity, as given in equation (2.8), for values of w_c in the range of 1500 to 2500 kg/m^3 (90 to 155 pcf), which covers concrete of normal density.

$$\begin{aligned} E_c &= (w_c)^{1.5} \times 43(f_c')^{0.5} \times 10^{-6} \text{ (N-mm units)} \qquad (2.8) \\ &= (w_c)^{1.5} \times 33(f_c')^{0.5} \qquad \text{(psi units)} \end{aligned}$$

For the commonly used density of 1900 kg/m^3, the relationship in equation (2.8) becomes $E_c = 3560 \times (f_c')^{0.5}$, which is equivalent to 0.75 times the value for dense concrete.

Equations (2.6), (2.7) and (2.8) yield very different values for E_c. If a value is available from tests on the particular lightweight concrete to be used then it should be used in preference.

2.3.5 Shrinkage and creep of lightweight concrete

Concretes containing lightweight aggregates have a higher capacity for moisture and so they are generally susceptible to larger initial and total shrinkage movement than those containing normal aggregate. The shrinkage for some concretes may be up to 40% higher. Creep of lightweight concrete also tends to be higher than that for normal concrete but not to the same extent as shrinkage.

Large variation of shrinkage and creep between concretes containing different fine and coarse aggregates preclude the presentation of meaningful guidelines. Characteristics for the particular concrete to be used should be obtained from the supplier, determined by tests or taken from published data.

The higher shrinkage and creep of lightweight concrete, compared with those of dense concrete, cause a larger long-term deflection. This, however, is more of a problem in reinforced concrete than in post-tensioned concrete, because in the latter it is possible to control deflection by a judicious combination of prestressing force and eccentricity.

The high creep also causes a greater loss in prestressing force. This has to be accepted and, therefore, the lightweight concrete requires a slightly higher initial prestressing force than would be required for a normal concrete member of the same section and carrying the same load. However, the smaller self-weight of lightweight concrete would require a smaller prestressing force. In practice the increase in the required prestressing force due to higher losses may be balanced by the saving due to lesser weight. The difference in prestressing force, if any, would be small and, if considered necessary, it should be possible to tailor the concrete section to make the best use of the tendons.

2.4 Post-tensioning tendons

In its early stages of development, prestressing was attempted by tensioning mild steel rods, which had a working stress of the order of only 140 N/mm^2 (20 000 psi). The problem in using such a low strength steel in a tendon was that a significant part of the prestress was lost as shrinkage and creep occurred in concrete and relaxation in steel. Concrete in compression may suffer a long-term strain of the order of 400×10^{-6} from shrinkage and creep, and prestressing steel would also undergo a corresponding strain. With a modulus of elasticity of 200 kN/mm^2, a strain of 400×10^{-6} is equivalent to a stress loss of 80 N/mm^2 in the tendon. In this instance, in mild steel with an initial stress of 140 N/mm^2 the loss of 80 N/mm^2 amounts to a 57% reduction of stress. A present-day high tensile tendon may have an initial stress of 1300 N/mm^2 (190 ksi) and a loss of 80 N/mm^2 in it is only 6%. The higher the initial stress in the tendon, the less significant are the shrinkage and creep losses. For this reason practical application of prestressing was held up until the production of high-strength steels.

Early prestressing tendons consisted of high-strength wires anchored singly or in multiples in suitable anchorage assemblies. Wires are still used in some post-tensioning systems but their most common use is in pretensioned precast products. In post-tensioned floors, *strand*—a rope comprising several wires—is almost universally used.

As stated in Chapter 1, the term strand can be rather confusing—it applies to the rope consisting of a number of individual wires wound together, it does not mean the individual wire comprising the rope. A tendon may consist of a single strand or several strands housed in the same duct (or sheathing). *Monostrand* tendons, consisting of a single strand, are used mainly in unbonded construction whereas *multistrand* tendons, for use in bonded applications, normally contain several strands. Multistrand tendons can be either circular or flat. The three tendons are shown in Figure 2.5.

The general properties of tendon steel are similar to those of rod reinforcement. It has a Poisson's ratio of 0.3, and a coefficient of thermal expansion of 12.5×10^{-6} per °C (7×10^{-6} per °F).

2.4.1 Strand

Strand, commonly in use in post-tensioned floors, is made from seven cold drawn high carbon steel wires. Six of the wires are spun together in a helical form around a slightly larger seventh straight centre wire. The strand is then given either a

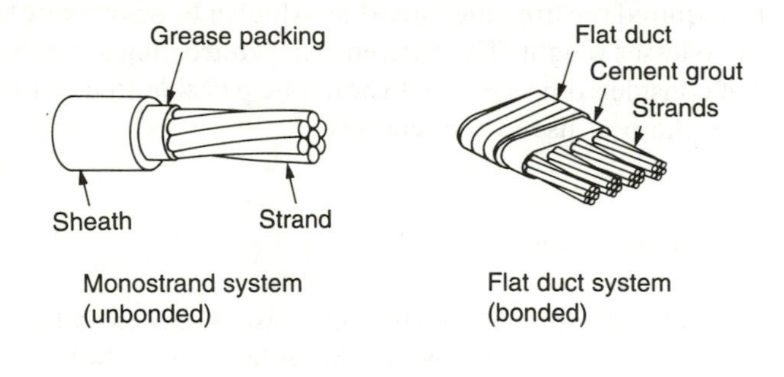

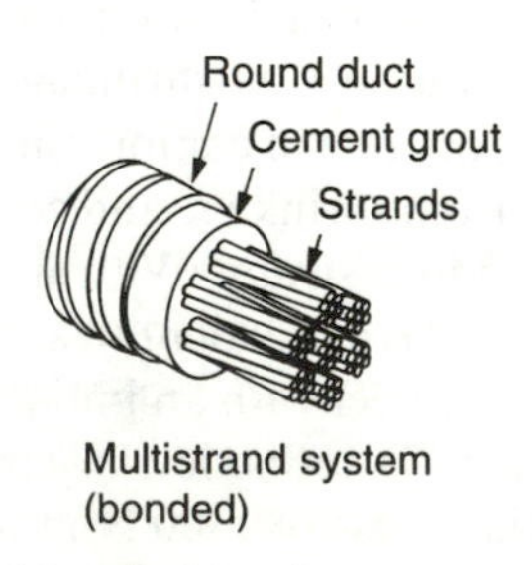

Figure 2.5 *Unbonded and bonded tendons*

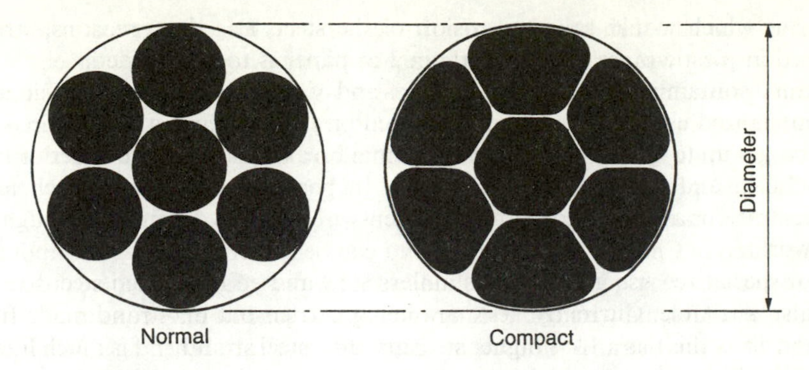

Figure 2.6 *Prestressing strand*

stress relieving treatment or it is run through a controlled tension and low temperature heat treatment process which gives it the low relaxation property. (Relaxation is discussed in some detail in Section 2.4.4.)

For a more compact type of product, the seven-wire strand is drawn through a die under controlled conditions of tension and temperature. In the process the individual wires are cold worked and compacted into the characteristic shape of the die-drawn strand. *Compact (or drawn) strand*, compared with normal strand, has a higher steel area for a given overall diameter and, therefore, it has a higher force to diameter ratio. Figure 2.6 shows the two types of strand in section.

Drawing a strand through a die increases its strength but reduces its ductility. The product is brittle and liable to failure without much warning. In order to avoid brittle and sudden failure the strand is annealed to increase its ductility.

For non-bonded application strand is impregnated with a protective fluid which penetrates to the centre wire, the strand is then coated with a corrosion resistant grease and finally a continuous polypropylene sheathing, of 0.75–1.00 mm minimum thickness, is hot extruded to cover the strand. The grease gives long-term protection and lubrication for ease of movement when stressing. The sheathing has a high impact resistance; it protects the strand from physical damage, and it prevents bond with the concrete.

In the early years of development of the unbonded system, strand was greased by hand and strips of waterproof paper were wound around the greased strand to prevent bond. The problem with this system was that the paper got damaged during handling, assembly of tendons, or during concreting by sharp aggregate or vibrators; this allowed the strand to develop local bond with concrete and led to difficulties with stressing. The next attempt consisted of pushing the greased strand through a plastic pipe; in theory this should have worked but subsequent exposure of such strand showed that water had found its way into the plastic pipe and caused serious corrosion of the strand. Water could have got in possibly while the tendon was stored on site, during concreting if the pipe got damaged, or while the tendon was waiting to be stressed and the anchorage to be protected. PVC was sometimes used to house the strand; when stressed, PVC may release chlorine which would cause corrosion of the steel. For these reasons, strand housed in plastic pipes or hand wrapped in paper is to be avoided.

Strand containing more than seven wires and of larger size is also manufactured for specialized use where large concentrated prestressing forces are required. Its size ranges up to 40 mm and it has a nominal breaking load of the order of 1000 kN (1.6 in and 225 kip approximately). In post-tensioned floors such force concentrations are not needed and the seven-wire strand is dominant throughout the world.

For specialized use galvanized, stainless steel and epoxy-coated steel strands are also available. Currently, tests are being carried out on strand made from carbon fibre; this has a 10% higher strength than steel strand and is much lighter in weight, but a reliable anchorage system suitable for commercial use is still to be developed. Other non-ferrous strands are also under development.

The ideal strand would consist of stable material, inert as far as possible against attack from chemicals in concrete and the environment, it would have a high strength, low relaxation, a low modulus of elasticity, high fatigue strength, a coefficient of thermal expansion similar to that for concrete, good fire resistance, not subject to brittle failure, and would be capable of being reliably and easily anchored. Quite a list, but many of the properties are already available in strand being produced for specialized use.

2.4.2 Corrosion of strand

Prestressing strand is susceptible to corrosion from the same chemical sources as rod reinforcement, such as oxidation and attack by chlorides. High-strength steel is also affected by hydrogen ions, which cause brittleness. Hydrogen is liberated by the chemical reaction of acids on iron and other metals which may be contained in the prestressing wires, and by the aluminium contained in high-alumina cements and in tendon grouts, where it was often added as an expanding agent.

The alkaline environment in the concrete, due to the presence of free lime, provides a very good protection to steel. The alkalinity is gradually lost through carbonation and then the steel becomes susceptible to attack. A small amount of hard oxidation also protects the steel and improves bond with the grout.

A rather dangerous phenomenon, characteristic of highly stressed steel, is *stress corrosion*, which causes a sudden fracture of the steel. Stress corrosion cracks result from the combined action of corrosion—from nitrates, chlorides, sulfides and other corrosive chemical agents—and high stress. Heat-treated wires are more prone to stress corrosion than drawn wires.

Strand, as used in post-tensioning, consists of small wires about 4 mm in diameter. Corrosion to a depth of, say, 0.1 mm represents a much greater loss of area in a 4 mm wire than it does in the case of a 20 mm reinforcement bar. For this reason, and to guard against stress corrosion, it is very important that prestressing strand is well protected.

Bonded tendons are safer in this respect, provided that the grout does not

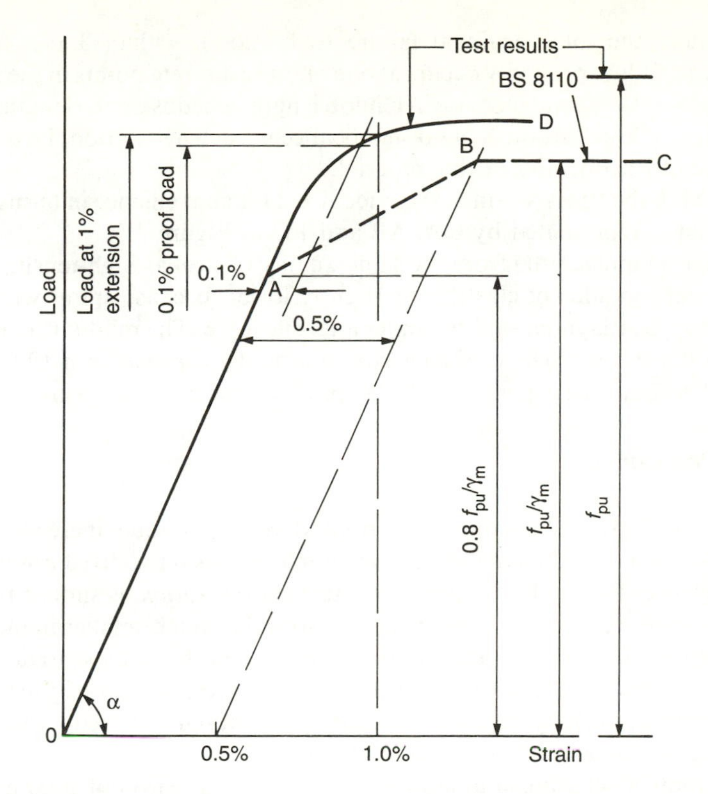

Figure 2.7 *Stress-strain curve for strand.* γ_m *is the partial safety factor for strength.*

contain any offensive chemicals and that no voids are left in the sheathing. Grouting should be carried out as soon after stressing as possible to avoid deleterious matter finding its way into the sheathing.

2.4.3 Stress-strain relationship

Curve OAD in Figure 2.7 shows the typical stress-strain relationship for strand. At low stress level the line is straight, then at some point it begins to curve which indicates onset of non-proportional strain. It then changes to a gentle curve, almost a straight line, leading up to the ultimate stress at failure.

The instantaneous modulus of elasticity is clearly defined by the slope of line OA, or tangent of angle α, but the point of onset of non-linearity is not clearly defined and, therefore, the stress-strain characteristic cannot be given in terms of the yield point which is used for many tensile materials. The load at 1% extension and the load at which 0.1% non-linear strain occurs (0.1% *proof load*) are used by manufacturers to control their process. High values of these two parameters indicate that the stress-strain relationship is linear in the normal range of stresses.

The total strain at failure may be of the order of 6% when measured in laboratory tests on 600 mm (2 ft) long samples subject to tension. ACI 318 and BS

8110 require the total strain at failure to be not less than 3.5%. In a real application, failure probably occurs at one or more discrete points in the length of a member and the actual increase in tendon length depends on the lengths of local yield zones. The 6% strain is of no significance in such a situation; however, it is indicative of the ductility of the material.

In BS 8110 the stress-strain curve is idealized for convenience in terms of three straight lines, represented by OA, AB and BC in Figure 2.7.

Strand is manufactured from the same quality of steel as high-tensile wire but its apparent modulus of elasticity is slightly lower, because spiral wires of the strand tend to straighten slightly under a tensile force. The modulus of elasticity (slope of the straight part of the curve) for strand varies between 190 kN/mm^2 and 200 kN/mm^2; the mean value of 195 kN/mm^2 is normally used in calculations.

2.4.4 Relaxation

Strand subject to a sustained tensile force undergoes a gradual increase in strain. Conversely, if a strand is stressed and anchored between two fixed points then it gradually loses force. The phenomenon, termed *relaxation*, is similar to that of creep in concrete, except that relaxation of strand is much smaller in magnitude than creep of concrete. The relaxation of stress in a particular strand depends on the initial stress, on ambient temperature and on the length of time. During manufacture strand may be given a heat treatment designed to reduce relaxation; these strands are called *low-relaxation strands.*

Relaxation is determined in laboratory tests over a period of 1000 hours (six weeks). This period is very short compared with the normal life expectancy of a building of 30 to 50 years. Extrapolating from laboratory tests, the relaxation of stress at 70% of the tensile strength at 20°C (68°F) maintained for a period of 500 000 hours (57 years) is expected to be of the order of 1.8% for low-relaxation (type 2) strands. For normal relaxation strands (type 1) the loss may be as high as 14%.

Relaxation rapidly increases with ambient temperature, as is evident from Table 2.6 which shows long-term characteristics for low-relaxation strand at an initial stress of 70% of strength. In ambient temperatures above 80°C (176°F) low-relaxation strand is not recommended, unless working loads are significantly reduced. At an ambient temperature of 30°C (85°F) low-relaxation strand would lose about 2.3% stress and in *normal relaxation strand* the loss would be over 20%.

In most buildings, temperature is controlled at a stable comfort level and strand is not subject to conditions which would cause high-relaxation loss. However, in buildings without air-conditioning, particularly in hot countries, it is important to use the appropriate loss figures.

Table 2.6 *Long-term relaxation and temperature at an initial stress of* $0.7f_{pu}$

Temperature	20°C	40°C	60°C	80°C	100°C
Relaxation	1.8%	3.5%	5.1%	7.5%	10.7%

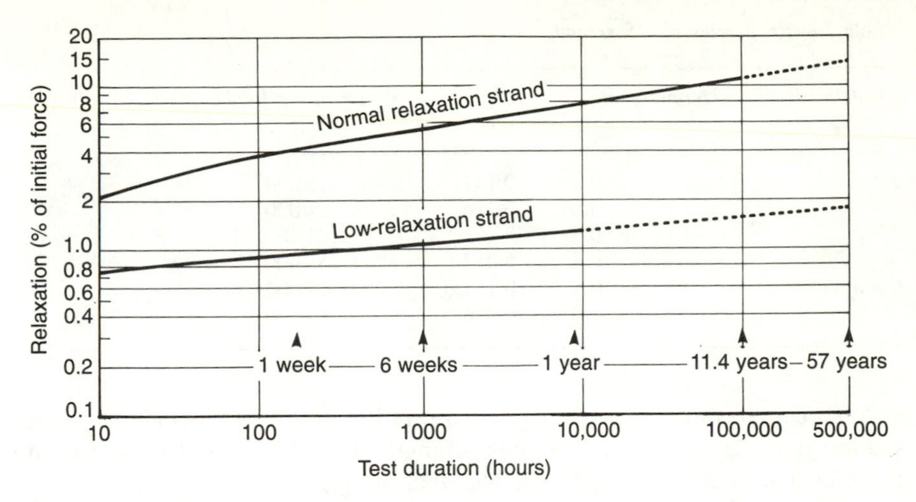

Figure 2.8 *Relaxation and time*

The relationship between time and relaxation plotted on a log-log scale is generally assumed to be linear. Figure 2.8 shows typical relationship for an initial load of 70% of breaking load at an ambient temperature of 20°C (68°F).

Figure 2.8 indicates that for normal strand the 500 000 hour (57 years) relaxation can be 2.6 times the 1000 hour value and for low-relaxation strand the ratio can be 1.6. For post-tensioned work, BS 8110 recommends values of 2.0 and 1.5 respectively.

2.4.5 Strand size and strength

Table 2.7 gives the data for British strand commonly used in post-tensioned floors.

It should be noted that for the same nominal diameter the strength and other properties of strand are very different for the three types and grades. It is important that strand is specified by type, diameter, strength and relaxation characteristics. Mixing of different strands in a member, particularly use of strands of the same diameter but of different types, should be avoided.

Table 2.7 *British low-relaxation strand*

Type	*Dia. mm*	*Area mm^2*	*Breaking load N/mm^2*	*Breaking load kN*	*0.1% proof load kN*	*Weight kg/m*
Standard	15.2	139	1670	232	197	1.090
	12.5	93	1770	164	139	0.730
Super	15.7	150	1770	265	225	1.180
	12.9	100	1860	186	158	0.785
Compact	15.2	165	1820	300	255	1.295
	12.7	112	1860	209	178	0.890

Table 2.8 *Properties of US strand*

Grade	*Nominal dia.* *in*	*Area* in^2	*Breaking load* *lb*
250	0.5	0.144	36 000
	0.6	0.216	54 000
270	0.5	0.153	41 300
	0.6	0.217	58 600
Compact	0.5	0.174	47 000
	0.6	0.256	67 440

American strand is very similar in size and strength. It is available in two grades: grade 250 which has a nominal strength of 250 000 psi and grade 270 with a nominal strength of 270 000 psi. The data are given in Table 2.8.

Wherever possible, particularly for strand which does not comply with British or American standards, strand characteristics should be obtained from the manufacturer. This also applies to its long-term relaxation and the effect of ambient temperature.

Strength of prestressing strand is seriously affected by a rise in temperature above 150°C; it falls almost linearly from 100% at 150°C to zero at 700°C. At low temperatures, the strength increases but ductility decreases. Typically, compared with the strength at 20°C, strand strength at −80°C is 5% higher and at −160°C it is 12% higher.

2.4.6 Transmission length

The length of bonded strand needed to develop a given force (usually taken as the initial prestressing force) in the strand is termed the *transmission length*. As discussed in Section 2.5.2, strand stressed from one end only is often bonded to the concrete at the other end. Except at bonded dead ends, the transmission length has no relevance in unbonded post-tensioning. In bonded applications, it indicates the distance beyond which the strand can be expected to have retained its force in case of damage.

Normal strand, with round wires, has a transmission length of about 30 times the strand diameter at a force of 70% of its breaking load when cast in concrete at a cube strength of 50 N/mm² ($f_c' = 6000$ psi). Compact strand has a smoother surface and a larger area of steel; its transmission length is 45 diameters. These lengths are for strand bonded in normal dense concrete; for lightweight concrete the lengths should be increased in proportion to the permissible bond stress. BS 8110 stipulates that for lightweight concrete the bond stresses should not exceed 80% of those calculated for normal-weight concrete; this amounts to an increase of 20% in transmission length for lightweight concrete.

2.5 Prestressing hardware

The prestressing hardware is available from a number of local and international manufacturers. Basically, it consists of anchorages (live, dead and intermediate, as discussed below), and sheathing for bonded systems. In order to protect the anchorage assembly from corrosion, manufacturers provide an integral protection system to suit their hardware. Grouting of the sheathing requires small tubes to be cast in the concrete which project at the top of the member and provide a path to the sheathing.

The following section gives typical details for the hardware, which would be of interest to the designer or the user. The sizes and descriptions are meant for general guidance only and it is recommended that actual data are obtained from the suppliers in the area.

Protection to an anchorage is required only at the live end, the dead end is cast in the concrete. The live anchorage is housed in a recess, or pocket, which is wide enough for the stressing jack and deep enough so that there would be adequate concrete cover to the assembly when the recess is made good. After stressing, the strand is cut off close to the face of the wedge using a disc cutter and the whole assembly is sprayed with a corrosion inhibitant. The assembly is then covered with a grease-filled cap and the recess made good with mortar containing a non-shrinking agent.

If a tendon is very long, then the losses at the far end may be high. Three options are possible: accept the high losses and use tendons at a lesser efficiency; provide live anchorages at both ends and stress the tendon from both ends; or use intermediate anchorages, called *couplers*.

Two-end stressing, while allowing strand to be used at a greater efficiency, uses two live anchorages (a live anchorage costs more than a dead anchorage) and requires more site operations. The choice between these alternatives would be made after considering the merits and demerits of each for the particular project.

2.5.1 Live anchorage

A few decades ago most manufacturers had their own patent devices for anchoring tendons. They included bars with threaded ends, enlarged ends of wire which passed through holes in thick plates such that the button sat on the plate, wedges, concrete male and female cones gripping a number of wires arranged in a circle, and a toothed conical wedge in a barrel. The cone and barrel system is now the most commonly employed device for post-tensioning floors and it is available from most manufacturers.

The anchorage casting is in direct contact with the concrete and is the component which transfers the prestressing force to the concrete. It is slightly conical in shape which allows strands to flare out for anchoring from their compact formation in the duct. The barrel is often an integral part of the anchorage casting in that the casting has sloping holes drilled in it, and machined

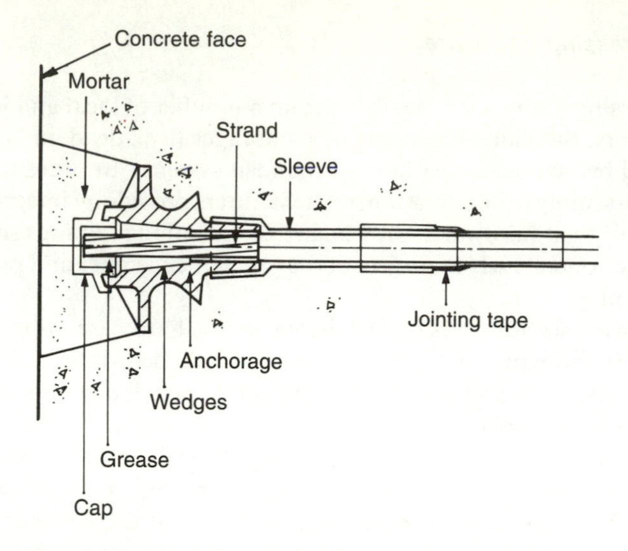

Figure 2.9 *Typical monostrand anchorage assembly*

smooth, in appropriate positions to take the wedges directly. Some systems use the casting to transfer the load to concrete and a separate heading block machined to accommodate the wedges.

Figure 2.9 shows a typical section at the end of a post-tensioned member containing a monostrand anchorage assembly, complete including the corrosion protection and Figure 2.10 shows a typical multistrand anchorage. Their representative dimensions are given in Tables 2.10 and 2.11.

Let us briefly look at the purpose of each component. A wedge (or cone or jaws) actually grips the strand; it is retained in a locked position by being pulled by the strand into a conical hole, either in a separate barrel or in a header block. The header block of a multistrand anchorage sits on the concrete, with a conical guide (sometimes referred to as a *trumpet*) which allows the several strands in a

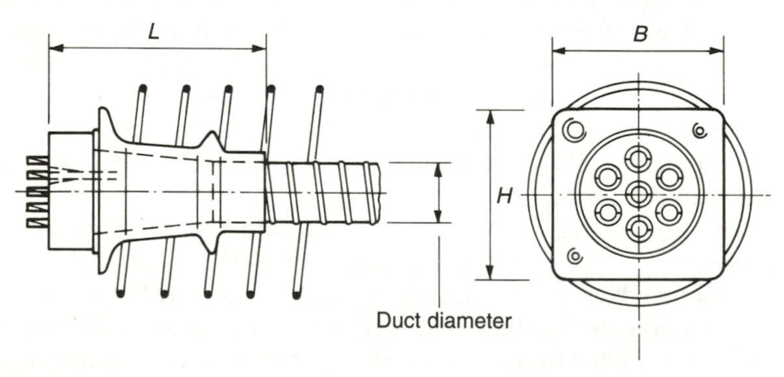

Figure 2.10 *Multistrand anchorage*

tendon to flare out to a spacing where they can be conveniently anchored. The guide may be either a casting or it may be fabricated from steel plate; it is cast into the concrete.

Some systems may have separate barrels into which the wedges fit. The barrels sit on a common thick bearing block which in turn bears on the concrete. This arrangement is now getting less common. Most systems have now discarded separate barrels; instead they use a steel block which has conical holes machined into it to receive the wedges and this block is seated on the guide.

As the strand is stressed, the wedge cone (consisting of two or three jaws) moves out of the barrel; at the required tendon force the cone is pushed into the barrel so that the strand is lightly gripped in the serrations of the cone. The hydraulic jack is then released; this action pulls the strand and the cone further into the barrel, by 6 to 8 mm (0.25 to 0.375 in) so that the strand is gripped tighter as a result. The 8 mm draw-in, of course, causes a loss in the prestressing force. The stressing operation is shown diagrammatically in Figure 2.11.

The mechanics of transferring the tendon force to concrete is worth a closer look as it brings to focus several important points which the designer should be aware of. There are three main components, or interfaces of components, whose integrity is essential for the reliability of the tendon.

Firstly, the wedge cone, which is made from particular grades of steel and is

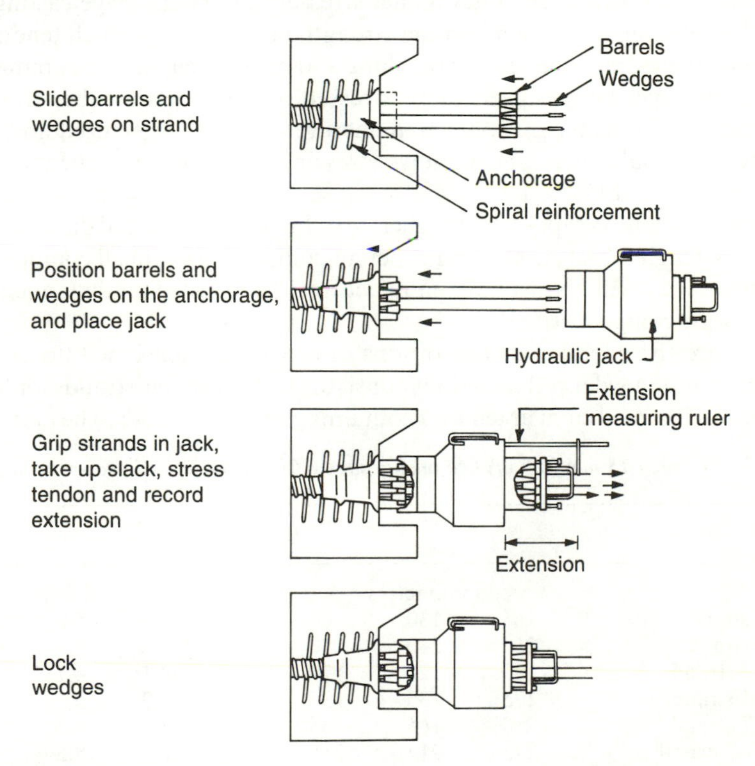

Figure 2.11 *Stressing of a tendon*

given a special heat treatment to harden its serrations. The cone wedge is longitudinally split into two or three pieces—depending on the manufacturer. If not split then, of course, it would not grip the strand. This perhaps is the most critical component in the anchorage assembly and most failures during stressing occur due to the serrations of the cone not gripping the strand. The serrations may shear if the steel is not of good quality, they may fail to bite into the strand if they have not been hardened properly, or they may fail because the cone angle is such that the strand does not come into contact with the whole of the toothed length in the cone. The importance of obtaining cones from a reliable source must be emphasized.

Secondly, the cone slides in the conical hole in the barrel (or the anchorage block) producing bursting stresses in the barrel cylinder; it is essential that the sliding surfaces are smooth, their slopes match exactly and that the barrel can withstand the bursting force with adequate margin of safety. It is unusual for a barrel, or a casting, to fail in bursting but all too easy to use a cone for the wrong size of strand—as shown earlier, strands of the same nominal diameter differ from each other in fractions of a millimetre in diameter.

Thirdly, the flanges of the anchorage casting should be strong enough not to fail and yet large enough not to overload the concrete. The two requirements are rather conflicting in that to reduce the bearing stress on concrete a large contact area is required, which generates higher stresses in the anchorage casting itself. This also sets a limit for the minimum strength of concrete at which tendons can be stressed. Bearing stresses approaching concrete strength are common and most anchorages are designed to be stressed at a minimum cube strength of 25 N/mm^2 ($f_c' = 3000$ psi). Some manufacturers offer a larger sized anchorage, usually to special order, which allows stressing at a concrete cube strength of 15 N/mm^2 ($f_c' = 1750$ psi).

Compact strand is capable of a higher force for a given nominal diameter and it generates a higher bearing stress in concrete, and, of course, in all components of the anchorage assembly. Not all manufacturers have tested all their anchorages for use with compact strand.

There are three tendon configurations in use in post-tensioned floors: single strand for unbonded use, flat tendons consisting of four or five strands for bonded use, and multistrand circular tendons containing up to 12 strands. The multistrand

Table 2.9 *Typical 13 mm (0.5 in) live anchorages (VSL Monostrand prestressing system)*

Type	$L \times B \times H$					
		millimetres			inches	
Single strand	65	130	60	(2.5	5.0	2.3)
Flat 5 strand	260	240	70	(10.2	9.5	2.8)
Multi 3 strand	180	120	120	(7.1	4.7	4.7)
Multi 4 strand	175	135	135	(6.9	5.3	5.3)
Multi 7 strand	210	165	165	(8.3	6.5	6.5)
Multi 12 strand	275	215	215	(10.8	8.5	8.5)

Table 2.10 *Typical 15 mm (0.6 in) live anchorages (VSL Monostrand prestressing system)*

Type	$L \times B \times H$					
		millimetres			*inches*	
Single strand	65	130	60	(2.5	5.0	2.3)
Flat 5 strand	277	240	70	(10.9	9.5	2.8)
Multi 3 strand	175	135	135	(6.9	5.3	5.3)
Multi 4 strand	210	150	150	(8.3	5.9	5.9)
Multi 7 strand	230	190	190	(9.1	7.5	7.5)
Multi 12 strand	320	250	250	(12.6	9.9	9.9)

tendons are designed for a maximum of 3, 4, 7, or 12 strands of 13 mm and 15 mm nominal diameter. Anchorages suitable for larger tendons are manufactured but for other uses because the magnitude of concentrated force available from them is not needed in post-tensioned floors. The anchorage sizes vary between manufacturers, Tables 2.8 and 2.9 show the typical. At the design stage, it is advisable to check the actual dimensions of the hardware available in the area.

It is not necessary to have seven strands in a 7-strand anchorage; it can accept any number up to a maximum of seven. Of course if the number of strands is less than the capacity of the anchorage then it is being used at a reduced cost efficiency.

2.5.2 Dead anchorage

General remarks about bearing stresses and anchorage dimensions (Section 2.5.1) are also applicable to dead anchorages where the same casting is used.

At the tendon end where access is not required for any operation after concreting, the anchorage assembly is cast in concrete. This assembly can be a simpler device because all that is required of it is securely to hold the end of the tendon and transfer the force to the concrete. It must, however, be reliable. If a wedge cone slips at the live end it can be replaced. If a dead anchorage fails then the choice is either to abandon that tendon or to cut the concrete out to gain access to the anchorage and try to replace the part that has failed. One may not be acceptable from the design point of view and the other would require an unscheduled site operation.

A pre-locked live anchorage can be, and often is, used at the dead end. Being inaccessible, it would be unwise to use the normal wedge cone for securing the strand, even if it is pre-locked; a higher grade wedge is preferred. Most of the prestressing systems offer a cheaper and more secure dead anchorage where a purpose-made barrel is swaged on to the strand end at their works. The swaged barrel bears on a steel plate with holes drilled for the strands to pass through. The plate transfers the load to the concrete. A design advantage of such a dead end is that the concrete is prestressed almost to the end of the floor.

In a variation on the above arrangement, Figure 2.12(a), the sheathing is also cut short to allow the strands to spread out, the strands are bonded over the

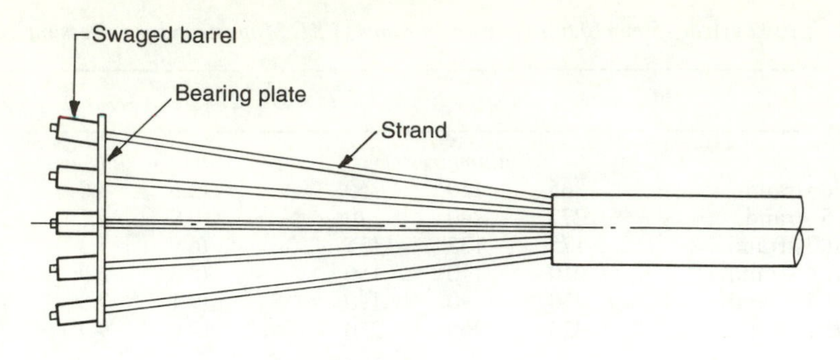

(a) Swaged barrel and plate

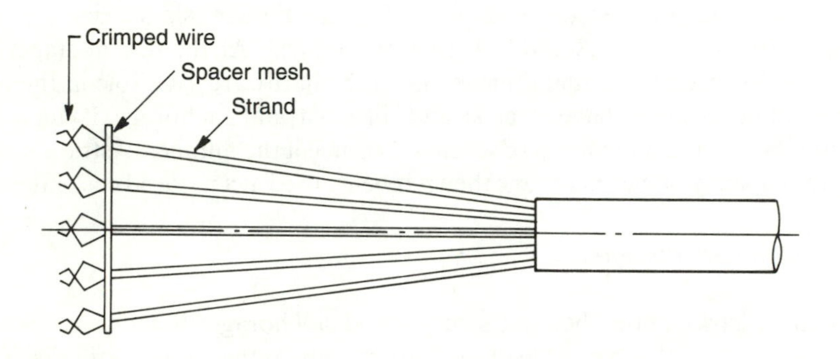

(b) Unravelled strand with crimped wire ends

Figure 2.12 *Dead-end anchorages*

exposed length. The overall length of this type of dead anchorage may be of the order of 400 mm (16 in). The concrete in this arrangement is not prestressed right to the end, but it is cheaper.

Where it is not essential for concrete to be prestressed right to the end, the sheathing may be cut short and the exposed length of the strand bonded without the swaged barrel and the steel plate, Figure 2.12(b). In order to reduce the bond length, the surface area of strand may be increased by unravelling the seven individual wires forming the strand. Sometimes the ends of the unravelled wires are also crimped to provide a mechanical anchorage. The length of strand bonded in concrete varies between different suppliers, but usually it is of the order of 900 mm (3 ft). The unravelled strand dead end is not recommended for unbonded use, because the grease and the lubricant are difficult to remove from the impregnated strand and their presence impairs bond.

Another method of saving on the number of dead anchorages is to loop the tendons where the dead ends would have been, as shown in Figure 2.13. The live

Figure 2.13 *Plan showing looped dead ends*

anchorages can be arranged either along one edge of a floor or along both edges.

2.5.3 Couplers and intermediate anchorages

In long lengths of slab, it is often convenient to cast the concrete in several operations. Each length may consist of several sections of different ages and the older one may be approaching the onset of shrinkage. It is desirable to apply prestress to each section in succession. This reduces the possibility of shrinkage cracks and, because shorter tendon lengths are now being stressed, the prestress losses are smaller.

The term *coupler* applies to a mechanical device which allows a new tendon to be coupled to the end of a tendon which has already been locked. An *intermediate anchorage* is a device which anchors a tendon in the middle of its length at a convenient construction joint and the same tendon continues past the intermediate anchorage. The latter is convenient only for single-strand systems and is used mostly in unbonded applications.

An intermediate anchorage, for use with unbonded systems, consists of a slotted plate which is dropped over the bare strand. The strand is then stressed and anchored using the standard cone and barrel. An unavoidable inconvenience is that the strand has to be threaded through the barrel, and, of course, the length of strand for the next part of the floor is stored in a coil near the intermediate anchorage. An ordinary jack would be almost impossible to use because the strand passes through a hole in its centre and threading this over a long length of strand is not practicable. A special twin-cylinder jack with an open throat for passing the strand through is needed for stressing at intermediate anchorages. An intermediate anchorage can be rather awkward to use and a coupler may be preferred in certain situations.

Couplers for multistrand tendons consist of an assembly with a number of holes into which the anchoring wedges fit and the same number of slots around the circumference. The strands are stressed and anchored in the normal manner using the wedges and then new strands with swaged barrels are placed in the slots. The whole assembly is concreted in with the next section. Approximate dimensions of typical couplers are given in Table 2.11.

Couplers for use with flat tendons are similar in principle. They are oval in

Table 2.11 *Multistrand couplers*

Nominal tendon size		*Length*		*Diameter*	
		mm	*(in)*	*mm*	*(in)*
3 × 13 mm	(3 × 0.5 in)	450	(18)	130	(5)
7 × 13 mm	(7 × 0.5 in)	550	(22)	170	(7)
12 × 13 mm	(12 × 0.5 in)	650	(26)	200	(8)
3 × 15 mm	(3 × 0.6 in)	500	(20)	150	(6)
7 × 15 mm	(7 × 0.6 in)	625	(25)	200	(8)
12 × 15 mm	(12 × 0.6 in)	725	(29)	240	(10)

shape, typically 330 × 110 mm (13 × 5.5 in); their length varies from 750 mm to 1250 mm (30 to 50 in) depending on the number and size of strands.

A monostrand coupler consists of a special anchorage casting with a female threaded projection. The strand is stressed and anchored in the normal manner. The end of the strand to be coupled passes through a male threaded block and the strand end is swaged in the manner of a dead anchorage. The block is then screwed into the projection of the special anchorage.

2.5.4 Sheathing and other hardware

Sheathing, or duct, for use in bonded post-tensioning is usually made from galvanized steel and is corrugated so that it is easy to bend to the required tendon profile while retaining a high radial collapse strength. Some manufacturers supply plastic sheathing. Sheathing for flat duct is rather difficult to manufacture with corrugations and sometimes it is supplied in rigid short lengths. Being only about 20 mm (0.75 in) deep, it can be bent to suit a tendon profile but, obviously, not as easily as a corrugated one.

Table 2.12 shows typical overall sizes for sheathing; the exact dimensions vary for different manufacturers. Normally, the inside area of a sheathing is not less than twice the area of the tendon to be housed in it. The clear inside dimension may be 5 mm to 7 mm (0.2 to 0.3 in) smaller than the overall diameter for round multistrand sheathing and two or three millimetres smaller for single strand and flat sheathings.

Table 2.12 *Typical sheathing diameters (outside)*

	13 mm	*(0.5 in)*	*15 mm*	*(0.6 in)*
Single strand	16	(0.63)	20	(0.80)
Flat 5 strand	70 × 19	(2.8 × 0.75)	70 × 19	(2.8 × 0.75)
Multistrand 3	45	(1.75)	50	(2.00)
Multistrand 4	50	(2.00)	55	(2.17)
Multistrand 7	60	(2.38)	67	(2.63)
Multistrand 12	72	(2.83)	87	(3.42)

As strands in curved profile are stressed, they tend to bunch up into the curve, thereby losing some of the eccentricity. Flat tendons are more suitable for use in slabs than their multistrand round equivalents, because of their smaller loss of eccentricity. For example, consider two tendons, each containing four 13 mm (0.5 in) strands, one in a flat sheathing and the other in a circular sheathing. For a given concrete cover, the height of centroid of steel above the sheathing soffit is:

Sheathing	*Radius*		*Bunching*		*Total*
Flat	9.5	+	2	=	11.5 mm (0.45 in)
Multistrand	22.5	+	9	=	31.5 mm (1.24 in)

However, a flat sheathing is more difficult to bend in a horizontal plane than a round sheathing. The latter should be preferred if the bonded tendons have to be curved horizontally, for example to avoid a hole in the floor.

Tendons need to be supported at regular intervals. Bonded tendons are more rigid than monostrand but they weigh more. The normal practice is to support bonded tendons at not more than 1.2 m centres and monostrand unbonded at 1 m (say 4 ft and 3 ft respectively).

The minimum radius of a bonded tendon is normally governed by the capacity of the sheathing to bend without suffering any damage. It may be larger than the 2.5 m (8 ft) normally considered a minimum for unbonded tendons. Curvature capacity data for the sheathing should be obtained from the supplier.

2.6 Equipment

The specialist equipment required for post-tensioning consists of the following items, not all of which would be required on site.

Stressing jack and pump
Swaging jack
Strand threading machine
Strand cutters or shears
Grout mixer and pump

The strand threading machine and the grouting outfit are required only for bonded tendons. Strand may be threaded into the sheathing either before placing the sheathing in position, or after it has been placed but before concreting, or after concreting. The pushing machine would, of course, not be needed on site if the tendon lengths are short enough to be delivered to site ready threaded.

Jacks are designed to grip the strand(s) either at the front or at the rear. In the latter case the design of jack is mechanically simpler but it needs an extra length of strand (of the order of one metre but check for the particular jack to be used) which is cut off after stressing. The older jacks did not have the facility for pushing the wedge cone forward after stressing and this used to be a manual operation; now almost all jacks automatically move the wedges forward into the barrel to lock the strand as part of the automated stressing operation. Jacks are equipped with facility for measuring extension of strand and are calibrated with their hydraulic pumps so that a direct reading of jacking force is displayed.

Apart from manoeuvrability with regard to the handling equipment and space available at a particular site, the important features of a jack are its force capacity and the stroke. Long tendons, if stressed with a jack of short stroke, may have to be stressed in stages. The process requires the anchoring cones to grip the strand and release it several times, which may weaken the serrations.

The sizes and weights of equipment vary in different systems but those for monostrand use are very similar in shape and size. These and their pumps are light enough to be handled manually. Flat tendons are usually stressed using the monostrand jack.

Multistrand jacks from different suppliers differ in shape, size and weight. Some are of a similar shape to the normal monostrand jack, but bigger, while others are much larger in diameter—up to 400 mm (16 in)—and shorter. A multistrand jack is much heavier than a monostrand jack, it may weigh as much as 300 kg (650 lb). It therefore needs crane time during stressing. Because of the wide variation in equipment from different sources, no details are given. It is recommended that particulars are obtained from specialists in the area.

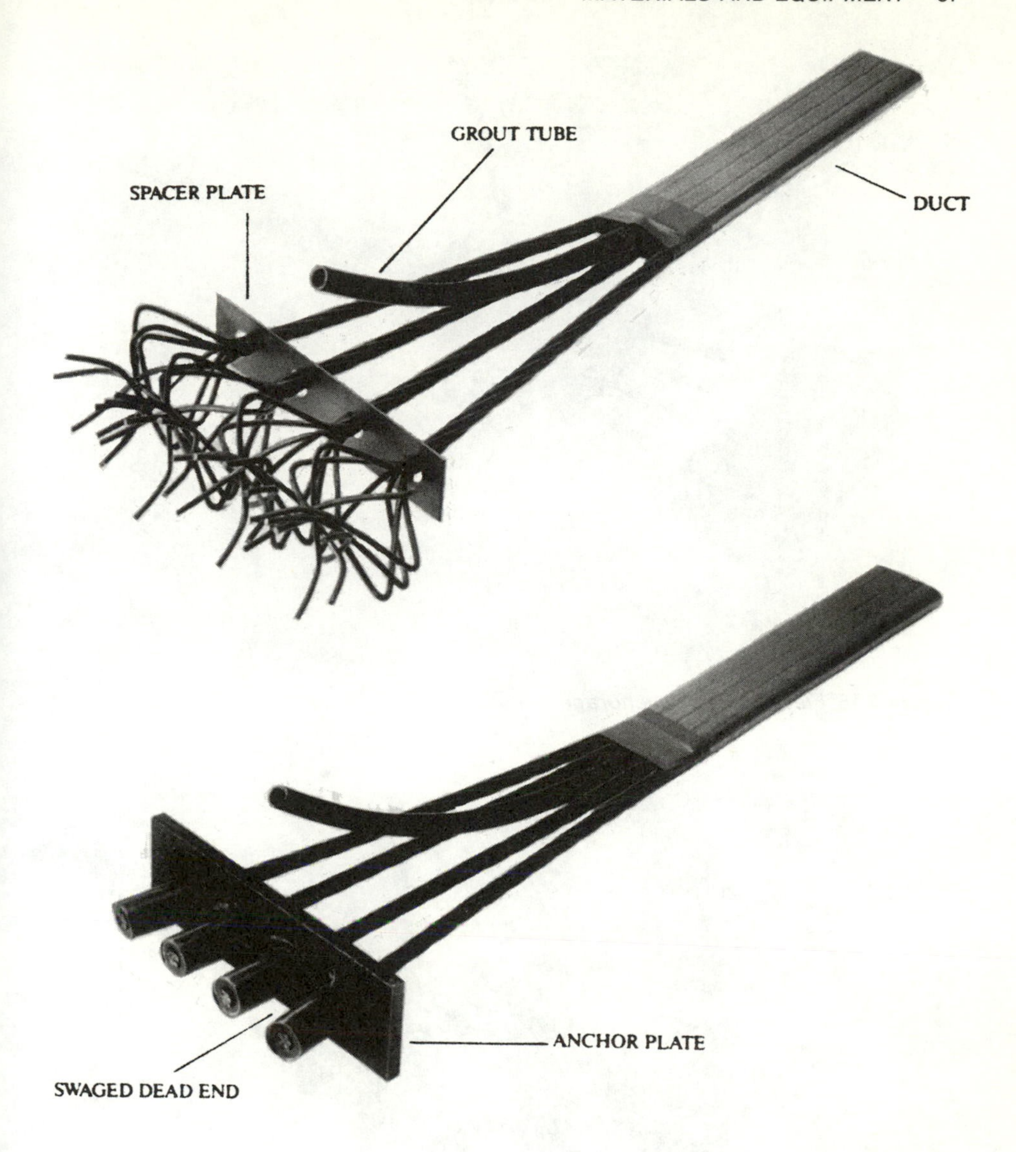

Figure 2.14 *Dead end anchorages*

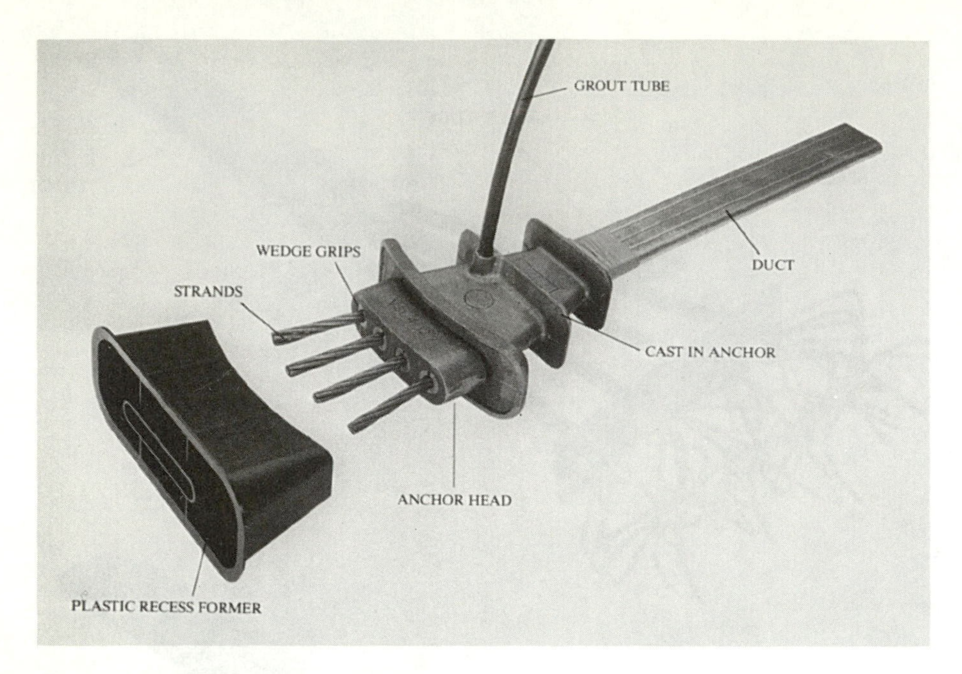

Figure 2.15 *Flat stressing anchorage*

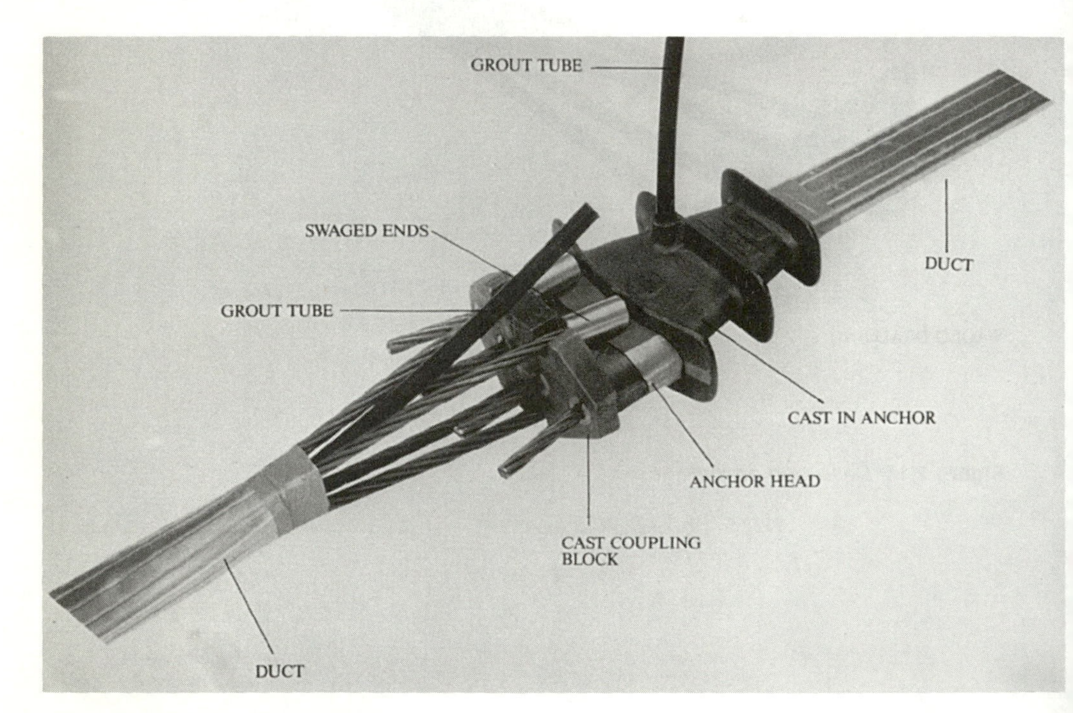

Figure 2.16 *Flat coupler*

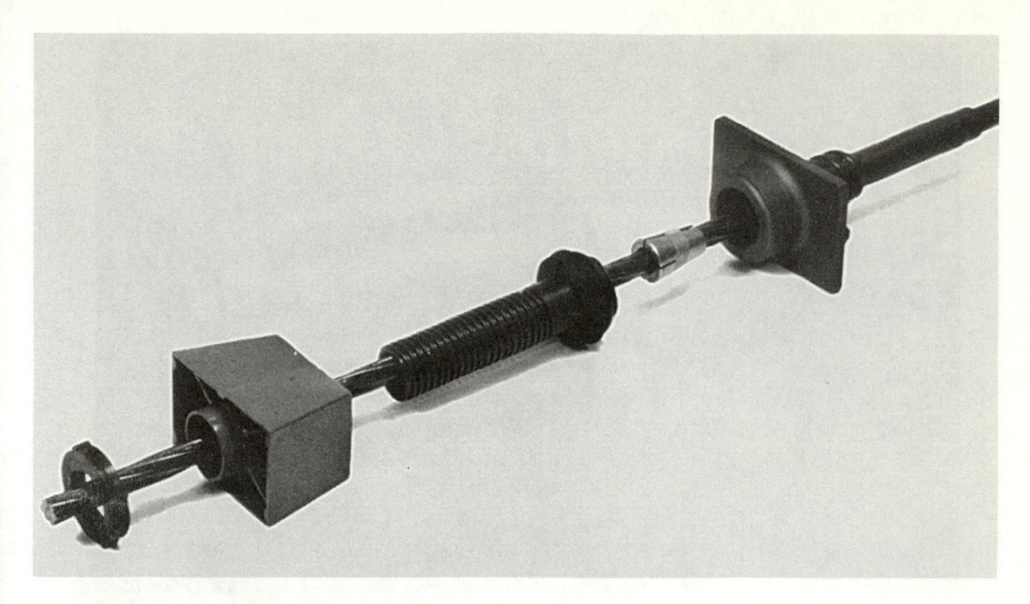

Figure 2.17 *Exploded view of monostrand anchorage*

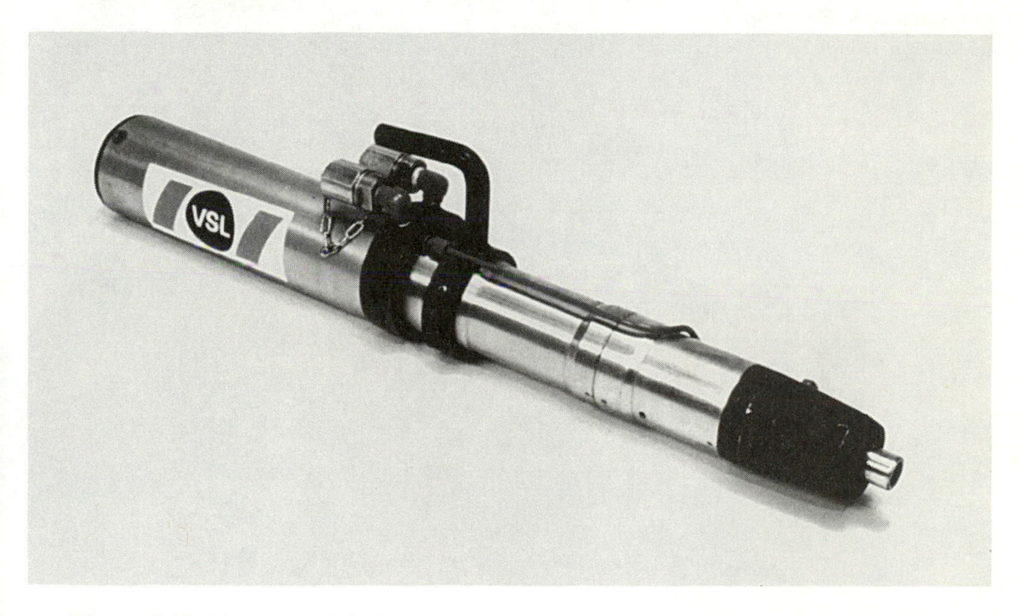

Figure 2.18 *Monostrand Jack*

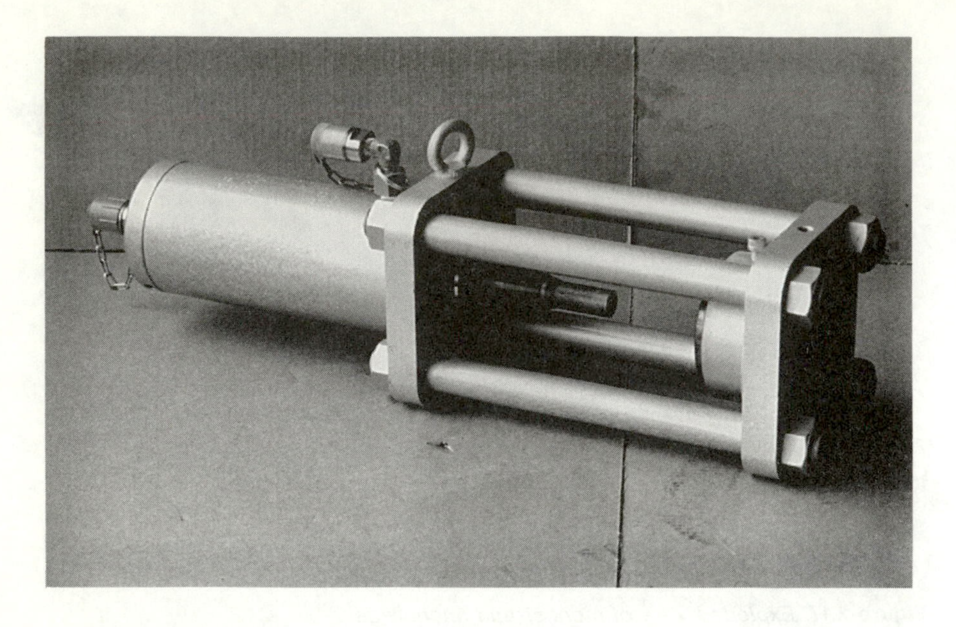

Figure 2.19 *Swaging Jack*

3 SLAB CONFIGURATION

A floor system consists of various arrangements of slab and beam elements, shown in Figure 3.1. These elements, and some of the commonly used arrangements, are discussed in this chapter. Possible applications of the various configurations are also suggested. These are meant only for general guidance; in fact, there is a considerable overlap in the applications.

3.1 General

Concrete floors have a variety of shapes. These can be summarized as:

- flat slabs, with or without drop panels
- ribbed and waffles floors
- beam and slab construction

In reinforced concrete the term *flat slab* is used to refer to a slab with or without drops and supported, generally without beams, by columns with or without column heads. It is understood that a flat slab will be designed spanning in two directions, supported on beam strips of the same depth as the slab. It may be solid or ribbed in both directions; the latter is called a *waffle* slab. In designing a flat slab, each panel is divided into column and middle strips in each direction, which are analysed and reinforced in accordance with the national standards.

In post-tensioned concrete, there is no generally accepted definition of a flat slab and the term is applied to any slab which has a flat soffit, with or without drop panels. The panels may be designed as column and middle strips (though this is not very common), as one-way spans supported on strings of strip beams, or as two-way spans supported on a grid of strip beams. A *strip* or *band* beam is a strip of slab in line with the columns which has been made strong enough to support the adjacent slab panels. A flat slab in post-tensioning does not describe any particular structural system in the same manner that a reinforced concrete flat slab does. The distinction is purely visual. The term *waffle* is also applied to post-tensioned floors with ribs running in two directions.

The beam and slab configuration needs a greater construction depth than other floor configurations, and is not generally used. The arrangement can be one-way spanning or two-way, and the design procedure is the same as that for other one-way spanning configurations.

From the structural design point of view, it is perhaps more meaningful to classify floors according to their internal structural system, such as:

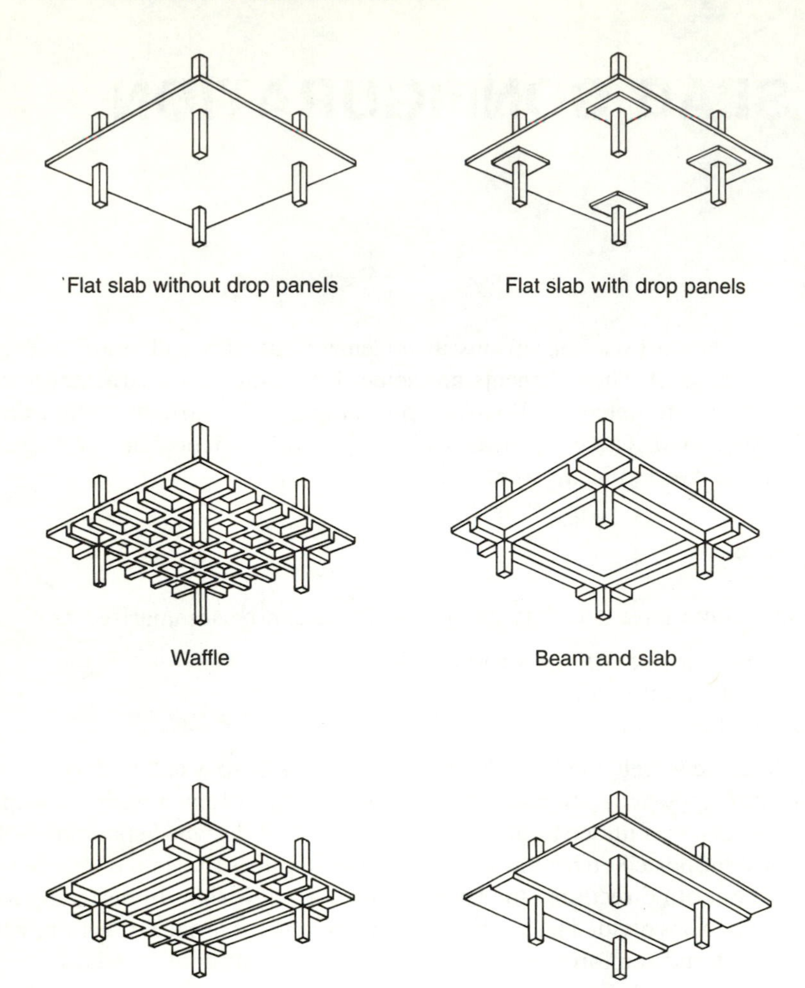

Figure 3.1 *Slab configurations*

- one- or two-way spanning
- solid or ribbed
- which elements are post-tensioned

Each combination of the above features has its merits and its uses; none can be said to be inherently better than the others in all circumstances and all locations. The choice between them is a matter of the requirements of the particular application, and of economy.

Figure 3.2 shows some of the typical tendon layouts associated with the various slab configurations. Diagrams (a) to (c) are suitable for one-way

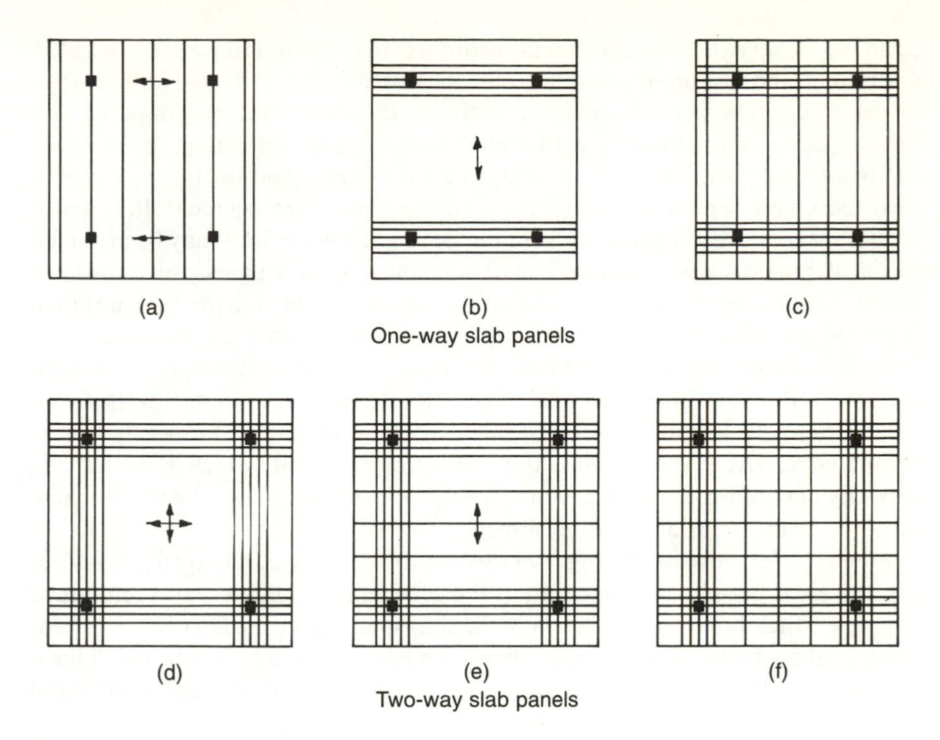

Figure 3.2 *Tendon layouts (arrows indicate reinforced concrete spans)*

spanning slabs, and (d) to (f) for two-way slabs. For further comments on the associated panel configurations see Section 3.3.

Tendons in a slab panel are either uniformly spaced across its width, or two or three tendons are grouped together and the groups distributed uniformly. A greater flexibility for making holes is achieved by keeping the anchorages uniformly spaced and then bunching two or three tendons together away from the anchorage points; this, of course, is not possible for short tendon lengths. Beam tendons can also be grouped together in a similar manner but the spacing of even the bunched tendons remains close and only small holes can be provided.

In Figure 3.2(a), only the slab is prestressed, it spans on either reinforced concrete beams or load-bearing walls; so the tendons are uniformly spaced and run in one direction only. This arrangement is used either where the beams are shallow and span short distances within the capacity of reinforced concrete, or where walls or deep reinforced concrete beams are acceptable; essentially it suits rectangular panels of about 2:1 aspect ratio.

In Figure 3.2(b) the slab is in reinforced concrete and spans on post-tensioned beam strips. The arrangement is suitable for approximately square panels, where strip beams are required and the span is too long for reinforced concrete. Some of the axial component of the prestress is absorbed by the adjacent reinforced concrete slab. It is necessary to assess the proportion thus lost from the beam and to check the serviceability stresses accordingly. In order to reduce the loss, a few additional tendons running parallel to the beams are often provided in the slab at its centroid. These tendons do not directly contribute to the strength of the slab but are useful in distributing load concentrations and controlling shrinkage.

Figure 3.2(c) is a combination of (a) and (b), the slab and the beams are both post-tensioned. It perhaps is the most commonly used arrangement. In order to reduce the loss of axial prestress from the beams, a few tendons may be provided in the slab as discussed above. The arrangement suits a panel aspect ratio of about 1:5 if the beam strips are to be contained within the slab depth; if not then the arrangement can be used with any aspect ratio up to a square panel.

Figure 3.2(d) shows the tendon arrangement where two-way reinforced concrete slab panels span on a grid of post-tensioned beams. Some of the beam axial prestress is lost in the slab. This arrangement is suitable for nearly square panels of relatively short spans, not exceeding say 8 m, which is within the capacity of reinforced concrete. This arrangement is found in waffle floors where only the beam strips are post-tensioned.

Figure 3.2(e) shows an arrangement where the slab spans in two directions but is prestressed in only one direction; in the other direction it acts as a reinforced concrete panel. It may be of use where the slab span is longer than convenient for reinforced concrete and some assistance from post-tensioning is required. This is a truly partially prestressed slab and a careful assessment of crack widths and deflection is required with this arrangement.

Figure 3.2(f) shows the arrangement for a two-way post-tensioned slab consisting of nearly square panels. It is suitable for solid slabs where a minimum construction depth is required. The arrangement is also used in waffle floors. It requires the tendon assembly sequence to be carefully worked out because the beam and the slab tendons are to some extent inter-woven. For reasons of difficulty in installation, this arrangement is not preferred.

3.2 Structural elements of a floor

A floor system essentially consists of slab and beam elements. A slab can be solid or ribbed, and one-way spanning or two-way; it may, or may not, have drop panels over the columns. A beam can be a downstand, an upstand or a strip. The various elements are discussed in this section.

3.2.1 Ribs in concrete floors

Ribbed floors are the most commonly used alternative to solid floors. They require less material and the reduced weight leads to further savings in the columns and substructure. Ribs running in two directions, of course, constitute a waffle floor. This section is equally applicable to one and two-way spanning floors, i.e., to solid and ribbed slabs and to solid and waffle slabs.

In reinforced as well as in post-tensioned concrete, it is the normal practice to avoid shear reinforcement in ribs. The section is changed near the support from ribbed to solid at a point where shear, or the support moment in the case of a continuous floor, is liable to be critical. This means that the ribs are almost never subjected to compressive stresses induced by applied loads. The discussion below, therefore, assumes that the rib is on the tension face of the floor.

At the ultimate stage of loading a ribbed floor is identical to a solid floor in both reinforced concrete and in post-tensioned construction as long as the depth of the compression block remains within the topping. From the serviceability design point of view, however, the structural function of a rib in reinforced concrete is markedly different from that in post-tensioned concrete.

First, consider a rib in reinforced concrete in the positive moment region of a floor. At midspan, where the moment is maximum, the section carries zero shear and the only function of the rib is to connect the topping concrete (the compression zone) and the rod reinforcement (the tension element). It does not directly contribute to the strength of the floor, and can be considered to serve as a separator between the compressive and tensile components and as a protection to steel against corrosion and fire. At any other point in the positive moment region, the rib carries some shear force corresponding to the change in moment but again it makes no direct contribution to the flexural strength. Therefore, in reinforced concrete there is no need to make the rib wider than strictly necessary to accommodate the reinforcement, provided of course that the shear force remains within its capacity.

Now consider a rib in a post-tensioned floor. At the time of stressing, the prestressing force induces a compressive stress in the rib. Later, when the slab is required to carry the imposed load, the flexural tension has to overcome this compression before the rib goes into tension. The rib effectively acts as a reservoir of compressive stress which it releases as and when required. Therefore, in post-tensioning the rib width, or rather the ratio of rib width to rib spacing, makes a direct contribution to the flexural strength.

How much compression should be stored in this manner? The magnitude of the applied moment represents the range of stress variation. At the optimum value of prestressing force in an ideal section, when fully loaded, the rib would have a tensile stress exactly equal to the permissible value. The concrete and the prestress would probably be most efficiently used if under permanent load the stress in the bottom fibre of the rib were equal to the permissible limit in compression and under applied loading the stress equalled the maximum allowed tensile value, though this may not be the cheapest design. In any case, such a solution is unlikely to be achieved in practice.

The self-weight of the concrete floor is normally balanced by the upward reaction from the tendon curvature; the prestress provided is often enough to balance all of the dead load and part of the applied load. Therefore the rib reservoir need not store any compression for self-weight of the floor and, in this particular discussion, self-weight can be considered to be of no importance. If the applied load is large then a correspondingly larger compression must be stored in the rib and this would need a wider rib, assuming that the depth is adequate.

Therefore, to a large extent the ratio of rib width to spacing depends on the magnitude of applied load.

If a post-tensioned ribbed section is found to be inadequate then two options are available—increase the rib width and/or the depth. In reinforced concrete, the rib width makes no direct contribution to its flexural strength, and so it is only the depth that can be increased.

The distinction between a rib and a beam is not clearly defined; it is of particular interest to the designer. The various national codes often specify different fire compliance and rod reinforcement content for the two. More concrete cover is required in a beam than in a rib, and a minimum amount of rod reinforcement must be provided in a beam while none, or a lesser quantity, is specified for a rib. Most of the standards also specify an upper limit for the spacing of ribs.

In a post-tensioned floor, it may be more economical to provide ribs at a wider spacing than the maximum specified, and design them as beams. In this case some additional rod reinforcement is needed in the ribs, but there would be fewer of them. The case is illustrated in Example 6.2.

BS 8110 specifies that ribs should be spaced at centres not exceeding 1.5 m ($\approx$ 5 ft) and their depth, excluding any topping, should not exceed four times their width. The minimum thickness of the topping should be one-tenth of the clear distance between the ribs but not less than 50 mm. The topping thickness is often governed by fire considerations.

ACI 318 specifies that ribs shall be not less than 100 mm in width; and shall have a depth of not more than $3\frac{1}{2}$ times the minimum width of rib. Clear spacing between ribs shall not exceed 800 mm.

There is no requirement for links in the ribs but some links are normally provided to support the tendons and to hold the bonded rod reinforcement in position.

3.2.2 Beams

In cross section, a beam can have its stem projecting below the soffit of the slab (*downstand*), or above (*upstand*), or it can be wide and shallow, in which case it is termed a *band* or *strip beam*, see Figure 3.3. Strip beams of the same depth as the slab are preferred, because the slab then has a flat soffit without any downstands.

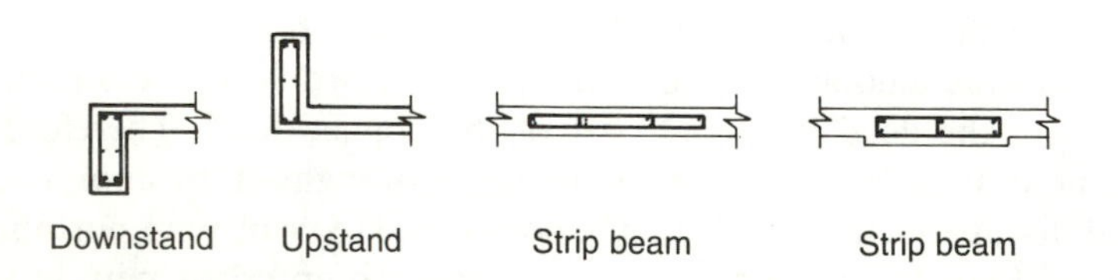

Figure 3.3 *Beam sections*

Downstand beams

Downstands are the traditional form of beams. Being deep and narrow, they are the most efficient in making use of the concrete and the steel. They also have a high moment of inertia compared to strip beams, and the flange is available for compressive stresses in the span. Therefore, a downstand beam has the least deflection compared with the other two shapes. Its high stiffness is very useful on a building facade which requires a tight control on deflections.

A flat soffit is generally preferred by the building occupants, and by the builders. Downstands are seen as intruding into the space below the slab soffit, space which could otherwise have been used for accommodating service runs.

Formwork for a downstand beam is more complicated than that for a strip beam, and stripping of formwork is time-consuming. Table forms and similar fast-tracking devices are more difficult to use with downstands.

Near supports the top of the beam is often congested with reinforcement and tendons, and care has to be exercised in detailing beams to ensure that enough space is available for compaction equipment.

For these reasons downstands are currently very much out of favour. They are used only where unavoidable, such as along the external edges for architectural reasons and in transfer structures, such as beams which are required to support the columns above.

National standards specify the flange width which should be taken as part of a beam. The width is required for calculating the moment of inertia of the section, which is needed for computing the stresses and deflections. BS 8110 gives:

for T-beams, $b_r + L_z/10$ or actual flange width if less,
for L-beams, $b_r + L_z/5$ or actual flange width if less.

where b_r = stem width
L_z = distance between points of zero moment, which can be taken as 0.7 times the effective span for continuous beams.

ACI 318 does not make any recommendation for the flange width of a post-tensioned T- or L-beam. Instead, the determination of an effective width of flange is left to the experience and judgement of the engineer. However, for reinforced concrete it recommends:

for T-beams, the least of $b_r + 8.d_s$ or $L/4$ or actual
for L-beams, the least of $b_r + 6.d_s$ or $L/12$ or actual.

where d_s = slab thickness
L = span length.

A method of constructing downstands, called the *shell beam* system, and shown in Figure 3.4, is gaining popularity in Europe. The system is designed for fast construction. It consists of preformed concrete U-sections which serve as permanent shuttering for the beams. The preformed units are able to span short distances, so they need fewer vertical props. The system, obviously, can be used only where downstands are acceptable.

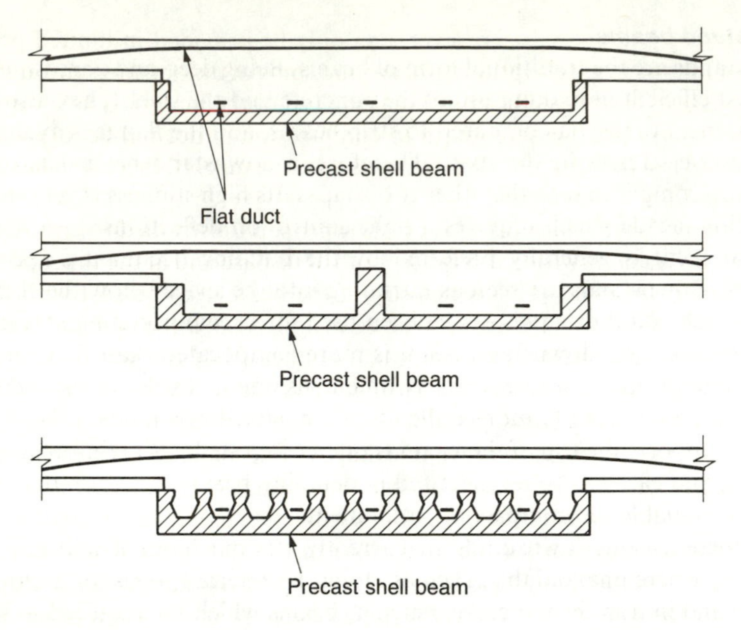

Figure 3.4 *Shell-beam arrangement*

The beam shell is prefabricated with holes in suitable positions for the slab anchorages where needed. The downstand beam can be either of the conventional deep and narrow shape, or it can be wide and shallow. The permanent formwork for the slab may consist of corrugated steel decking units as commonly used in steelwork construction, or of precast pretensioned thin planks, say 50 mm (2 in) deep. These also need some propping.

The precast units are usually pretensioned, made by the long line method. Attempts have been made to make structural use of these units by incorporating a sufficiently high prestress to carry most of the dead and superimposed loads. This, however, can only be of a limited success, if the units are not to be too heavy for transporting and handling. Also, the prestress from the precast unit is not transferred to the *in situ* concrete and shrinkage of the *in situ* concrete may cause a differential slip at the interface. Therefore, fully composite action between the precast unit and the *in situ* concrete is difficult to develop; provision of shear links between the two concretes, however, does help in this respect.

The shell-beam system provides a permanent formwork requiring a minimum of propping and it is quick. It can be used with advantage where downstands are acceptable and precasting facilities are easily available.

Upstand beams

Upstand beams avoid the main problem with downstands—intrusion into the space below the slab soffit, and they allow the soffit formwork to be flat and uncomplicated. Upstands, however, can only be provided in limited positions, such as along external elevations and around some of the openings through floors.

An upstand is not as efficient in making use of the materials as a downstand because it only has the narrow rib in compression in normal flexure whereas a downstand has the adjacent slab acting as a wide flange. It, therefore, has a higher deflection than a downstand.

The vertical forms for upstands must be properly restrained to ensure that the beam width remains within tolerance, both at its base and at the top, and that the beam remains vertical. This is not easy if the beam is to be cast with the slab, because diagonal bracings are difficult to install. Also, if the upstand is cast with the slab then the vertical pressure causes upheaval of the wet concrete in the adjacent slab; to prevent this, narrow shutters are required on top of the slab. Upstands are usually constructed after the slab concrete has hardened. In this case, steps must be taken to prevent leakage of fine material from the joint which would otherwise leave a weak construction joint.

In spite of their disadvantages, upstands are often useful at vertical ducts where service runs coming out of the duct close to the slab soffit would not allow a downstand and the span is too long or the support too narrow for a strip beam.

For cantilevers, the upstand is the preferred shape of the beam rather than the downstand, because the flange at the bottom of the section is in compression. In this case, the expressions given in 3.2.2 for the flange width apply, but L_z can be taken as twice the cantilever span provided, of course, that the flange continues beyond the cantilever root.

Band or strip beams

A strip beam, being wide and shallow, has the least effective depth of the three alternatives. It, therefore, uses steel in the most inefficient manner of all, but it does not need any additional concrete or special formwork.

A strip beam is usually wider than the column, so that it becomes necessary to provide a reinforcement cage at the support to transfer the load from the beam to the column.

The rod reinforcement and tendons are easy to assemble, the concrete can be compacted with the same ease as in the slab and there is much less chance of a tendon being damaged by a vibrator. A strip beam is, therefore, the easiest and fastest to construct and is much favoured even though it needs more tendons and possibly more rod reinforcement.

Most national standards specify a minimum amount of bonded rod reinforcement and links to be provided for the entire length of a beam. If beam strips are analysed as integral parts of a slab panel then it is perhaps not necessary to provide links over the whole of the strip length; in Australia links are provided only near the columns in such cases.

If the beams are of such a size that shear, or support moment, is critical and increasing the beam depth is not a viable option then drop panels or prefabricated shearheads may be used, though the former option is best avoided.

3.2.3 Drop panels

A drop panel, shown in Figure 3.1, is used to enhance the shear strength of a strip beam, or to increase its flexural strength over the support section. Drop panels, however, make the erection and the removal of formwork rather cumbersome. Easily removable forms, essential for fast-tracking, become more involved and assembly of reinforcement in the drop panels can be a time-consuming operation if a prefabricated cage cannot be dropped in position.

Wherever possible, drop panels should be avoided and use of a proprietary shearhead considered instead. Shearheads are dealt with in detail in Chapter 10.

3.3 Panel configuration

The elements of a floor panel were individually considered in Section 3.2. This section discusses combinations of post-tensioning and reinforced concrete in one and two-way spanning floors.

3.3.1 One-way span

A one-way slab requires tendons running only in one direction—along its span. In the other direction sufficient rod reinforcement is provided to prevent shrinkage cracks, to distribute any load concentration, and to satisfy the requirements for the ultimate condition and those of the relevant standards.

One-way slabs are often preferred because they are easier to construct and because they need fewer beams; they are also easier to design and detail. A particular advantage is that of much reduced congestion of tendons and reinforcement near the columns.

For a one-way slab spanning on beams, there are four possible arrangements of beams and slabs using reinforced concrete and post-tensioning:

- reinforced concrete slab on reinforced concrete beams
- post-tensioned slab on reinforced concrete beams
- post-tensioned slab on post-tensioned beams
- reinforced concrete slab on post-tensioned beams

The first case, where the slab and the beam are both in reinforced concrete, is outside the scope of this book. A post-tensioned floor spanning on reinforced concrete strip beams is frequently used. The slab normally spans in the longer direction; if the shorter span is within the capacity of reinforced concrete then this arrangement should be considered. It would have the merit of using the two techniques to their best advantages.

Reinforced concrete beams generally have a span-to-depth ratio of the order of 20, whereas the ratio may be about 40 for a post-tensioned slab. This suggests that the arrangement is suitable where the column spacing in the two directions is in the ratio 2:1. In fact, a range starting at 1.5:1 may be worth investigating.

Below the 1.5:1 ratio of long to short span length, it is normally better to adopt the third of the above configurations and use post-tensioning for the beam strips as well as the slab. Alternatively, the slab may be spanned in two directions.

The last configuration, that of post-tensioned beams and a reinforced slab, is rather rare. It would be used in special circumstances only, such as when a short span slab is supported on a bridging beam.

In one-way floors where beams are post-tensioned, the axial component of the beam prestress spreads into the adjacent slab, so that the full prestressing force is not available in the beam section. This does not affect the equivalent load of the tendons or the ultimate strength of the beam but it does matter in serviceability calculations.

Downstand beams, or strip beams slightly deeper than the slab, can be used in any of the situations mentioned above. The deeper beam will be better able to cope with the loads from the slab panel and may be more economical than a beam within the slab thickness, though it is slower to construct—see Section 3.2.4.

3.3.2 Two-way span

A two-way spanning slab makes use of the concrete in both directions. It is shallower than a one-way slab because it carries only a proportion of the load in each direction. However, being shallower, the tendons work at a smaller internal lever arm, and, therefore, a two-way slab may need more tendons than a deeper one-way slab carrying the same applied load. A two-way slab is often used where construction depth is to be minimized. It is also used where some control needs to be exercised on the distribution of the slab load in the two directions; the principle is described below.

In a rectangular slab panel supported on four sides, the load is transferred in each of the two directions; the proportion of load depends on the elastic properties of the slab in the two directions, and on the span lengths. More load goes along the shorter span and in the stiffer direction. At a span ratio of approximately 2:1 most of the load goes along the short span and the panel effectively acts as one-way spanning. Therefore, two-way spans are most useful for span ratios in the range of 1:1 to about 1.5:1; above this the advantage is gradually lost. Two-way post-tensioning of a floor is likely to be used where the ratio of span lengths in the two directions is in the range 1.0 to 1.5 and the slab loading is heavy, such as in a warehouse, or where both the spans are long.

In a two-way reinforced concrete solid slab the proportion of load transferred in the two directions is determined by the ratio of span lengths, the proportion along span L_1 being $L_2^4/(L_1^4 + L_2^4)$. Apart from considerations of continuity, very little can be done to change this ratio.

In a post-tensioned floor much more control can be exercised over the load distribution by the arrangement of the tendons. Each curved tendon can be seen to exert a vertical load on the concrete by virtue of its curvature, usually in a direction opposite to the applied load. At the support the tendon curvature is reversed and it sheds its load onto the supporting beam. Therefore, if desired, the tendon profile in one direction can be made deeper so that the curvature is more acute, with the result that much more load is carried along that direction. The same result is achieved if more tendons are provided in one direction. This is an extremely useful tool, available only to post-tensioning.

Extending the idea further, in a partially prestressed floor it is possible to run the tendons in one direction only. The upwards force resulting from the tendon curvature can be deducted from the total applied load, and the remaining load distributed in proportion to the slab geometry in the two directions. Using this arrangement, a greater proportion of the load will go along the direction of the tendons than would have been possible in a simple reinforced concrete slab. Such a slab acts as post-tensioned in one direction and as reinforced in the other. In determining the proportion of load distribution in the two directions it should be noted that the moments of inertia of the slab in the two directions will be different, because the reinforced concrete section will crack whereas the post-tensioned section will remain intact. The arrangement is particularly useful where beams in one direction are capable of carrying a larger proportion of the load than those at right angles; the tendons can then be used to distribute the load to suit the beams.

A two-way slab may be designed as reinforced in both directions, post-tensioned in one direction only or post-tensioned in both directions.

The first combination, that of a two-way slab in reinforced concrete in both directions, is rather rare in a generally post-tensioned floor. Its occurrence is likely to be incidental rather than by choice, such as where the framing geometry of a post-tensioned floor leaves a panel somewhere, which is too small for post-tensioning. If the adjacent beams are post-tensioned then attention should be given to the spread of the axial component of the prestress into the slab.

Each of the three slab arrangements can be used with beams in reinforced or post-tensioned concrete. If the slab span is long enough to require post-tensioning then, in all probability, the beams will also need to be post-tensioned, unless for some reason they are downstand and of sufficient depth to be designed in reinforced concrete. In the latter case, the reinforced concrete beams will absorb some of the prestress from the slab, leaving it deficient in the axial component, unless provision has been made in the design for this spread.

The combination of a reinforced concrete slab and post-tensioned beams is also likely to be used only in very specialized circumstances. It is feasible but attention must be given to the spread of prestress into the slab.

A solid flat slab is suitable for domestic and office buildings with approximately equal spans in the two directions. It is economical with spans not exceeding about 10 m (35 ft). If drop panels are provided then the solid flat slab can be used for longer spans, say up to about 12 m (40 ft). Lightweight concrete may be more suitable in the latter case than the normal dense concrete.

Where heavier loading and/or long spans—up to 20 m (65 ft)—are needed, a waffle slab may be a good choice. Its possible applications include warehouses, industrial buildings and public halls.

Two-way slabs with downstand beams can be used in either of the above situations, provided of course that the downstands are acceptable.

3.3.3 Solid or ribbed

A solid slab can, from a theoretical point of view, be considered as a particular case of a ribbed floor where the ribs are touching together. It forms the largest reservoir of compression for a given depth and often a solid floor is used where a minimum section depth is the main criterion and its additional weight is not significant. Self-weight, of course, becomes of increasing significance as the span length increases.

Formwork for a solid slab consists of a simple flat surface and vertical edge boards. A solid floor is heavier than a ribbed one and the props must be designed for the loading, though the weight of wet concrete is rarely critical for floors of the normal span length. Formwork for a ribbed floor requires the flat platform and the vertical edge boards but it additionally needs proprietary or purpose-made formers shaped to the profile of the ribs, which must be secured in position. Stripping of soffit shutters for ribs, obviously, requires more work and sometimes the arrises get damaged. These need careful repair because their tensile strength is often relied on in post-tensioning.

For a solid slab the rod reinforcement, tendon supports and tendons are assembled on a flat platform without any other physical impediment, and the concrete can be placed and compacted with confidence in the open area; the only precaution to be taken is against the vibrators damaging or displacing the tendons.

Assembly of rod reinforcement, tendon supports and the tendons is cumbersome in ribbed or waffle floors, because these are housed within the confines of the rib. Some nominal rod reinforcement is also needed in the shape of links to maintain the tendons in the specified profile.

Concrete has to be placed and compacted in narrow ribs which may already be congested with reinforcement and tendons. At times it is difficult to find room for a poker vibrator, which makes it more difficult to achieve a good compaction of concrete in the narrow ribs. Thus, while a ribbed floor may have a lesser concrete quantity, the ease with which concrete can be deposited and compacted in a solid floor may make the latter more attractive. In high-rise buildings, on the other hand, pumping large volumes of concrete to the upper floors may be an expensive operation and so ribs may be preferred.

Considering fire, a much larger surface area is exposed to the high temperature at the soffit of a ribbed floor than that of a solid floor. Also, the solid floor has a larger mass and a larger capacity as a heat sink. Therefore, in a fire the rate of rise of temperature is slower in a solid floor. For this reason the fire requirements are more onerous for ribs, in that more concrete cover is prescribed to reinforcement and tendons.

The amount of work involved in the design of a solid floor is slightly less than that for a ribbed floor. Because shear is rarely critical in a solid floor, no calculation is required for the point where the section should be changed to solid, and the moment of inertia remains constant which simplifies the computation of deflection.

In summary, the main advantage of a solid slab is the ease of construction, because of the easier formwork and reduced site labour needed. It is particularly useful where the self-weight of the floor is not a major consideration and is the ideal solution for spans under about 10 m (33 ft) carrying heavy loads. A solid floor is also likely to be chosen where a minimum construction depth is the main criterion, though it may not be the most cost effective solution. For longer spans, the self-weight of a solid slab will probably be a considerable disadvantage—at least in residential and office buildings.

A ribbed floor makes use of the concrete and the steel in the most efficient manner, and is likely to be a good economical choice if construction depth is not a major consideration.

3.4 Span-to-depth ratio

One of the major advantages of post-tensioning is that the floor can be made shallower than it would be in reinforced concrete. A reinforced concrete floor will have a larger deflection than a post-tensioned floor (with equal load and depth). The reason is that, to a large extent, the deflection can be controlled in a post-tensioned floor, though this would not make an efficient use of the prestress.

The minimum possible depth for any type of floor depends on a number of factors, such as concrete strength, span length, intensity of loading, whether the section is solid or ribbed, and rib proportions in the latter case. In general, a solid slab is likely to give a shallower depth than other types.

The question arises as to how shallow a post-tensioned floor can be. The usual practice is to quote span-to-depth ratios which are based on actual practical designs. It is, perhaps, worth devoting a few paragraphs to the theoretical aspects in considering the minimum depth. Apart from the practical considerations of buildability and economy, the minimum depth of a floor is determined from two design criteria—strength and deflection.

Consider the flexure of a floor required to carry a uniformly distributed load w on a span of length L. Assume that the section remains uncracked.

$$\text{Moment} = \text{stress} \times \text{section modulus}$$
$$wL^2 \quad \alpha \quad \sigma . D^2 \tag{3.1}$$

where σ = flexural stress
D = section depth

Expression (3.1) indicates that for a constant uniformly distributed *total* load, depth is directly proportional to span. However, the floor self-weight increases in proportion to the depth, and therefore to the span length, with the result that, the total load being constant, the capacity available for the imposed load reduces. Therefore, for a constant *applied load*, the increase in depth must be more than linear. Exactly how much, depends on the ratio of applied load to self-weight.

Considering the deflection δ,

$$\delta \quad \propto wL^4/D^3 \tag{3.2}$$
$$\delta/L \propto wL^3/D^3$$

The above relationship shows that, assuming the depth to increase in proportion to the span, the deflection is also proportionally higher. In other words, if the total load is constant, and span and depth are doubled then deflection is also doubled, i.e., δ/L remains constant. Some standards, BS 8110 for example, stipulate two criteria for deflection; that it must not exceed a prescribed proportion of the span length and that it must not exceed a given absolute value. For a constant deflection, depth should be proportional to $L^{4/3}$.

Evidently, the relationship between span and depth is a complex one and cannot be expressed simply as linear. In BS 8110, for reinforced concrete, the complexity is simplified by specifying two different span-to-depth ratios, above and below 10 m (33 ft). In the case of post-tensioned concrete the situation is even more complex, because the net load causing deflection depends on the tendon geometry, which can be used to control deflection to some extent.

A number of span-load combinations were designed for minimum possible depths, at a maximum long-term deflection of 20 mm for simply supported solid slabs. The results are shown in Figure 3.5, where *minimum depths* for spans ranging from 6 m to 20 m (20 ft to 65 ft) are plotted for imposed loads of 2.5, 5.0, 10.0 and 15.0 kN/m^2 (nominal 50, 100, 200 and 300 psf). It is possible to obtain a different set of curves from those shown in Figure 3.5 by choosing other values for the design criteria (concrete strength, continuity, etc.), but the plotted values are considered as approaching the minimum.

It is unlikely that a 20 m (65 ft) span slab would be constructed in solid concrete. The curve has been extended to this span only to indicate a base line starting point for depth in the design of long span ribbed or waffle floors. An actual design will almost certainly have to include economy as one of the governing parameters which would require a deeper section than indicated in Figure 3.5. Economy, being dependent on the relative costs of material and labour, varies from one location to the next, so that it is not possible to give any meaningful guidance on economical depths.

Note that Figure 3.5 is based on the following assumptions:

- $f_{cu} = 40$ N/mm^2 ($f_c' = 4600$ psi), normal concrete
- Tendon centroid 35 mm (1.375 in) above slab soffit
- Serviceability stresses limited to 2.3 N/mm^2 tension and 13.33 N/mm^2 compression (330 and 1950 psi respectively)

The span-depth ratios in Figure 3.5 vary from 32 to 54. Caution is needed in using a very slender slab. Initial stresses at jacking and final serviceability stresses may both approach their respective permissible limits, deflection under certain conditions may be excessive, or the floor may have an undesirably low natural frequency, though the initial stresses, and the creep deflection, can be controlled to some extent through application of prestress in two stages.

Table 3.1 shows the suggested range of span-depth ratios for different floor configurations. The lower and the upper values in a range correspond approximately to applied loads of 10 and 2.5 kN/m^2 (200 and 50 psf) respectively.

The span-depth ratios for the beams may be more important in defining the overall construction depth of a structure. Beams are normally heavily loaded

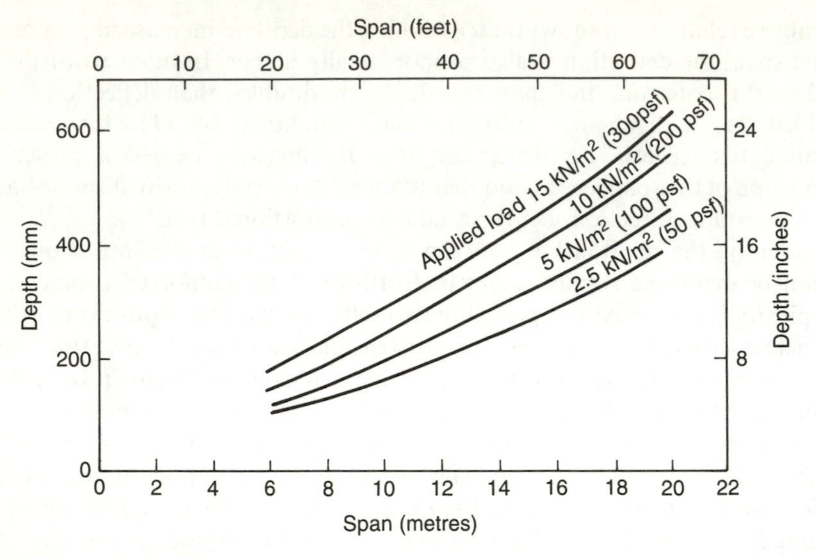

Figure 3.5 *Minimum depth–applied load for solid slab*

Table 3.1 *Span-depth ratios for floors*

Floor type	*Span-depth range*
One-way solid	30 to 45
Ribbed slab	25 to 35
Solid flat slabs	35 to 45
Waffle floors	20 to 30
Beams	13 to 33

Table 3.2 *Span-depth ratios for beams*

Total load		*Live/Dead load ratio*		
kN/m²	*ksf*	*1.5*	*1.0*	*0.5*
150	3.1	13.3	14.0	14.5
100	2.1	16.3	17.0	17.8
75	1.6	18.8	19.5	20.3
50	1.0	23.0	23.5	24.5
25	0.5	—	31.5	33.0

and, therefore, the average prestress level is higher in beams than in slabs. The maximum load that a given beam section can carry depends on the prestressing force, among other factors. An upper limit on the prestressing force is imposed by the intensity of the dead load on the beam, if the initial stresses are not to exceed the specified limits. This is also true of slabs but it is of significance only in highly stressed members, such as beams. For this reason, simple span-depth ratios are more difficult to give for beams. Ratios in the range of 13 to 33 can be tried; the lower figure corresponds to a total load of 150 kN/m^2 and the higher to 25 kN/m^2 of the plan area of the beam, with live-to-dead load ratios between 0.5 and 1.5. Table 3.2 gives the values.

The depth of a continuous slab or beam, or a two-way slab, can be smaller than that of a single span. It is not possible to produce simple span-depth ratio curves for continuous members and two-way slabs because of the large possible variation in the ratios of adjacent span lengths in the case of continuous floors and in the ratio of side lengths for two-way floors. If a minimum depth is the requirement in such a floor then a number of trial calculations would have to be carried out, perhaps starting with a depth of 0.9 times that for a simple span of the same length.

Figure 3.6 *New Port III – USA*

Figure 3.7 *Monostrand floor post-tensioning*

4 PLANNING A STRUCTURE

This chapter deals with some of the factors to be considered when a post-tensioned structure is being planned, before the process of calculations is begun. It includes discussion of a few design-related topics, such as, the difference between the behaviour of columns for reinforced and post-tensioned concrete floors, the longitudinal shortening of the floor and its interaction with the vertical elements, and the spread of prestress into adjacent zones which have a lesser, or no, prestress. The relative merits of the alternatives—reinforced or post-tensioned, bonded or unbonded, etc.—have been discussed in Chapter 1 and are not repeated here.

In planning the structure of a building, the factors to be considered include:

- buildability
- shape and position of shear walls
- column spacing
- one- or two-way spanning
- floor depth
- solid or ribbed
- column size
- construction joints
- economy
- adaptability

Some of the above have also been discussed in Chapter 1 and are not repeated. Economy is often the most important consideration, but it is listed last, because the cost of a structure depends on decisions resulting from the consideration of the other points, and economy summarizes the values of these decisions. Depending on the type of building, the cost of the structure may vary from 10% of the development cost for a prestige building, to 80% for a utilitarian project. This reflects the relative importance of the structure and its direct cost.

4.1 Design objectives and buildability

The structural design of a building is guided by certain design objectives; the obvious ones concern structural integrity and robustness, while others pertain to buildability and the intended use of the building—see Table 4.1.

The possible objectives for a high-rise building of a predominantly vertical

Table 4.1 *Design objectives and post-tensioning*

Objective	*Effect of post-tensioning*
Design	
Minimum structural depth	Allows high span/depth ratio
Column free space	Allows long spans without excessive self-weight
Low self-weight	Minimum depth, possibly in lighweight concrete
Small column size	Low floor weight leads to reduce column moment
Economy	Economical above 8 to 10 m span (26 to 33 ft)
Low lead time	Can be faster than steelwork and reinforced concrete
Design effort	May take slightly more time
Construction	
Simpler formwork design	Flat soffit floor designs easily achievable
Quick formwork turnaround	Formwork released early
Quick assembly of steel	Minimum steel quantity. Tendons in long lengths
Fast concreting	Less concrete quantity, uncongested steel
Minimum backpropping	Self supporting at early age
Low labour requirement	Less material quantity, easier assembly of steel
Crane time	Less material, crane may be needed for jacking
Site presence	Specialist operatives needed
In use	
Limited deflection	Provides initial camber and control
Low maintenance	No microcracks, slow deterioration
Possible small holes	Steel spaced much further apart than in an RC floor
Large holes	Need careful design study and specialist operatives
Future refurbishment	Less structural depth and flat soffit permit maximum flexibility for renewal of services
Demolition	Needs study of structure and specialist knowledge

aspect, and those for a low-rise building of large plan area, may be similar but the means employed for achieving these, and the emphasis required on particular operations, differ. For example, while it is desirable to construct the frame in as short a time as reasonably possible, in a tall building this implies a quick turn-around floor construction cycle whereas in a large plan building the construction of successive floors will probably be carried out in stages, so that the time taken in completing a floor has a lesser importance. For the same objective the planning of site operations is also quite different for the two types of buildings.

The designer should recognize the buildability requirements for the project and make adequate provision for these in the design.

The main objectives usually include a minimum of structural presence, fast construction, and reliability in service. The advantages of post-tensioning have already been discussed in Chapter 1; however, it is of interest to see how post-tensioning helps in achieving the objectives set out in Table 4.1. Some of the items are self-explanatory and are not discussed further.

Lead times for the procurement of materials are part of the overall building program. Building elements on long lead times have two fundamental disadvantages. Firstly, the material has to be ordered some time before it is delivered, which requires financing in advance. Secondly, the design has to be frozen a long time before the material is incorporated into the building. Often the tenants' requirements are not fully known until a late stage in the building program, and it becomes necessary to incorporate late changes in the design. If this involves materials on a long procurement period then it becomes difficult to keep costs under control. Structural steelwork often has a long delivery period. In reinforced concrete construction it is a question of how quickly the reinforcement supplier can bend and deliver the rod reinforcement—only the small diameter bars can be bent on site. In post-tensioning, strand and hardware can be stored on site and the requisite lengths of strand cut from the reel as and when needed. For last minute holes, the tendons can simply be deviated out of the way. Post-tensioning, therefore, offers the shortest lead time and maximum flexibility for accommodating late design changes.

The ease with which a building can be refurbished in the future is important in measuring its serviceable life. A building where services cannot be renewed, because of insufficient storey height, or service zone interruptions such as by downstand beams, cannot be easily refurbished and the only alternative may be to pull it down.

In order to design an economical structure, facilitate its construction, and enable its later refurbishment, the designer should consider the following points. However, the desire to achieve optimum efficiency of design should be tempered by the essential need for simplicity of installation.

- The positioning of the columns is important in post-tensioned concrete, as it is in other forms of construction. A regular rectangular grid is likely to be more economical than one in which the span lengths vary to the extent that reversal of span moments is possible. In the authors' opinion, the preferred span ratio for a panel would be in the range 1:1.3 to 1:1.6, and the outer spans of a continuous floor would be about 80% of the inner spans in length.
- Keep the formwork simple. Avoid downstand beams, column heads and drop panels. In ribbed and waffle floors, reduce the number of moulds by spacing the ribs apart. Purpose-made moulds may prove more economical than the proprietary ones.
- Provide construction joints to suit the construction sequence wherever possible.
- Keep the tendon layout simple. Avoid interleaving of tendons. Space tendons as far apart as the design permits, to allow holes to be cut after construction.
- Bunch tendons in twos and threes. This saves on the number of supports and is quicker to install.
- Avoid congestion of steel. This usually means that the structural members should not be reduced in size to the theoretical minimum.
- Detail rod reinforcement to allow ease of assembly.
- Use pre-assembled reinforcement cages and standard mesh in preference to loose bars which have to be tied on site.

4.2 Restraint from vertical elements

In a concrete structural frame, the vertical elements, such as columns and walls, are usually designed monolithic with the floor. When such a floor is stressed, part of the tendon force is resisted by the vertical elements and is lost as far as the axial stress in the floor concrete is concerned. The axial prestress in the floor equals the difference between the tendon force and the force resisted by the vertical elements. This does not affect the magnitude of the equivalent balancing load, because the force in the tendon is not affected. In fact, because the average stress on the concrete is now lower than it would have been if the vertical elements had no stiffness, elastic and creep losses in prestress are also lower; therefore, the tendon force is marginally higher.

For a normal floor supported on columns, the difference between the tendon force and the compression in the concrete may be of the order of 1 to 2%. This is usually ignored, because this order of accuracy is insignificant considering the uncertainties in the properties of the materials and the tolerances. However, the loss can be quite significant if stiff walls or large columns are monolithic with the floor. Such conditions are likely to occur in the lower storeys of a tall building and where retaining walls running full, or partial, height of the columns are cast monolithically with the frame. In such a case, the serviceability tensile stresses may exceed the permissible values in the floor if the loss is not allowed for in the calculations. The vertical elements themselves may get overloaded.

Consider a single span beam framing into two columns, Figure 4.1(a). For simplicity, assume that the columns are equal in size, are fully fixed at the bottom, and are free to rotate at the top. If a different set of conditions is desired then the effective column height H can be adjusted accordingly. If the columns provided only vertical support and the floor were free to slide horizontally, then there would develop a gap of width δ between the beam and one of the columns, Figure 4.1(b), because of the shortening of the beam due to the axial component of the prestress. In order for the gap to close, shears P_v are needed as shown in Figure 4.1(c).

$$\begin{aligned} \delta &= \text{shortening in length } L \\ &= P_s L / A_c E_c \end{aligned} \tag{4.1}$$

where P_s = Axial prestressing force in concrete, not the tendon force
$= P_t - P_v$
P_t = Tendon force
P_v = Force resisted by the vertical elements

Each column must deflect a distance $\delta/2$ under force P_v.

$$\begin{aligned} \delta &= 2P_v H^3/(3E_c I_v) = P_s L/A_c E_c \\ P_v &= 1.5P_s(LI_v)/(A_c H^3) \\ &= P_s/K_r \\ &= P_t/(K_r + 1) \end{aligned} \tag{4.2}$$

where $K_r = 1.5H^3A_c/(I_vL)$ (4.3)

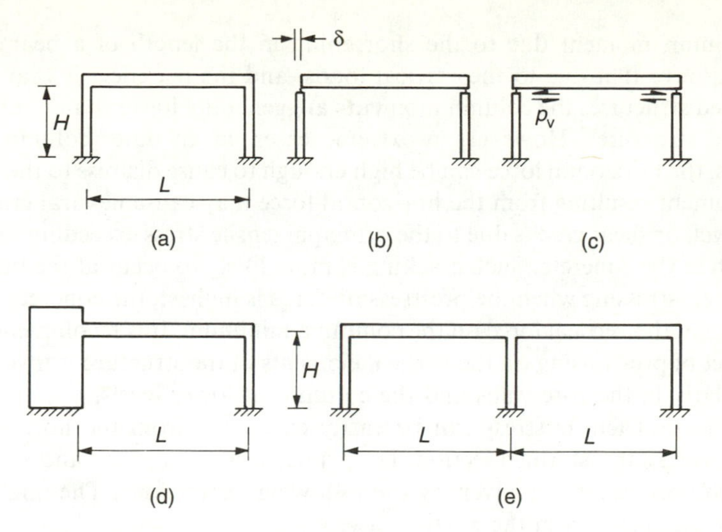

Figure 4.1 *Loss of axial prestress due to column stiffness*

I_v = Moment of inertia of the column
A_c = Cross sectional area of the beam

K_r is a non-dimensional factor which relates the deformation of a beam in direct tension to that of a column in bending. The modulus of elasticity E_c cancels out, assuming that the columns and the floor have the same concrete.

For the frame shown in Figure 4.1(d), where the beam is supported on a very stiff shear wall at one end and on a column at the other end, only the one column deflects and therefore

$$K_r = 3H^3 A_c/(I_v L) \tag{4.4}$$

In the case of two equal bays of length L each, see Figure 4.1(e), the centre column remains unaffected. The frame is, therefore, equivalent to Figure 4.1(d) and Equation (4.4) is applicable. If the two span lengths differ only by a small amount, so that the deflection of the centre column can be neglected, then Equation (4.4) can be used, taking an average value for L.

Equations (4.2), (4.3) and (4.4) are fairly simple to apply to a frame consisting of one or two bays. For longer frames, it becomes necessary to solve a set of simultaneous equations; further complexities arise if the columns differ in stiffness. For a uniform prestressing force over the whole length of the member, the phenomenon is analogous to a drop in temperature—a problem which most frame analysis programs are capable of solving. The required hypothetical temperature drop is given by Equation (4.5).

T = temperature drop
$$= P_t/(A_c E_c . C_t) \tag{4.5}$$

where C_t = thermal coefficient for concrete

The column moment due to the shortening in the length of a beam acts in opposition to that due to the vertical loads, and the net effect is that in post-tensioned structures the column moments are generally lower than in reinforced concrete structures. However, in extreme cases, in an outer column of high stiffness, the horizontal force can be high enough to cause distress to the column. The moment resulting from the horizontal force may cause flexural cracks at a floor level, or shear cracks due to the principal tensile stress exceeding the tensile strength of the concrete. Such cracking is more likely to occur at the time of, or soon after, stressing when the prestressing force is highest, the concrete strength lowest, and the vertical force on the column a minimum. It is recommended that the effect of prestressing on the vertical elements of the structure is investigated, particularly in the core walls and the columns at lower levels.

The flexural tensile stress can be easily calculated from the normal elastic theory, using a transformed section. The principal stresses σ_1, σ_2, and α, the angle at which they occur, are given by the following expressions. The angle α is in radians, measured from the x-axis.

$$\begin{aligned} \sigma_{1,2} &= 0.5\sigma_y \pm 0.5(\sigma_y{}^2 + 4\tau^2)^{0.5} \\ \alpha &= 0.5 \tan^{-1}(2\tau/\sigma_y) \end{aligned} \tag{4.6}$$

where σ_y = the vertical stress
τ = the shear stress

The above equations are based on elastic theory and are not compatible with the ultimate state shear analysis in BS 8110. The distribution of elastic shear stress over the concrete section is normally parabolic, being maximum at the section centroid and zero at the edges.

From the above discussion it is obvious that the loss in the axial component of the prestressing force can be minimized by positioning the stiff vertical members, such as shear walls and cores, near the middle of the floor area, or in such a manner that they offer the least resistance to the shortening of the floor.

Rectangular cores and channel-shaped shear walls, while being very efficient for resisting lateral forces, can absorb a significant proportion of the prestressing force if they are poorly positioned. The worst case is that of two rectangular cores located at either end of a long floor. Staircases framed in columns can have a very high stiffness, as they may form a vertical truss triangulated by the flights. Figure 4.2 shows some of the preferable and the undesirable arrangements.

A similar loss in the axial component of the prestressing force occurs in suspended ground floors due to the stiffness of the basement retaining walls at corners, and where the tendons runs along the wall. Away from the corners, the external walls offer very little resistance; they can be treated in the same manner as the columns, and the loss in prestress can be calculated using either Equations (4.2) to (4.4), or the temperature drop analogy.

Where the service cores and shear walls cannot be located in structurally desirable positions, the loss can be reduced by delaying the connection between the vertical element and the floor. A gap (or infill) strip, say one metre wide, is left

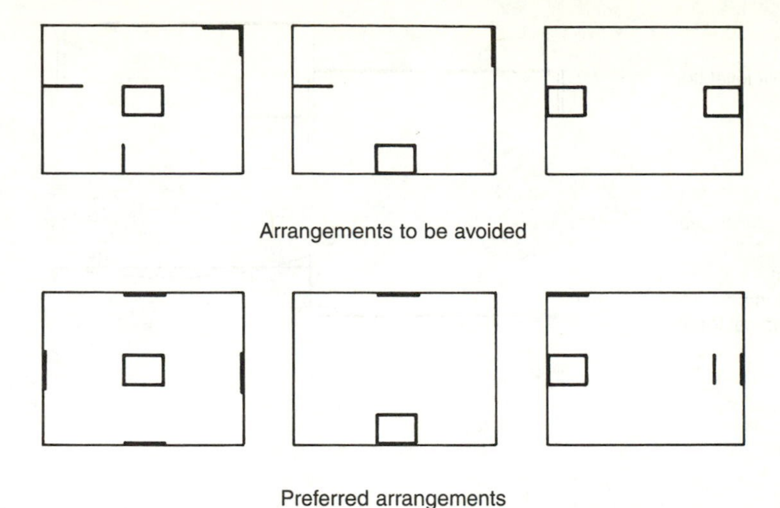

Figure 4.2 *Core and shear wall arrangements*

unconcreted between the stiff vertical element and the slab, for as long a period as practical. This requires the slab edge to be supported for an extended period. The gap strip is also useful in reducing the effect of creep at movement joints.

It is also possible to design a sliding connection between the wall and the floor, where the slab is vertically supported by the wall but it remains free to slide laterally. After a suitable period the sliding joint is made ineffective.

Suitable joints and filler strips are shown in Chapter 12.

The lateral forces resulting from the restraint offered by the vertical elements may have a rather unexpected effect in split-level car parks, such as that shown in Figure 4.3(a). The floor would normally be designed as simply supported, taking local account of the negative moment which would develop at the centre column, shown in Figure 4.3(b). The lateral restraint offered by the external columns sets up the system of forces indicated in a simplified manner in Figure 4.3(c), which produces tension on the inside face of the upper floor-column connection, opposing the tension from the normal vertical loads. On the next floor down, the tension is on the outside face, adding to that from the vertical loads. Depending on the magnitude of the lateral forces, the upper floor may crack near the centre column on the *underside*, and the lower floor on the *topside*—in the first case because no tension is anticipated on the underside and in the second case because the tension on top is larger than anticipated.

4.3 Dispersion of the prestressing force

In a post-tensioned floor, the anchorages are normally located along the slab edge. Immediately behind an anchorage, the force from each tendon is concentrated over the area of concrete in direct contact with the anchorage; the zone between

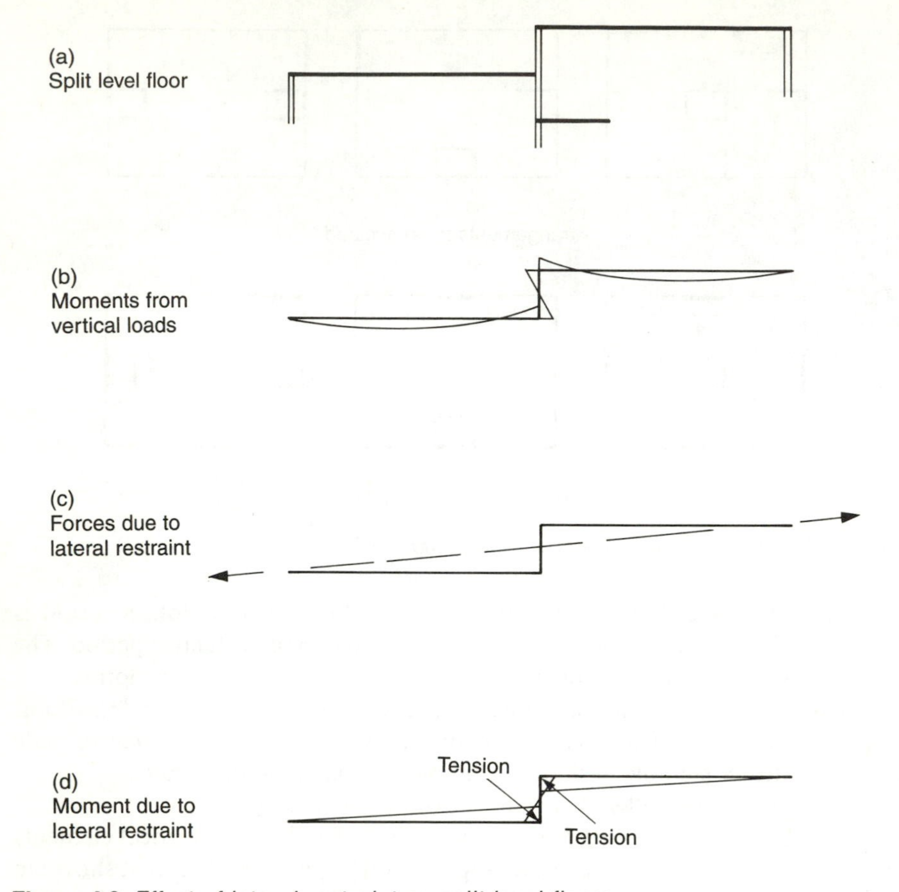

Figure 4.3 *Effect of lateral restraint on split-level floors*

two adjacent anchorages has no prestress. The prestressing force disperses through the concrete, and, a certain distance away from the anchorages, the slab can be assumed uniformly stressed.

ACI 318 limits the spacing of anchorages to a maximum of six times the slab thickness. BS 8110 does not specify any such limit. In a 400 mm (16 in) deep floor slab, in theory the anchorages may be spaced at as much as 2.4 m (8 ft) centres. If the anchorages are spaced this far apart, then the unstressed wedges of concrete may not be able to support the design load; they may be particularly vulnerable to damage from concentrated loads. In order to reinforce these areas, the normal practice is to provide a mesh of rod reinforcement in a strip along the edge, of width equal to the anchorage spacing.

A different problem arises, from the same cause, in post-tensioned beams. Near the anchorages, the beam carries the full prestressing force; the force disperses into the adjacent concrete, which is likely to have a lesser, or no, prestress. Therefore, away from the anchorages, the beam section has lost some of the axial prestress. In some cases the loss may be large enough to result in the serviceability tensile stresses exceeding the permissible values.

It is worth noting that only the axial component of the prestress is dispersed in this manner; the equivalent balancing load, being a function of the tendon force, remains unaffected.

This loss must be allowed for in the serviceability state calculations for the beam. The severity of the problem can be reduced by providing a few tendons in the slab, at its centroid level, running parallel to the beams. Such tendons also enhance the capacity of the floor to support load concentrations by increasing the effective width of the slab.

Traditionally, a concentrated force is assumed to disperse at an angle of 45°. In fact, the stress on a plane, at right angles to the line of force and at a given distance from the point of application, is a maximum on the axis of the force and quickly drops away either side of it, see Figure 4.4. The total force acting on a length of the plane bounded by the 45° angle, i.e. between the lines $x/c = -1$ and $x/c = +1$, lines is approximately equal to the applied force; this is the basis of the commonly assumed 45° dispersion.

$$\sigma_y = \text{average stress on mid-plane} = P/2c$$

Figure 4.4 shows the stress at the middle of an infinitely long slab strip subject to concentrated loads. The stress σ_y along the 45° line, at $x/c = 1$, is less than one-ninth of that on the axis ($x/c = 0$). The diagram can be used to assess the stress at any point due to the axial component of prestressing force in a tendon.

In the case of a post-tensioned slab, with anchorages at a spacing S, the stress diagrams from the adjacent anchorages overlap, and it is reasonable to assume a uniform distribution a certain distance away from the edge. Figure 4.5 shows the stress patterns in an isolated long strip of unit thickness and width S, at distances $S/4$, $S/2$ and S from the anchorage. The values shown in diagrams (a), (b) and (c) should be multiplied by the average stress P/S to get the actual stress at a point. The isolated strip is not a true representation of a section of a post-tensioned slab because of the discontinuity at the edge, but it does indicate that the stress becomes nearly uniform at a distance S, suggesting a dispersion at a gradient of 1:2, i.e. 27° instead of the traditionally assumed 45°.

4.4 Column moments

The structural size of a column is determined by the combination of direct load and moment resulting from the frame action of a floor. The moment being highest at the floor-column junction, this section often governs the column size.

In a structural frame with a post-tensioned floor, there are two factors which reduce the magnitude of the moment at the floor-column junction, compared with a frame containing a reinforced concrete floor. First, the upward acting equivalent load from the curved tendon reduces the net downward load on the beam or slab, which directly reduces the moments. Second, the axial prestress

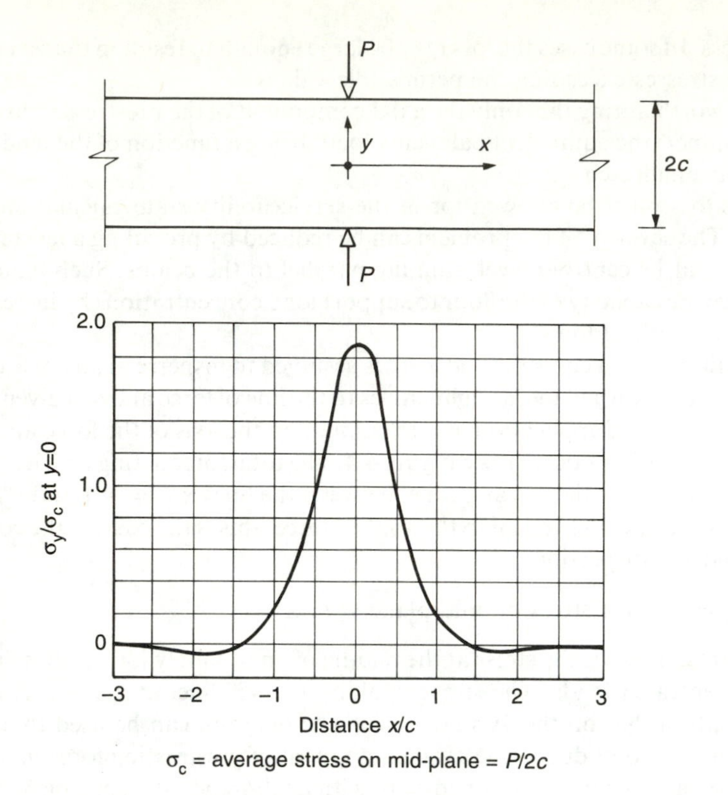

Figure 4.4 *Dispersion of a concentrated load (after Timoshenko and Goodier, 1951)*

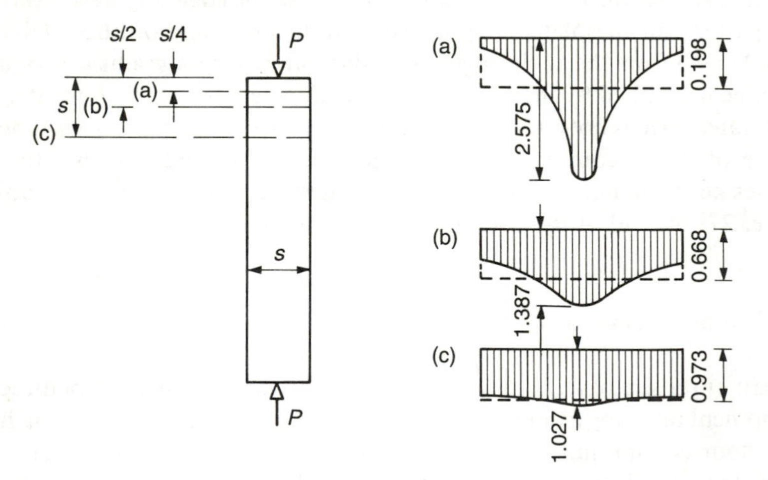

Figure 4.5 *Stress in a strip under concentrated loads*

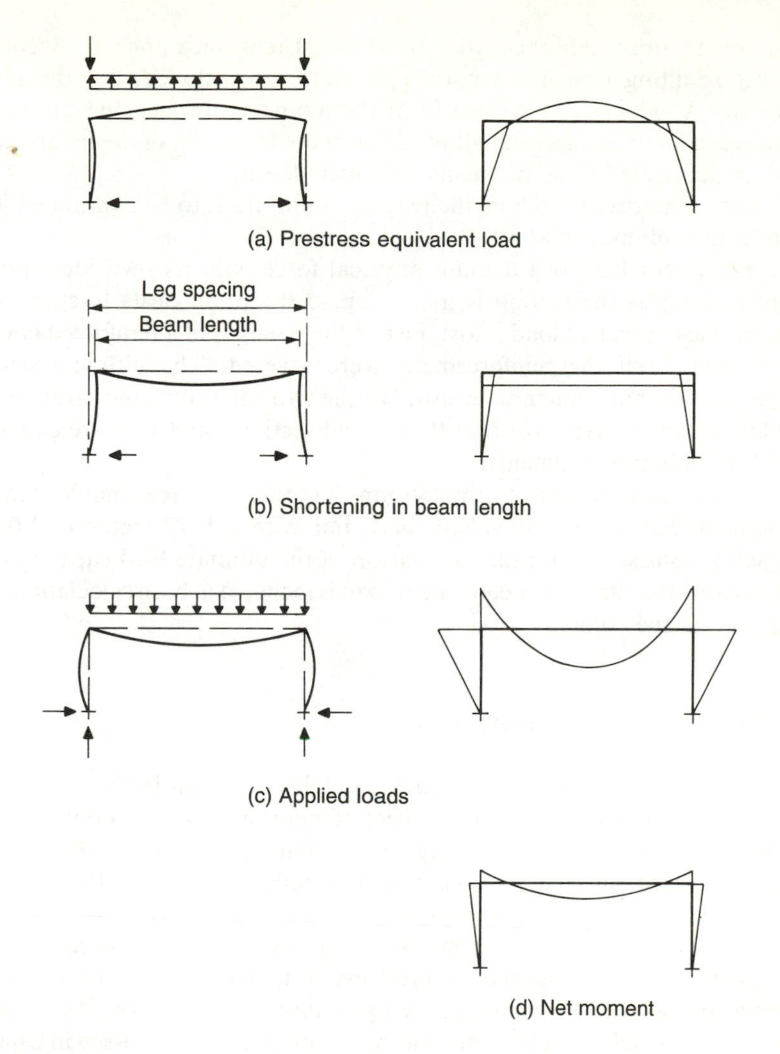

Figure 4.6 *Effect of prestress on column moment*

and shrinkage strains in the floor cause a slight shortening in the length of the floor, which pulls the columns inwards, inducing moments opposite to those caused by the normal loads. The moments in the floor itself are also affected in a corresponding manner; the former is automatically included in serviceability calculations, and the latter may be included in the final stage calculations if desired. Figure 4.6 illustrates the points. The left-hand sketches in Figure 4.6 show the loads and the exaggerated shape of the deflected frame, the right hand sketches show the corresponding moment diagrams. Figure 4.6(d) is the net moment, and a comparison with the moment in Figure 4.6(c) illustrates the reduction in column moment.

The columns are normally designed for ultimate strength, with factored loads and moments. In checking the strength of a post-tensioned floor, the secondary moments resulting from the tendon curvature are included, but the tendon eccentricity is *apparently* excluded from the moments. In fact, the effect of the tendon eccentricity is implicitly allowed for in the lever arm of the tendon and it would be duplicated if the moments also included it.

The question arises as to how the tendon curvature is to be considered in the design of the columns, if at all.

The equivalent load is a definite physical force with its own identity; it is present as long as the tendon is intact, up to the point of its fracture. Once fractured, the equivalent load is lost, just as the strength of a reinforced concrete beam would be if the reinforcement were severed. The ultimate strength requirements do not contemplate either of the two situations; the assumption is that plastic hinges have formed at the critical sections, and they are capable of resisting the ultimate moments.

In calculating the moment on the columns it is, therefore, reasonable to include the moment due to the equivalent load, but with a load factor of 1.0. The approach is consistent with the calculation of the ultimate load capacity of the beam, because the effect of the equivalent load is included in both calculations—for the beam and the column.

4.5 Movements in a concrete floor

In reinforced concrete, shrinkage causes a slight reduction in the length of the member and creep has the long-term effect of increasing the deflection of a floor. Post-tensioned concrete also undergoes shortening due to shrinkage and its deflection is also affected by creep; additionally, because of the axial pre-compression, a further shortening occurs because of elastic strain and creep. Therefore, deformation is higher in post-tensioned than in reinforced concrete, because of the axial component of the prestress. Both reinforced and post-tensioned concretes are affected by changes in temperature and humidity.

The factors, which affect the deformation of concrete, are discussed in Chapters 1 and 7; this section deals with the relative magnitudes of the movement caused by each.

Consider the deformations of a typical post-tensioned slab panel. Assume the following for this exercise:

Modulus of elasticity	$E_{ci} = 22\ \text{kN/mm}^2$
Creep coefficient C_c	$= 2.5$
Thermal coefficient C_t	$= 8 \times 10^{-6}/°\text{C}$
Shrinkage strain	$= 250 \times 10^{-6}$
Average prestress	$= 2\ \text{N/mm}^2$
Temperature change	$= 10\ °\text{C}$
Length of member L	$= 10\ \text{m}$

The changes in the length of the member are:

Elastic shortening – $2.0 \times 10/22$	=	-0.91 mm (0.036 in)
Creep – 2.5×0.91	=	-2.28 mm (0.090 in)
Shrinkage – $10000 \times 250 \times 10^{-6}$	=	-2.50 mm (0.098 in)
		-5.69 mm (0.224 in)
Thermal $\pm 10 \times 10 \times 8 \times 10^{-3}$	=	± 0.80 mm (0.031 in)
Change in humidity, say	=	± 1.00 mm (0.039 in)
		± 1.80 mm (0.070 in)

For a 10 m long slab, stressed to an average of 2.0 N/mm^2, the long-term shortening of its length would be 5.69 mm (0.224 in). To allow for the average humidity condition, deduct 1.0 mm from this figure. Then the change in length, including the humidity and the temperature changes, would be -4.69 ± 1.80 mm, i.e., the shortening in a 10 m length could vary from -6.49 to -2.89 mm (-0.256 to -0.114 in). The total shortening, 6.49 mm, represents 0.065% of the length.

It is worth noting that the above factors, except elastic shortening and creep, also affect reinforced concrete. In this example, the elastic shortening and creep account for approximately 50% of the movement. A reinforced concrete member of the same length would have only 50% of the shortening.

If the member was stressed to an average compression of 6 N/mm^2 (900 psi), as a beam can be, the elastic shortening, creep and humidity changes would be proportionately larger. Then, the shortening in a 10 m length could vary from -12.9 to -9.3 mm (-0.51 to -0.37 in).

In a reinforced concrete member of the same length, there being no axial stress, no shortening in length occurs due to the elastic and creep phenomena, and the effect of humidity is also smaller. Therefore, the variation in length may be of the order of -2.5 ± 1.0 mm, smaller than that for post-tensioned concrete.

Creep, shrinkage and temperature change depend on a number of environmental factors, as discussed in Chapter 2, and, therefore, the difference between a reinforced and a post-tensioned member would vary in accordance with the local conditions.

The expected range of movement in a post-tensioned floor should be assessed for each project. The architectural and the structural designs of the building must allow for the larger movement expected in a post-tensioned concrete floor, else the concrete may crack somewhere. This is of importance at the expansion joints where the relatively larger gap between two columns, or the sliding joint, may be unsightly if exposed on elevation. The expansion joint in the floor may need a cover strip. The filler in the joint must be sufficiently resilient to withstand the large movements.

4.6 Crack prevention

Most of the factors, which can contribute to the cracking of a floor, have been discussed in this and in other chapters. The subject is, however, considered important enough for post-tensioned concrete to be summarized here.

A concrete member can develop cracks through lack of care of the concrete at an early age, and through overloading. Additionally, in a post-tensioned member, the potential sources of cracks are:

- restraint from formwork between two downstand faces
- restraint from formwork at a hole in the floor
- restraint from stiff vertical members
- insufficient axial prestress if its dispersion has not been taken into account
- construction joints, where the concrete may lack in tensile strength
- brick or block wall panels erected early and tight against the columns

In order to prevent cracks attributable to formwork, its design should allow removal of some strategic narrow strips as soon as the concrete has hardened. Formwork inside a hole should be removed as soon as practical.

Re-entrant angles, such as at the corners of a hole or when a floor is stepped in plan, are potential points of stress concentration. It is a good practice to provide some rod reinforcement at these points—straight bars at 45° across the re-entrant angle are the common form of reinforcement. The amount of reinforcement is a matter of judgement; two 10 mm diameter bars top and bottom should be sufficient at holes up to about one metre long.

Restraint from the vertical elements to free movement of the floor, and the dispersion of prestress into adjacent concrete which has a lower average prestress, or is not prestressed, should be allowed for in the calculations.

At construction joints which are not in compression due to prestress, it is prudent to provide sufficient rod reinforcement for the total tension which may develop, assuming that the concrete has no tensile strength. Intermediate anchorages, of course, need bursting cages on the stressed side.

A partition consisting of brittle material and built hard against the structural frame containing post-tensioned floors may itself develop cracks. A partition trapped hard between two downstand beams may also show signs of distress. The erection of such partitions should be delayed to allow as much movement of the floor to take place as is permitted by the construction programme, and soft joints should be provided to allow shortening of the frame.

The restraint to movement due to the stiff vertical elements can induce horizontal thrusts of such magnitude that the restraining elements themselves become overloaded. As discussed in the Section 4.4, the effect of prestressing a floor is to reduce the magnitude of column moments. However, in particular circumstances the horizontal thrust on a column may produce the more critical condition for design. In this case, the column would tend to crack on the opposite face to that expected for the normal vertical loading, viz, on the inside face of a single bay. Most vulnerable in this respect are the columns in the lower storeys of tall buildings, columns in the basement where the retaining wall extends only to part of the height, and the end columns in long floors particularly if one end is restrained by a stiff core. The column forces should be calculated and allowed for in the design. In the case of low retaining walls, it may be possible not to connect the wall and the columns.

4.7 Tendon profile

The geometry and equivalent loads of the various tendon profiles used in post-tensioned floors are discussed in detail in Chapter 5.

Economic considerations often require that the prestressing force be the minimum which satisfies the serviceability conditions. Any deficiency in the strength of a member can be made up by adding bonded reinforcement; therefore, the level of prestress tends to be determined by the serviceability requirements.

In a simply supported member the amount of prestress is easily determined, because there are no redundancies. The problem of determining an economical tendon profile is more difficult in continuous members. The process requires that a profile be guessed initially and then refined in design by repetition. The design and refinement process is easier if the initially chosen profile is a reasonable approximation to the final profile.

Assuming that the final tensile stress is the criterion, an obvious goal is to so tailor the prestress that the final tensile stress equals the permissible value at all critical sections. A good starting point is for the product of the prestressing force and the eccentricity to follow the shape of the of the dead load moment diagram for the continuous member. An even better pattern to follow is that of the stress diagram corresponding to the dead load moment diagram, because this allows for the different section moduli at the top and the bottom of the section; it is, however, too complicated to calculate the stresses just for the initial shape of a profile.

In following the dead load moment diagram, advantage can be taken of the property that it is the sag of the parabola which is important, not its absolute positioning. Parabolas having the same sag are equivalent, and they all produce the same equivalent moment diagram after including the secondary moments. Therefore, in each span, the sag of the tendon profile should represent the sag of the parabola of the dead load moment diagram. This is tantamount to balancing a fixed proportion of the total load. The problem of optimizing the prestressing force can, therefore, be transformed into a consideration of how much load should be balanced by the tendon curvature. In order to balance a constant load w_e in spans of differing length, the ratio $P.e_p/L^2$ should be constant.

In order to vary the equivalent load in a span, either the prestressing force can be changed, or the eccentricity. If the tendons are placed at maximum eccentricity, then the tendon force must vary between spans, requiring a different number of tendons in each span. Alternatively, the same tendons may run through all spans and the eccentricity may vary from span to span. It is, of course, possible to use a combination of the two.

In continuous members of varying span lengths, or differing loads, it is often economical to curtail the tendons to suit the prestressing requirements. Figures 4.7(a), (b), and (c) show the possible tendon curtailments for providing a larger

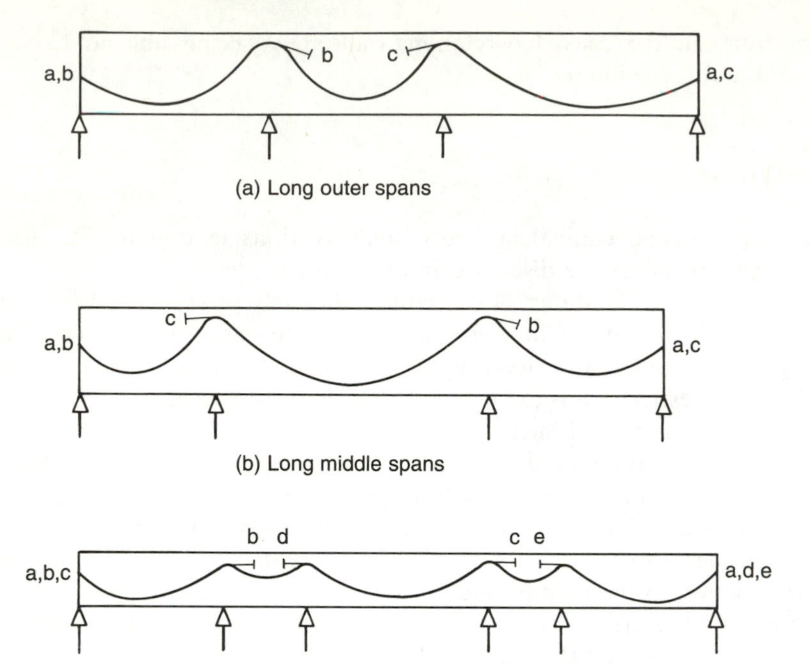

Figure 4.7 *Curtailing the tendons*

prestressing force in the external spans, in the internal spans, and in three alternate spans. In Figures 4.7(b) and (c), some of the tendons can run through all of the spans. It is desirable to avoid the complication of locating live anchorages away from the slab edge; this is a restriction which excludes the possibility of providing optimum prestress in all possible cases.

As a curved tendon is stressed, the strands tend to bunch up in the sheathing into the curve. Therefore, the centroid of the tendons is not coincident with the centre of the sheathing, which results in a reduction in eccentricity. The reduction is normally of the order of 9 mm (0.35 in) for multistrand tendons and 2 mm (0.08 in) for flat tendons used in floors. A useful approximation is to assume the tendon centroid at the third point of the sheathing diameter.

4.8 Access at the live end

Ideally, alternate anchorages would be stressed from opposite ends; this would distribute the effect of friction losses, and so provide an approximately uniform prestress in all the spans. In some locations, however, access may not be available to one or both of the slab edges for stressing the tendons. Two options are available in this situation.

- If access is available along one edge of the slab, then all of the live anchorages

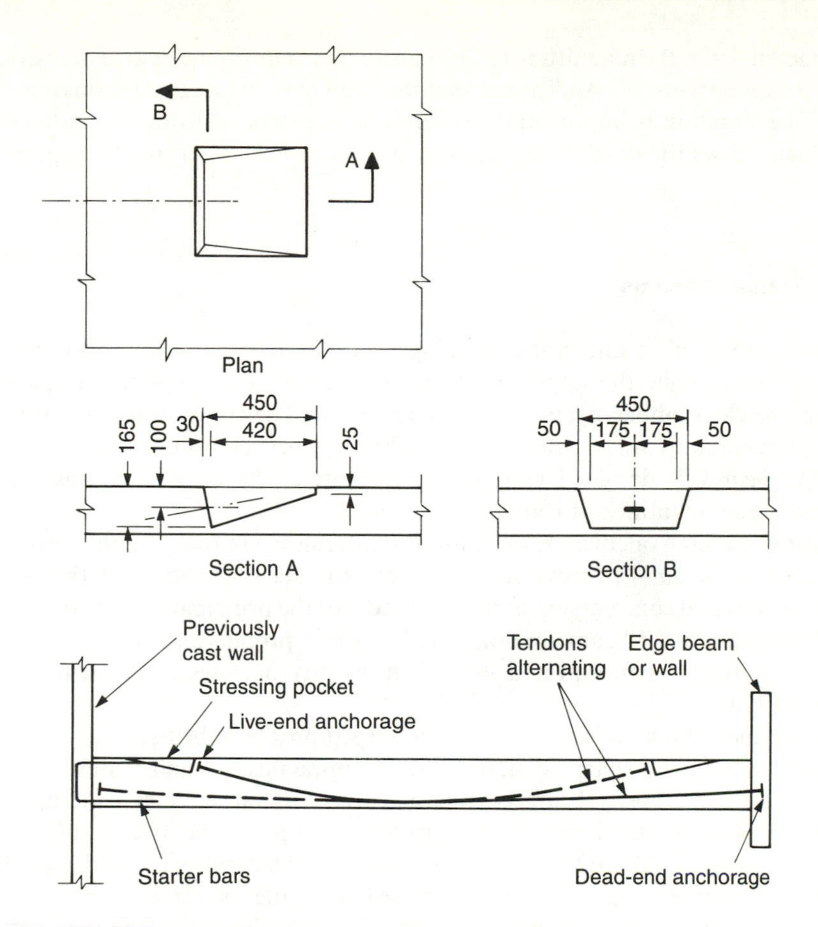

Figure 4.8 *Live anchorage for stressing from slab surface*

are located along this edge. The disadvantage of this arrangement is that the prestressing force is lower at the far end. If the far span is shorter than the others, then this layout is, in fact, preferable to that of alternate anchorages being stressed from opposite ends.

- If the loss in prestress from the above arrangement cannot be accepted, or if access is not available at either of the two ends, then the live anchorages can be placed in pockets, away from the support centreline, as shown in Figure 4.8. Sufficient space must be available behind the pocket for stressing the tendons. The dimensions shown in Figure 4.8 are typical for a 4-strand flat tendon.

It should be appreciated that such a tendon does not transfer any of the shear force to the support, and that the concrete between the anchorage and the support is not stressed. The unstressed strip must, therefore, be designed in reinforced concrete.

When stressed, the elastic shortening, shrinkage and creep between the stressed and the reinforced lengths induce tensile stresses in the unstressed concrete strip; the magnitude of this tension depends on the degree of restraint of the support system. Additional reinforcement may be required for this tension.

The situation is improved if the tendons are stressed from alternate ends, which allows the dead anchorages to be placed nearer, or at, the support.

4.9 Transfer beams

Most modern office and hotel buildings contain a column-free lobby at the ground floor, while the upper floors have columns at a much closer spacing. Often, the clear lobbies extend over several floors. The concentrated loads from the upper columns are transferred to suitable supports below through a series of *transfer beams*. If designed in reinforced concrete, the transfer beams would require large quantities of rod reinforcement.

Post-tensioning offers a very economical alternative in resisting the flexure of the transfer beam. Prestress also enhances the shear capacity of the beam, because the axial compression at the ends reduces the principal tensile stress. The tendons in a transfer beam are normally harped in profile (discussed in Chapter 5), the equivalent concentrated load then directly opposes the load from the upper column.

Sometimes, the span is too short for a harped profile to be used as it requires the tendon to have a sharp bend, needing an impractically small radius. The size of such a short transfer beam is governed by shear rather than bending. The tendon, in such a short beam, may be profiled as a parabola. The load from the upper column would not be opposed directly, but the parabola would provide a worthwhile moment opposing that from the load. If the span is too short for even a parabolic profile then straight tendons may be provided, so placed as to induce no, or very little, tension at the top.

The tendons in a transfer beam must be stressed in several stages, as the upper floors are constructed and their dead weight is imposed on the transfer beam. If fully stressed in one operation, the compressive stress at the bottom, and perhaps the tensile stress at the top, is very likely to exceed the safe value, possibly resulting in failure. The high level of prestress in a transfer beam requires particular care in its demolition. An uncoordinated removal of dead load, without a corresponding reduction in prestress, may cause an explosive failure.

A variation on the transfer beam is the cantilevered cap on the top of the core in a high-rise building, from which the lower floors are hung. In this system, one edge of a floor is supported on the core and the other by the hangers. The problem with reinforced concrete, or steelwork, in this situation is that of deflection. As each floor is constructed, the cap deflects and the hangers extend, thereby increasing the deflection of the previously completed floors relative to the core. Post-tensioning of the cap has the advantage that the deflection of the cap can be controlled, and it may even be possible to cause the cap to deflect upwards at each stressing stage, compensating for part of the elongation of the hanger.

4.10 Durability

For the normal building, which is not expected to house any deleterious substances, the most vulnerable constituent is the steel. Durability consideration, therefore, requires adequate protection of the tendon and the rod reinforcement to guard against their corrosion. The concrete cover is expected to provide this protection.

A building in post-tensioned concrete can suffer a greater loss of strength than in reinforced concrete, paradoxically because post-tensioning is more efficient. A floor contains a much larger quantity of steel if reinforced than post-tensioned, and the main reinforcement may consist of 20 or 25 mm diameter bars, when the strand consists of 4 or 5 mm diameter wires. Depending on the loading and span length, the spacing of strand in a post-tensioned slab may be two to four times the spacing of rod reinforcement in a reinforced concrete slab carrying the same imposed loading but 50% deeper. If reinforcement is spaced at 200 mm centres then the loss of one bar per metre width amounts to a reduction in strength of 20%; a post-tensioned floor may have strands at 600 centres and a loss of one strand is tantamount to a reduction of 60% in its strength. A corrosion loss of one millimetre in the diameter of a 20 mm reinforcement bar represents a 10% reduction of area, whereas it is equivalent to a 44% reduction in a 4 mm wire.

The strand tendons in a post-tensioned concrete member need a better quality of protection than rod reinforcement does in reinforced concrete. Post-tensioned elements normally remain crack-free under permanent loads, and they can even be designed to be tension-free under normal load conditions if required. The crack-free cover provides a better protection to steel than that given by the cover in a reinforced concrete member, which must be cracked.

The degree of protection provided by the concrete cover depends on the permeability of the concrete, the depth of cover, the chemical alkalinity of the immediate environment and the exposure conditions. The permeability of concrete can be minimized by having an adequate cement content, a low free water-cement ratio, by good compaction, and by sufficient hydration of the cement through proper curing.

Many processes of deterioration of concrete occur in the presence of water. Therefore, a structure should be so designed and detailed that water and moisture have a minimum access to concrete. Where unavoidable, the detailing should ensure that the water does not pool on the concrete surface or get trapped in contact with the concrete. Cracks can provide an access route for the moisture, and measures should be taken to avoid formation of cracks by the generation of heat, or due to shrinkage. Concrete is more likely to deteriorate if the sections are thin.

Car-parking floors and access areas to the buildings are normally exposed to the weather; they may also occasionally come into contact with deleterious chemicals. An enclosed building is likely to have a controlled environment, and its floors are not expected to undergo freezing or come into contact with injurious chemicals.

The strand used in unbonded post-tensioning is provided an additional protection by the extruded sleeve and grease packing. This protection is, however, not available to the anchorages.

4.10.1 BS 8110 requirements

Where concrete may be subjected to freezing and thawing, air entrainment should be considered. For 28-day strengths below 50 N/mm^2 (7000 psi), BS 8110 recommends the following air content by volume of the fresh concrete at the time of placing.

7% for nominal 10 mm maximum aggregate ($\frac{3}{8}$ in)
6% for nominal 14 mm maximum aggregate ($\frac{9}{16}$ in)
5% for nominal 20 mm maximum aggregate ($\frac{3}{4}$ in)
4% for nominal 40 mm maximum aggregate ($1\frac{1}{2}$ in)

With regard to the exposure conditions of a concrete structure, BS 8110 recognizes five categories of environment, as shown in Table 4.2.

For normal and lightweight concretes with 20 mm aggregate, BS 8110 recommends the minimum depth of cover to all steel, including stirrups, and the concrete contents shown in Table 4.3.

4.10.2 ACI 318 requirements

Where the concrete is subject to freezing and thawing cycles, ACI 318 recommends air entrainment in normal and lightweight concretes in the proportions given in Table 4.4. The specified tolerance in the air content is 1.5%. A moderate exposure is one where in a cold climate the concrete will be only occasionally exposed to

Table 4.2 *Exposure conditions*

Environment	*Exposure condition*
Mild	Concrete protected against weather or aggressive conditions
Moderate	Surfaces sheltered from severe rain or freezing whilst wet, concrete subject to condensation, concrete surface continually under water Concrete in contact with non-aggressive soil
Severe	Surfaces exposed to severe rain, alternate wetting and drying or occasional freezing or severe condensation
Very severe	Surfaces exposed to sea water spray, deicing salts, corrosive fumes, severe freezing whilst wet
Extreme	Surfaces exposed to abrasive action, e.g. sea water carrying solids or flowing water with pH ≤ 4.5 or machinery or vehicles

Table 4.3 *Cover for durability*

Exposure	*Nominal cover, mm Normal weight concrete*				*Nominal cover, mm Normal weight concrete*			
Mild	20	20	20	20	20	20	20	20
Moderate		30	25	20	45	40	35	30
Severe		40	30	25		50	40	35
Very severe		50	40	30		60	50	40
Extreme			60	50			70	60
Maximum water/cement ratio	0.60	0.55	0.50	0.45	0.60	0.55	0.50	0.45
Minimum cement content, kg/m^3	300	325	350	400	325	350	375	425
28-day cube strength, N/mm^2	35	40	45	50	25	30	35	40

Table 4.4 *Air content for frost resistant concrete*

Maximum size of aggregate		$f_c' \leq 35$ *N/mm*^{2} (5000 *psi*)		$f_c' \geq 35$ *N/mm*^{2} (5000 *psi*)	
mm	*in*	*Severe exposure*	*Moderate exposure*	*Severe exposure*	*Moderate exposure*
9.5	$\frac{3}{8}$	7.5 %	6.0 %	6.5 %	5.0 %
12.5	$\frac{1}{2}$	7.0 %	5.5 %	6.0 %	4.5 %
19.0	$\frac{3}{4}$	6.0 %	5.0 %	5.0 %	4.0 %
25.0	1	6.0 %	4.5 %	5.0 %	3.5 %
37.5	$1\frac{1}{2}$	5.5 %	4.5 %	4.5 %	3.5 %

moisture prior to freezing, and where no de-icing salts are used.

The ACI 318 requirements for special exposure conditions for concrete using normal density and low density aggregate are given in Table 4.5. The minimum cement content of concrete exposed to freezing and thawing in the presence of de-icing chemicals is specified as 252 kg/m^3 of concrete.

Table 4.6 shows the ACI 318 recommendation for the nominal minimum concrete cover to all steel (including ties) for prestressing tendons and rod reinforcement, ducts, and anchorage ends. For concrete members exposed to earth, weather, or corrosive environments, and in which the permissible tensile stress is exceeded, the minimum cover should be increased by 50%.

4.11 Fire protection

The cover concrete acts as an insulation to the tendons and rod reinforcement, delaying the rise of temperature of the steel in case of fire. The strength of steel is affected when exposed to high temperature; the prestressing strand is more vulnerable in this respect than high-strength rod reinforcement. Typically, the

Table 4.5 *Concrete for special exposure conditions*

Exposure condition	*Maximum water-cement ratio, normal aggregate*	*Minimum f'_c for low density aggregate (N/mm^2)*
Concrete intended to have low permeability when exposed to water	0.50	25
Concrete exposed to freezing and thawing in moist condition	0.45	30
For corrosion protection for reinforced concrete exposed to deicing salts, brackish water, seawater or spray from these sources	0.40*	30*

Note: If minimum concrete cover shown is increased by 10 mm ($\frac{3}{8}$ in), water-cement ratio may be increased to 0.45 for normal density concrete, or f_c' reduced to 30 N/mm (4500 psi) for low-density concrete.

Table 4.6 *Minimum cover (from ACI 318)*

Concrete cast against and permanently exposed to earth	70 *mm*	(3 *in*)
Exposed to earth and weather		
Slabs	30	($1\frac{1}{4}$)
Other members	40	($1\frac{1}{2}$)
Not exposed to earth or weather		
Slabs	20	($\frac{3}{4}$)
Main steel in beams	40	($1\frac{1}{2}$)
Ties in beams	20	($\frac{3}{4}$)

strand strength drops from a 100% at 150°C to nil at 700°C; the drop in the strength of the high-strength rod reinforcement is from a 100% at 300°C to 5% at 800°C. The depth of cover is, therefore, of a greater importance in post-tensioned than in reinforced concrete.

The fire resistance of a member is affected by a number of factors, including:

- depth of concrete cover
- disposition and properties of reinforcement and tendons
- the type of concrete and aggregate

Table 4.7 *Minimum dimensions and cover.*

Material and element		*Fire period (hours)*					
		0.5	*1.0*	*1.5*	*2.0*	*3.0*	*4.0*
Dense concrete							
Simply supported beams	width	100	120	150	200	240	280
	cover	25	40	55	70	80	90
Continuous beams	width	80	100	120	150	200	240
	cover	20	30	40	55	70	80
Simple plain soffit floors	thickness	75	95	110	125	150	170
	cover	20	25	30	40	55	65
Continuous plain soffit floors	thickness	75	95	110	125	150	170
	cover	20	20	25	35	45	55
Simply supported ribbed floors	thickness	70	90	105	115	135	150
	rib width	80	110	135	150	175	200
	cover	25	35	45	55	65	75
Continuous ribbed floors	thickness	70	90	105	115	135	150
	rib width	70	75	110	125	150	175
	cover	20	25	35	45	55	65
Lightweight concrete							
Simply supported beams	width	80	110	130	160	200	250
	cover	25	30	45	55	65	75
Continuous beams	width	80	90	100	125	150	200
	cover	20	25	35	45	55	65
Simple plain soffit floors	thickness	70	90	105	115	135	150
	cover	20	20	30	35	45	60
Continuous plain soffit floors	thickness	70	90	105	115	135	150
	cover	20	20	25	30	35	45
Simply supported ribbed floors	thickness	70	85	95	100	115	130
	rib width	75	90	110	125	150	175
	cover	20	30	35	45	55	65
Continuous ribbed floors	thickness	70	85	95	100	115	130
	rib width	70	75	90	110	125	150
	cover	20	25	30	35	45	55

Note: the cover refers to the tendon or main reinforcement, and the thickness refers to the depth of the topping concrete.

- conditions of end support
- size and shape of the member
- stress level

Rapid rates of heating, large compressive stresses or high moisture content (over 5% by volume) can lead to spalling of concrete cover at elevated temperatures, particularly for thicknesses exceeding 40 mm to 50 mm. Such spalling may impair performance by exposing the steel to the fire or by reducing the cross-sectional area of the concrete. Concretes made from limestone aggregates are less susceptible to spalling than concrete made from aggregates containing a high proportion of silica, e.g., flint, quartzites and granites. Concrete made from lightweight aggregates rarely spalls. Acceptable measures to avoid spalling are:

- an applied finish by hand or spray plaster, vermiculite, etc.
- the provision of a false ceiling as a fire barrier
- the use of lightweight aggregates
- the use of sacrificial tensile steel

A light welded steel mesh fabric, placed within 20 mm of the concrete surface, is often used as sacrificial reinforcement to prevent spalling. A problem arises if the durability considerations require a cover of more than 20 mm to all steel.

The behaviour of a structure, when exposed to fire, depends on the availability of the alternative support route for the floor loads; the condition of the end support is, therefore, significant. Adequacy of an alternative route requires that the steel is detailed so that, in the case of the primary route becoming ineffective, the load can be sustained through the alternative route, albeit at a reduced load factor.

The rate of rise of temperature of steel embedded in concrete depends on the ratio of surface area exposed to the fire and concrete volume, and on the mass of the concrete. The surfaces exposed to fire are the soffit for the slab, and sides and soffit for beams and ribs. Downstand beams and ribs, having a larger surface area to volume ratio, are normally required to have more cover than a slab with a flat soffit. A tendon located at the corner of a member receives heat from two surfaces, and is, therefore, more vulnerable than a tendon away from a corner. National standards usually specify the minimum concrete cover and the minimum size of a member for various fire periods. Some national standards also give limiting dimensions for the concrete members, which would generally ensure a satisfactory structural behaviour of the member; these are discussed in Chapter 3. Requirements of BS 8110 with regard to the fire protection are given in Table 4.7. In the case of downstand beams and ribs where the width, measured at the tendon level, is greater than the minimum, the covers shown in Table 4.7 can be reduced by 5 mm for every 25 mm increase of width; the reduction in cover is limited to 15 mm for dense concrete and 20 mm for lightweight concrete, and in no case should the resulting cover be less than that specified for a flat soffit floor of the same fire resistance period.

4.12 Minimum and maximum prestress

Utilizing only the tensile strength of concrete, it is possible to satisfy the serviceability requirements for a lightly loaded short-span floor without providing any prestress. Some bonded steel would, of course be required to comply with the ultimate strength requirements. If such a member is subjected to an increasing load, there is a sudden increase in the deflection as the concrete cracks, because the moment of inertia of the section drops suddenly from that of an uncracked section to that of a cracked section.

With a small amount of prestress, it is conceivable for the cracking stage to be the critical condition rather than the serviceability or the ultimate states. Such a member would fail in an explosive manner without warning.

At the other extreme, a member with a high average prestress is also likely to fail in an explosive manner, without much warning, if it is subjected to an increasing load. In this case, the ultimate strength is governed by the compressive strength of the concrete rather than the tensile strength of steel. This state is likely to be approached if the size of a member is reduced to the minimum possible.

Both of the above extremes are undesirable. Failure of a properly designed member should occur by the yielding of the steel—tendons and rod reinforcement, so that ample warning is given by the increasing deflection.

No minimum limit is specified in BS 8110 for the average prestress in concrete. The upper limit is considered to be satisfied if the ultimate moment of resistance exceeds the moment necessary to produce a flexural tensile stress in the concrete at the extreme tension fibres equal to $0.6(f_{cu})^{0.5}$. In this calculation the prestress may be taken as the value after all losses have taken place.

ACI 318 stipulates a minimum average prestress of 1.0 N/mm^2 on the gross concrete section after all losses, and it limits the maximum area of steel—tendons plus rod reinforcement as discussed in section 8.2.

4.13 Additional considerations for structures in seismic zones

It is well known that earthquakes impart very large horizontal loads to structures in addition to the vertical gravity loads. The primary objective of seismic-resistant design is, therefore, to provide adequate horizontal load resistance in the frame members, together with moment resistance and ductility in the joints. Seismic resistance is normally provided by a beam-column frame, often strengthened by shear walls. Although not forbidden by earthquake codes, the use of flat slabs, without beams, as part of a seismic moment resisting frame is quite rare, especially in areas of very high seismic risk. Normally the only requirement for the slabs themselves, therefore, is that they be able to act as horizontal diaphragms, distributing the seismic forces across the structure. This function can usually be fulfilled without difficulty by slabs designed for normal vertical gravity loads. For this reason, a detailed account of seismic-resistant design is outside the scope of this book. Some issues relating to the planning of the structure, with particular reference to floors, are discussed in this section, while detailing requirements are briefly discussed in Section 12.6. For a fuller treatment of seismic-resistant design, the reader is referred to specialist texts (Dowrick, 1987; Key, 1988; Naeim, 1989).

The response of a building to seismic loads can be significantly affected by its structural configuration both in plan and in elevation, making it important that seismic behaviour is fully considered at the conceptual design stage. As a general rule, desirable aspects of buildings in seismic zones are simplicity, regularity and symmetry. These properties help to ensure that the seismic forces are distributed

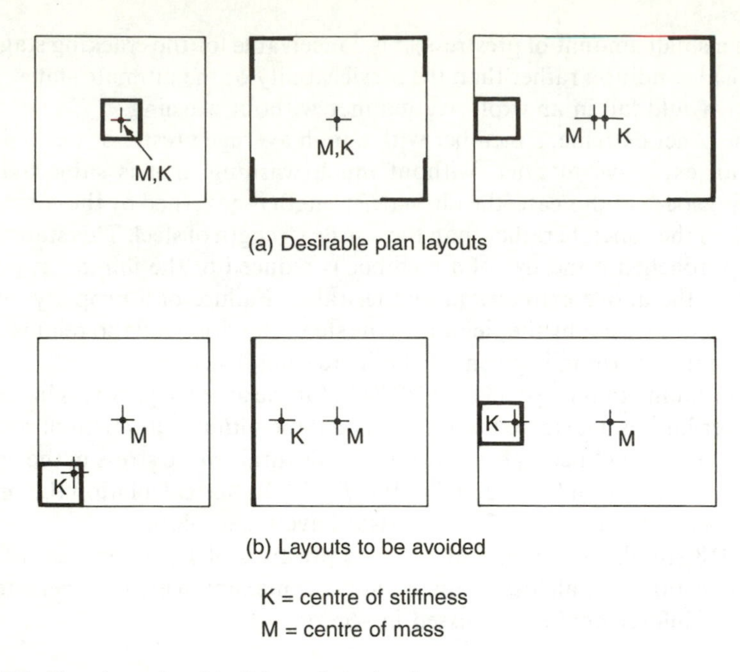

Figure 4.9 *Plan layouts of buildings in seismic zones*

evenly, whereas any variations in mass or stiffness distribution are likely to lead to an increase in the dynamic response.

Provision of horizontal load resistance in a building can be achieved by using a moment-resisting frame (i.e. one where the beam-column joints are designed to withstand the seismically-induced moments), by providing bracing, by the use of shear walls, or by some combination of these methods. The use of shear walls is popular and effective, but can be inconvenient in office buildings, since a high density of shear walls may interfere with the need for large, unobstructed areas. In locating the horizontal load-resisting elements, care should be taken to ensure that the centres of mass and stiffness of the building remain approximately coincident. Any eccentricity between the mass and stiffness centres will result in the generation of additional torsional moments and displacements under earthquake loading. Figure 4.9 shows some desirable plan layouts of horizontal load-resisting elements for seismic areas, and some which should be avoided.

Irregularities in the plan shape of a building should also be avoided where possible. For example, the L-shaped building shown in Figure 4.10 has two problems. Firstly, the lateral stiffnesses of the two legs of the building are substantially different, resulting in differential movements between the two legs, and hence very large stresses at the interface. Secondly, as discussed above, the layout causes an eccentricity between the centres of mass and stiffness, which is likely to result in significant torsional motion. Very similar problems are likely to be encountered in buildings which are tapered in plan.

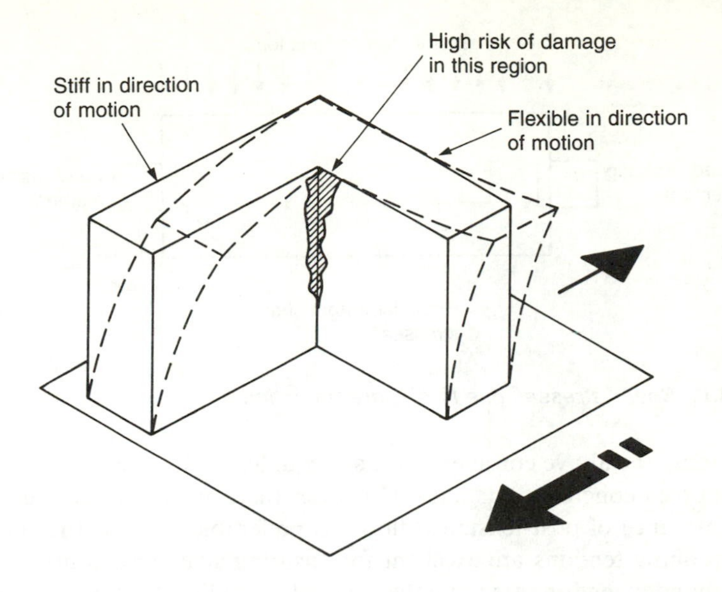

Figure 4.10 *Seismic movement of building with irregular floor plan*

Similarly, the vertical configuration of the building should be kept as regular as possible. Ideally, floors should have the same plan configuration at each level, with no setbacks, since the sudden change in stiffness at the level of the setback causes a concentration of load at that level. Floors should also be equally spaced as far as is practicable. Storeys which are higher than the others in a building are normally less stiff, and so sustain very large displacements, causing a *soft storey* failure. If, as is common, it is desired to have a high ground storey, the columns must be very carefully designed to ensure they have adequate strength and stiffness to resist such a failure.

In resisting the horizontal forces caused by an earthquake, it is important that all the load-resisting elements are securely tied together, ensuring good composite behaviour and providing additional redundancy. In many instances, this requires the floor slabs to act as diaphragms, distributing the horizontal loads across the structure. Post-tensioned floors can be regarded as rigid diaphragms, causing the loads to be distributed between the horizontal load-resisting elements in direct proportion to their stiffnesses.

The diaphragm action causes horizontal shears to be set up in the floor; usually these will be small and can be neglected, but occasionally configurations may be used which give rise to quite high shears, such as that shown in Figure 4.11. Even when the average shear is low, care should be taken to ensure that openings in the floor are adequately reinforced, are not so numerous or closely spaced as to cause loss of diaphragm action, and do not interfere with the attachment of the floor to walls or frames.

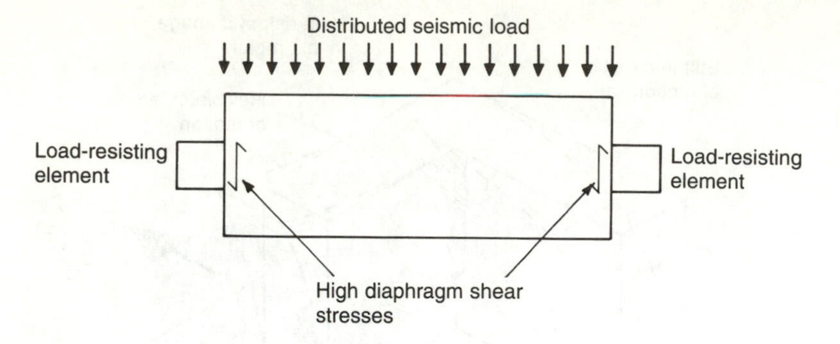

Figure 4.11 *Shear stresses due to diaphragm action*

In general, the above considerations are equally applicable to post-tensioned and reinforced concrete floor slabs. However, there are some respects in which the performance of post-tensioned floors is preferable. Firstly, the continuous post-tensioning tendons are excellent for ensuring structural continuity. Also, when unbonded tendons are used, they are able to redistribute peak forces along their length, and so have a lower risk of failure than bonded reinforcement. The only additional issue to be considered for post-tensioned floors is the location of the anchorages. These should be spaced as evenly as possible so as to avoid congestion and stress-raising in highly stressed zones. Anchorages should be positioned as far as possible from potential plastic hinge locations.

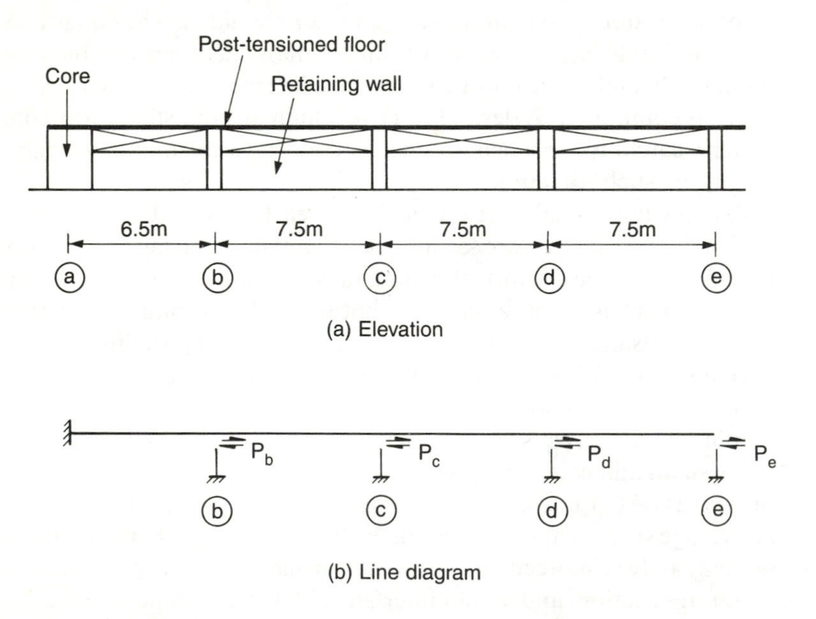

Figure 4.12 *Example 4.1*

Example 4.1 Restraint from columns

Figure 4.12 shows part of a building with post-tensioned floor. At one end is a stiff core, and a short retaining wall is cast monolithic with the columns in all the bays. Calculate the thrusts which would develop in the columns when the floor is stressed. Assume that the core is infinitely stiffer than the columns, the columns are pinned at top, and are fixed 1.5 m below the centroid of the floor. The other relevant data are:

Tendon force, constant from (a) to (d) = 1000 kN
Tendon force from (d) to (e) = 1200 kN
Cross sectional area of the associated floor(A_c) = 0.50 m^2
Moment of inertia of each column (I_c) = 0.01 m^4
Effective column height (H) = 1.50 m

Solution

Let δ_{12} represent change in length between points 1 and 2, and P_x the thrust on column x. For convenience, use kN/m^2 as the units for E_c the modulus of elasticity of concrete.

In an unrestrained floor, the elastic shortening will be:

$$\delta_{ab} = 1000 \times 6.5/(0.5 \times E_c) = 1300/E_c$$
$$\delta_{ac} = 1000 \times 14.0/(0.5 \times E_c) = 28000/E_c$$
$$\delta_{ad} = 1000 \times 21.5/(0.5 \times E_c) = 43000/E_c$$
$$\delta_{ae} = 1200 \times 7.5/(0.5 \times E_c) + \delta_{ad} = 61000/E_c$$

In the presence of the column thrusts, at each column position the sum of floor extension and column deflection must equal the above shortenings. Therefore

$$\text{at b, } (6.5P_b + 6.5P_c + 6.5P_d + 6.5P_e)/(A_cE_c) + P_b.H^3/(3E_cI_c) = \delta_{ab}$$
$$\text{at c, } (6.5P_b + 14P_c + 14.0P_c + 14.0P_e)/(A_cE_c) + P_c.H^3/(3E_cI_c) = \delta_{ac}$$
$$\text{at d, } (6.5P_b + 14P_c + 21.5P_d + 21.5P_e)/(A_cE_c) + P_d.H^3/(3E_cI_c) = \delta_{ad}$$
$$\text{at e, } (6.5P_b + 14P_c + 21.5P_d + 29.0P_e)/(A_cE_c) + P_e.H^3/(3E_cI_c) = \delta_{ae}$$

E_c cancels out from the two sides of the above equations. Substituting values of A_c, I_c and H gives:

$$125.5P_b + 13.0P_c + 13.0P_d + 13.0P_e = 13000$$
$$13.0P_b + 140.5P_c + 28.0P_d + 28.0P_e = 28000$$
$$13.0P_b + 28.0P_c + 155.5P_d + 43.0P_e = 43000$$
$$13.0P_b + 28.0P_c + 43.0P_d + 170.5P_e = 61000$$

Solving the above simultaneous equations gives the following column shears.

$P_b = 44.6$ kN $P_c = 102.1$ kN $P_d = 173.1$ kN $P_e = 294.0$ kN

5 TENDON PROFILES AND EQUIVALENT LOADS

In post-tensioned floors, the tendons seldom run in straight lines. They are normally curved in the shape of shallow parabolas; occasionally, a profile may consist of short parabolas joined by straight lengths.

In order to calculate the effect of a tendon on the structure, it is necessary to know its height along the length of the member. Tendon heights at various points are also needed later for their assembly on site. The geometry of the tendon profiles is discussed in this chapter, the idea of equivalent loads is introduced, and equations defining the various profiles in common use, and their equivalent loads, are given. The application of the equivalent load is discussed in Chapter 6.

5.1 General

As an example of a tendon profile containing the various shapes in general use, consider the continuous beam shown in Figure 5.1(a). It consists of three continuous spans with a cantilever at one end. The loading consists of predominantly concentrated loads on the cantilever and the next span 1–2, and of uniformly distributed loads on the other two spans 2–3 and 3–4.

Figure 5.1(b) shows a possible tendon profile for such a beam. In the cantilever and in span 1–2, where the loadings are predominantly concentrated, the profile consists of straight lines; the triangular shape in span 1–2 is called *harped*. In spans 2–3 and 3–4, where the load is uniformly distributed, the profiles are parabolic. The profile is symmetrical in span 2–3 and its lowest point is obviously at midspan. In span 3–4, the profile is unsymmetrical, and the length b, the distance from the support to the lowest point, is not known in this case.

Figure 5.1(c) shows the equivalent load corresponding to the profile shown at (b). A tendon cannot be bent at sharp angles, all corners in the profile are rounded off with short parabolas, as over supports 2 and 3, and at midspan 1–2. Without these short curves the profile would look like Figure 5.1(d) whose equivalent load is shown in Figure 5.1(e). At preliminary stages of design, the profile is often simplified to the form shown in Figure 5.1(d).

The profile in Figure 5.1(b) can be broken down into the three basic elements

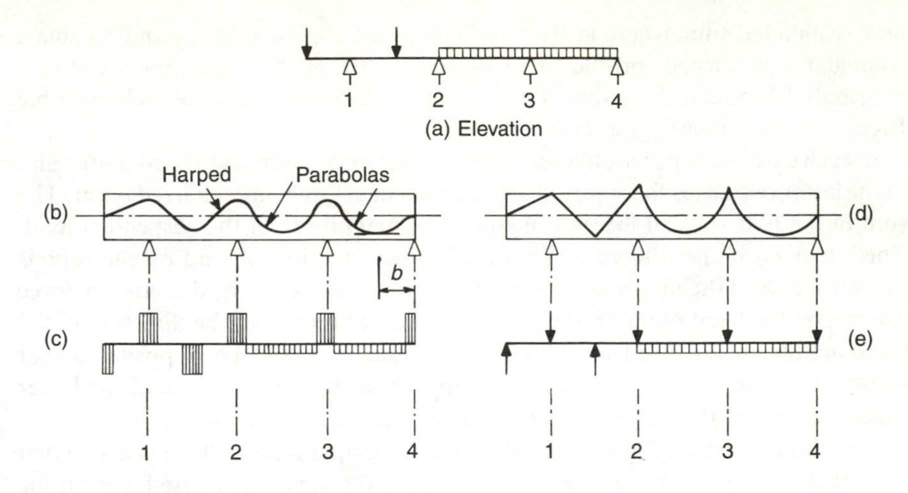

Figure 5.1 *Tendon profile in a continuous beam*

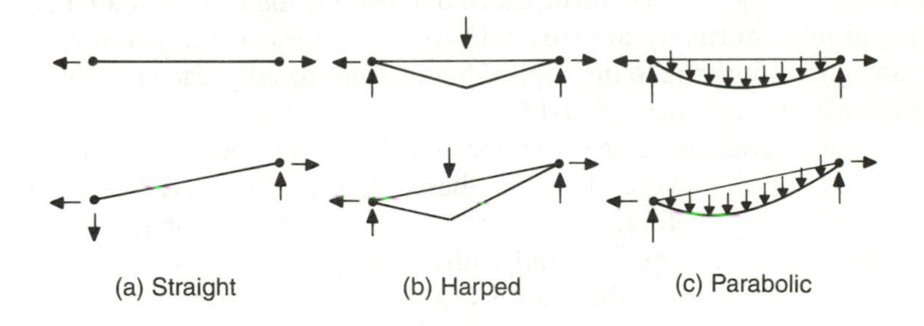

Figure 5.2 *Basic tendon profiles*

shown in Figure 5.2—a straight line, a triangle (harp) and a parabola. The upper diagrams show symmetrical profiles and the lower diagrams show their unsymmetrical forms.

5.2 Equivalent load

Most of the computer software packages for the design of post-tensioned floors have the capacity to calculate tendon geometry, equivalent loads and secondary moments from a simple input. Nevertheless, an understanding of the equivalent loads generated by different profiles and their combinations is very useful in the design of post-tensioning.

In order to appreciate the action of a tendon it is helpful to consider it as a length of rope strung between two fixed points representing the end anchorages. In the absence of any other load, and assuming the rope to be weightless, it would remain in a straight line as shown in Figure 5.2(a). If a concentrated load were now suspended somewhere in the middle then the rope would sag and assume a triangular, or harped, profile as in Figure 5.1(b). If the load were uniformly distributed between the support points then the rope profile would resemble Figure 5.1(c), which is parabolic.

In each case the rope would exert a horizontal force on the two fixed points and in the latter two cases there would also be a vertical force on each fixed point. The sum of the two vertical forces will equal the magnitude of the suspended load. These loads are also shown in Figure 5.2. The direction of load on the rope is downwards and the arrows at the two fixed points indicate the direction of force exerted by the fixed point on the rope; they do not represent the direction of the force acting on the fixed points which, of course, would be opposite to that shown. The upper diagrams have the supports at the same level and the lower ones at different levels, but this does not affect the basic load pattern.

From the above analogy it is evident that a unique load pattern is associated with each profile. A straight tendon does not exert any transverse force on the concrete member, a harped tendon exerts a concentrated force and a parabolic tendon, of the basic $y = Ax^2$ form, exerts a uniformly distributed load. In fact, the tendon profiles normally approximate to the shapes of the bending moment diagrams corresponding to the applied loads. Additionally, each profile exerts an axial force along the member axis.

It is often convenient to see a tendon profile as an imposed bending moment diagram. It should, however, be remembered that the tendon represents a line of compression and, therefore, the bending moment diagram is on the compression face of the member, i.e. opposite and a mirror image of the convention in concrete design where the moment diagram is drawn on the tension face.

A composite profile, a combination of one or more of the basic profiles, corresponds to a load distribution which is a combination of the individual load patterns for the components of the profile. For example, the profile of a rope with a straight length near one end and a parabolic shape for the remaining length corresponds to a uniformly distributed load along the parabolic length and no load along the straight portion, Figure 5.3.

Applied loads are sometimes triangular in shape, such as on a beam supporting a two way slab; this load pattern is associated with a cubic curve of the form $y = A.x^3$. However, the cubic profile is almost never used for tendon drape, it being sufficient to use a parabola corresponding to a uniform loading.

In calculations, the prestressing force in a tendon and its profile can be considered in two alternative ways, see Figure 5.4.

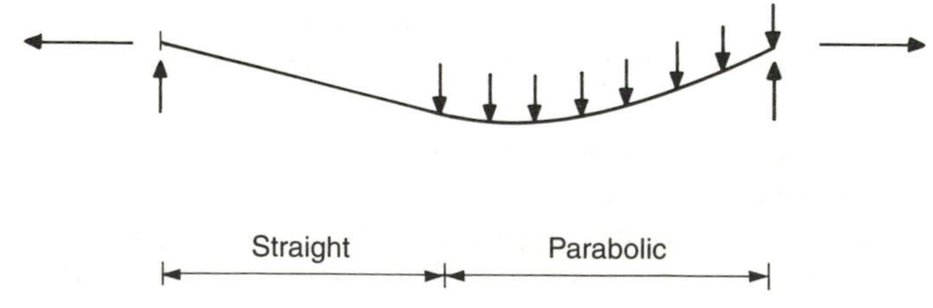

Figure 5.3 *A composite profile*

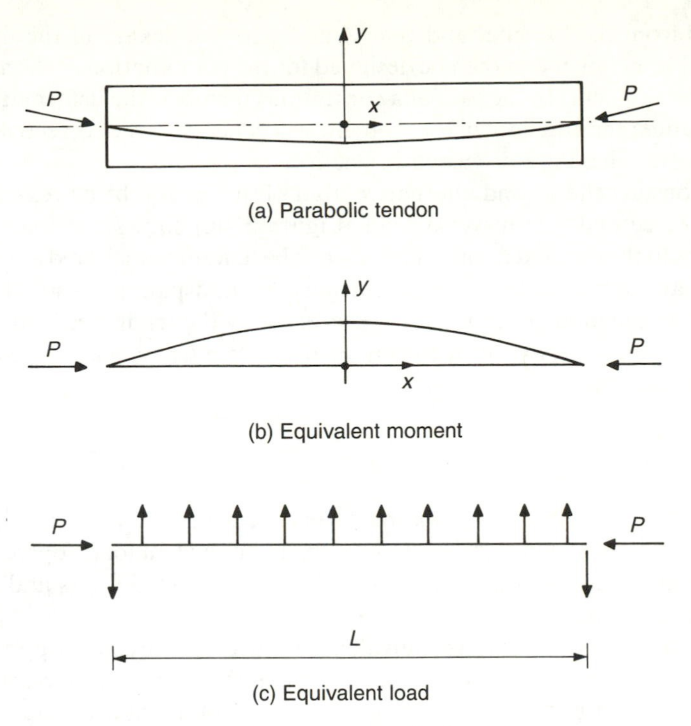

Figure 5.4 *Equivalent alternatives for a parabolic tendon*

- either as an axial force and a moment diagram represented by the product of the prestressing force and its eccentricity at each point along its length, Figure 5.4(b)
- or, as an axial force and an equivalent load acting at right angles to the member along its length, Figure 5.4(c). The equivalent load will, of course, give rise to the bending moment diagram shown in Figure 5.4(b).

Consider the first alternative, that of a tendon profile being represented by an axial force and a moment diagram. Let y_m be the tendon ordinate at midspan, measured from the section centroid.

Then $y = Ax^2 + y_m$, and the moment at x is given by

$$M_x = Py = P(Ax^2 + y_m) \tag{5.1}$$

The diagram for the equivalent moment M_x is shown above the beam centreline in Figure 5.4(b) following the normal convention used in concrete design where the diagram is drawn on the tension face. Note that by convention the eccentricity, denoted by e, is taken to be positive when the tendon is located below the section centroid. However, in this case the standard geometric convention is being followed where positive is upwards, and hence the symbol y_m is used rather than e_p.

In the case of a simply supported span this moment must be superimposed on moments from the loading, and that is as far as the flexure of the member is affected. The member can then be designed for the combination of the axial force and the net moment. In the case of a continuous member, the deformation of the member under the influence of the moment will generate some corrective forces at the supports. These are discussed in Section 5.3.

Now consider the second alternative, that of the tendon being represented by an axial force and a transverse load. Figure 5.4(c) shows the load which is equivalent to the stressed parabolic tendon. The tendon is anchored at the section centroid at each end and the eccentricity at midspan is $-y_m$. For static equilibrium, the moment at midspan produced by the tendon eccentricity must equal the moment produced by the uniformly distributed *equivalent load* w_e, due to tendon curvature.

$$M = -Py_m = w_e . L^2/8$$
$$w_e = -8Py_m/L^2 \qquad (5.2)$$

Both the above alternatives are mathematically correct and either can be used to analyse a post-tensioned member. However, the equivalent load approach is very much the favoured method in the design of post-tensioned floors and is further discussed in Chapter 6.

A tendon is draped so that its equivalent load acts in a direction opposite to the dead and superimposed loads, i.e., the equivalent load is arranged to act upwards on a span in a normal floor. It then balances part of the design load; hence the equivalent load is also termed the *balanced load* and the analysis associated with this approach is known as the *load balancing method*.

In Figure 5.1, diagram (d) shows the equivalent loads for the tendon profile (b). Diagram (b) represents the practical shape of the tendon profile as it might be used, and its equivalent load diagram represents the true loading which the tendon exerts on the continuous beam. Most of the computer programs are designed to work with the true equivalent load diagram, such as that shown in Figure 5.1(d). However, at the initial design stage and for manual calculations, it is expedient to use a simplified profile and its simpler equivalent load diagram, see Figures 5.1(c) and (e). The inaccuracy resulting from the simplification is negligible in most cases. It is worth noting that the profile in diagram (c) projects outside the outline of the beam; this is quite acceptable, knowing that the curves, to be introduced later to round the corners off, will bring the profile within the desired envelope of the required concrete outline allowing for the necessary covers.

5.3 Secondary moments

Consider a two-span continuous post-tensioned beam. Ignore the self-weight of the beam. Before prestressing, the beam soffit is in contact with the three supports, Figure 5.5(a). When post-tensioned with a straight eccentric tendon, a uniform moment is induced along the length of the beam. If the beam were not held down at the support, it would deflect upwards, creating a gap δ between the

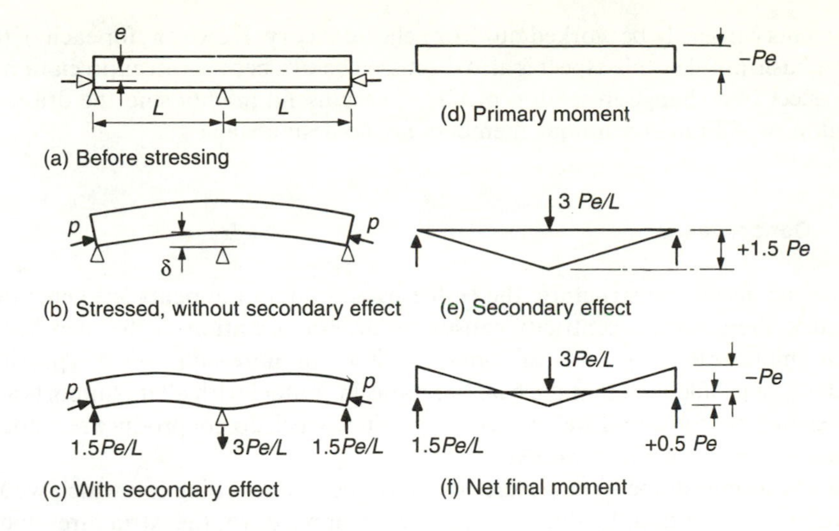

Figure 5.5 *Secondary forces and moments*

middle support and the beam soffit, Figure 5.5(b). The bending moment due to the tendon eccentricity, Py_x at any point x, is called the *primary moment*.

In order to maintain contact with the middle support, a force is generated between the support and the member, which produces a bending moment diagram along the member of such a shape that the gap δ closes, Figure 5.5(c). This restoring force is called the *secondary force* and its corresponding moment the *secondary moment*. Its value at a point x is given by $M_x - Py_x$, where M_x is the moment produced by the prestress on the indeterminate structure at point x.

The primary system is internal to the beam structure, in that it causes flexure of the beam but has no direct effect on the support reactions; the shears are balanced by the tendon slopes over the length of the beam and there is no residual shear. Flexure of the member due to the primary system may generate secondary restoring forces as described above, and it is these secondary forces and their corresponding secondary moments which amend the support reactions.

In the above example, the two-span beam has only one redundancy and, therefore, only one secondary force is generated. In an indeterminate structure the number of secondary forces equals the number of indeterminancies, and the secondary moment is the moment due to all such secondary forces acting simultaneously.

Obviously, the net effect of a tendon is the sum of the primary and the secondary moments, Figures 5.5(d) to (f). Effectively, the virtual position of the tendon differs from its actual position. This virtual position is referred to as the *line of pressure*.

The secondary moments, being caused by the concentrated forces at the support points, are always linear, varying uniformly over the length of a span; the secondary shear forces are constant over the span length. The secondary moments can easily be worked out from elastic theory. However, for each of the three basic profiles, this aspect is also discussed below, because an appreciation of the effect of a change in tendon profile is very useful in choosing the drape of tendon profiles in continuous members at the design stage.

5.4 Concordance

In an indeterminate structure, the restoring secondary moments are generated because the tendon eccentricity causes flexural deformation of the member in such a manner that it would lose contact with one or more supports. It is possible to devise a profile such that the beam remains in contact with all its supports and no secondary moments develop. Tendon profiles which do not produce secondary moments are termed *concordant*.

If a tendon is draped in the exact shape of the moment diagram which would develop if a certain loading pattern were applied to the structure under consideration, then no secondary moments will be generated, because the drape is already based on a shape which includes the effect of indeterminancy. The loading need not represent the actual loading to be applied. For example, in a three-span slab, no secondary moment results if a tendon is profiled in the shape of the bending moment diagram due to a single point load on any of the spans. Of course, such a profile is unsuitable if the beam carries a uniformly distributed load and it would be much better if the profile were parabolic, representing the uniform loading.

5.5 Tendon profile elements

The three elements of tendon profiles (the straight line, the harp and the parabola) are discussed below. Geometrical equations defining the curves and their equivalent loads are given where appropriate. In each case it has been assumed that the tendon slope θ at the anchorage is small so that

$$\sin\theta \approx \tan\theta \approx \theta, \quad \text{and} \quad \cos\theta \approx 1$$

5.5.1 Straight tendon

Straight tendons on their own are most commonly used in ground slabs. In post-tensioned suspended floors a short straight length is usually provided immediately behind a live or a pre-locked dead anchorage, the end from which the tendon is to be stressed; a short straight length may also be provided to bridge any gap between two curves. A straight tendon does not have any load shape directly associated with it but it may be useful for transferring shear between adjacent supports of a continuous member.

Two types of eccentric straight tendons are considered below—running parallel to the member axis with a constant eccentricity, and with the eccentricity varying linearly along the member length.

A straight eccentric tendon, running parallel to and below the axis of the member, generates a constant primary moment of magnitude Pe, where P is the prestressing force and e the eccentricity, see Figure 5.5(d). For the tendon profile shown, the secondary force acts downwards on the beam and upwards on the middle support, pulling the two together. Therefore, under applied loading, the middle support itself has its load reduced by the amount of the pull and the two outer supports have a corresponding increase in their reactions. The result is a transfer of shear from the middle support to the outer supports. In this case a load of $3Pe/L$ is transferred from the centre support, half the value to each of the two outer supports. If the tendon is placed above the neutral axis then the transfer of load will be in the opposite direction, i.e., from the outer supports to the centre one.

For Figure 5.5,

$$\begin{aligned} R_2 &= -3Pe_p/L \\ R_1 &= R_3 = -R_2/2 = +1.5Pe_p/L \end{aligned} \tag{5.3}$$

For unequal spans, L_1 and L_2,

$$\begin{aligned} R_2 &= -1.5Pe_p(L_1 + L_2)/L_1L_2 \\ R_1 &= +1.5Pe_p/L_1 \\ R_3 &= +1.5Pe_p/L_2 \end{aligned} \tag{5.4}$$

In these equations, the reaction acting upwards on the beam, and downwards on the support, is taken as positive. Eccentricity e_p is positive when the tendon is below the section centroid.

Now consider the case of a two-span beam with end anchorages at the section centroid but the tendon raised at the centre support, Figure 5.6. Note that the eccentricity e in this case is negative. Tendon eccentricity varies linearly along the span length. The shape of the primary moment diagram is that of the tendon profile, and its magnitude at each point equals the product of the prestressing force and the eccentricity. The top of the beam is in compression and the bottom in tension, the beam tends to deflect downwards. The centre support, however, does not allow any deflection at that point and it exerts an upward force which, in turn, generates downward reactions at the two outer supports. A secondary bending moment is thereby induced, exactly opposite to that produced by the eccentricity of the tendon. The two opposing moments, the primary and the secondary, cancel each other out and as a result the member has only the axial force acting on it, and no net bending moment at all.

There is, however, a transfer of load from the outer supports to the centre support. If the tendon at the centre support is below the section centroid then the transfer of load is from centre support to the outer ones.

For Figure 5.6,

$$\begin{aligned} R_2 &= 2Pe/L = 2P\tan\theta \\ R_1 &= R_3 = -R_2/2 = -Pe/L \end{aligned} \tag{5.5}$$

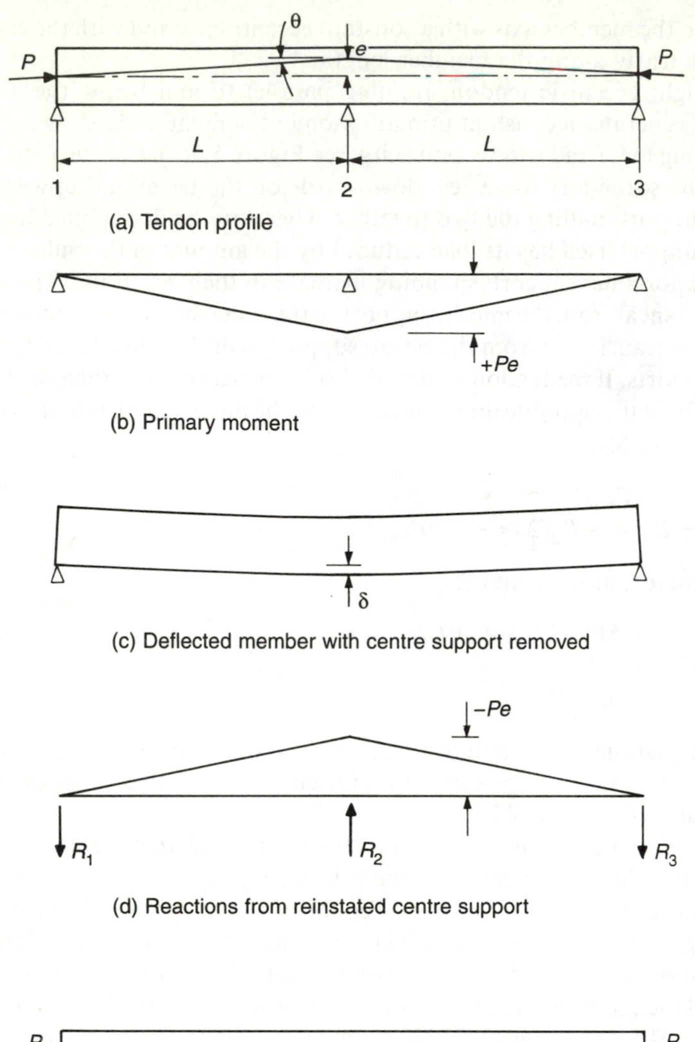

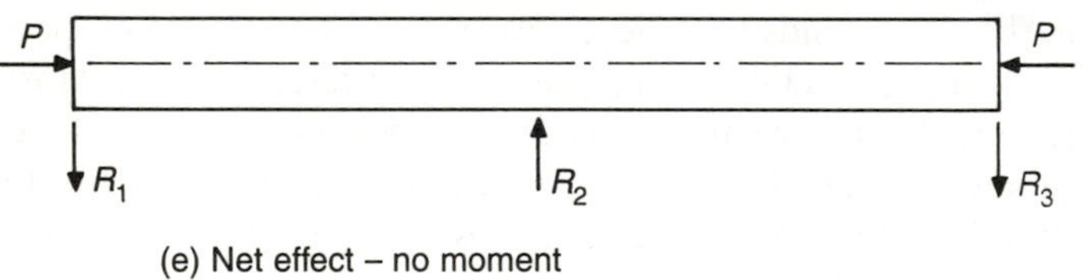

Figure 5.6 *Straight tendon in continuous spans*

If the two spans are unequal, of lengths L_1 and L_2 as shown in Figure 5.7 then the net result still amounts to an axial force and no moment. The reactions, indicating the transfer of load in this case are:

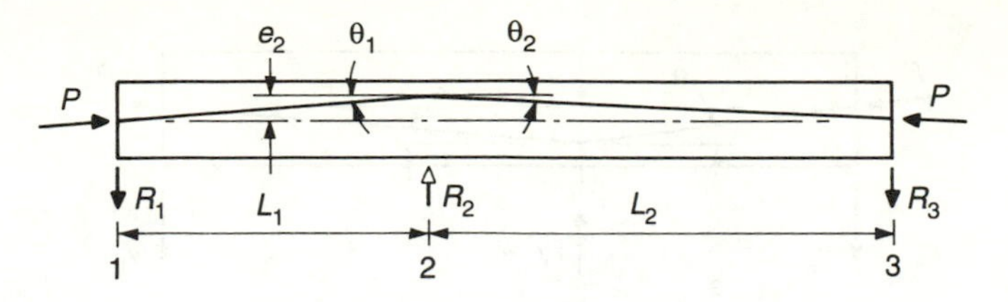

Figure 5.7 *Straight tendon over two unequal spans*

For Figure 5.7,

$$\begin{aligned} R_1 &= -Pe_2/L_1 = -P\tan\theta_1 \\ R_3 &= -Pe_2/L_2 = -P\tan\theta_2 \\ R_2 &= -(R_1 + R_3) = Pe_2.(L_1 + L_2)/L_1L_2 \end{aligned} \tag{5.6}$$

where θ_1 and θ_2 are the angles of tendon slope in spans L_1 and L_2 respectively.

This property of a straight tendon, that *it can be quite arbitrarily displaced at an internal support but the tendon remains equivalent to an axial load only*, is very significant. It effectively means that a straight line profile, such as that shown in Figure 5.7, can be superimposed on any other profile (harped, parabolic, etc.) in a continuous member without affecting its equivalent load. *Conversely, in the case of a continuous span with the tendon draped as a harp or a parabola, the eccentricities at the interior supports can be ignored in calculating the equivalent load, provided that the total sag of the profile is taken in the calculations.* This is another very useful property and will be utilized in further discussions.

It is important to remember that the secondary reactions R_1, R_2 and R_3 in Equations (5.3), (5.4), (5.5) and (5.6) are actual physical forces which act on the supports. They are the corrective forces resulting from the eccentricity of prestress. These secondary forces and the corresponding secondary moments are to be added to other external forces and moments, such as those resulting from applied loads, to arrive at the net moments and shear forces.

5.5.2 Harped profile

A harped profile gives rise to an equivalent concentrated load. This profile is suitable for members which carry dominant concentrated loads, such as transfer beams where a column cannot be carried down to its foundation and must be supported by a beam, or a slab which carries a set-back facade above. The primary system in this case consists of the triangular moment diagram representing Pe_p, and the associated equivalent point load W and the shears V_a and V_b in the beam.

In Figure 5.8(a) the tendon is at the section centroid at both ends. In Figure 5.8(b) at one end the tendon is at the section centroid but at the other end it has an eccentricity e_r (negative) and the span eccentricity is e_m (positive). For the general case of Figure 5.8(b),

$$V_a = P\tan\theta_1 = +Pe_m/a$$

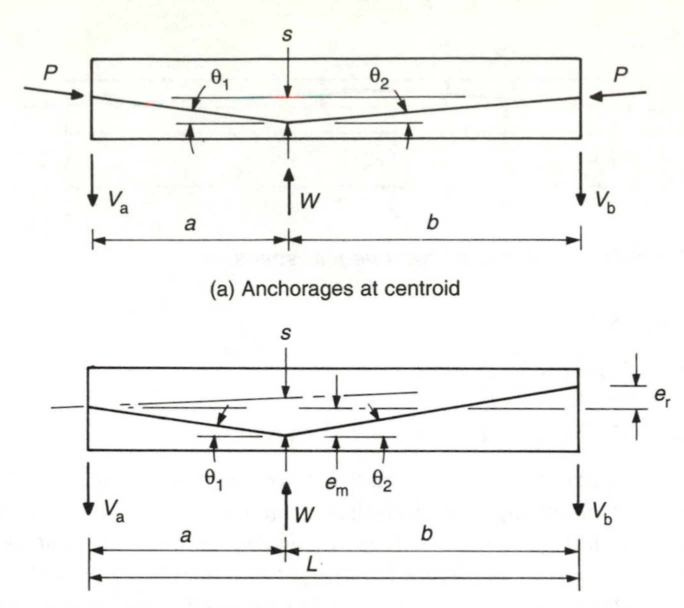

Figure 5.8 *Harped tendon*

$$V_b = P \tan \theta_2 = +P(e_m - e_r)/b \tag{5.7}$$
$$W = -V_a - V_b = -PsL/ab$$

where s = the sag of the profile
L = span length = $a + b$

Sag s is measured below the straight line joining the two ends of a profile and its value is negative in the normal drape where the tendon is lower in the span than at the supports.

The *equivalent* concentrated load W depends on the total sag of the profile, and is independent of e_r the eccentricity at individual supports. Therefore, Figures 5.8(a) and (b), in which both tendons have the same total sag s, represent the same equivalent load. However, in (b) there is a moment at one end and the support eccentricity affects the values of θ_1 and θ_2. If the shears V_a and V_b are calculated for the two diagrams, their values will be found to differ.

It is, of course, not possible to provide a sharp kink in a tendon as the harped profile implies; the tendon, in fact, is arced with a radius of about 2.5 m (8 ft) and, therefore, the reaction is a distributed load over the length of the arc, but it is convenient to think in terms of a concentrated load. The short curve is in practice treated as a parabola.

5.5.3 Parabolic profile

Most of the suspended floors in buildings are designed for a uniformly distributed load which corresponds to a parabolic profile.

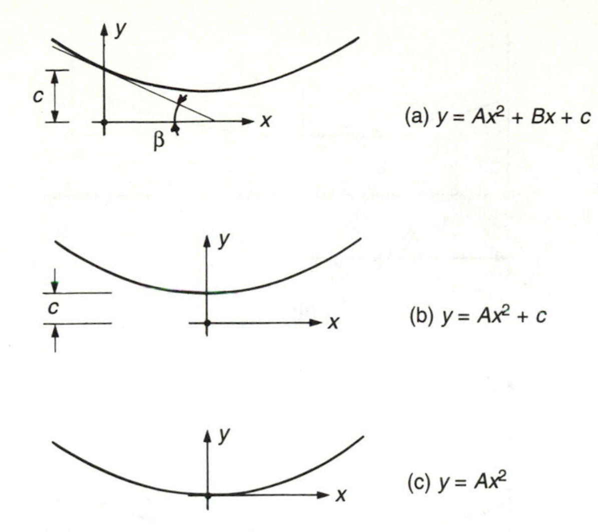

Figure 5.9 *Parabolas*

A second degree parabola is represented by the general Equation (5.8a). In this equation A represents the curvature, so that the smallest radius for the curve is $1/(2A)$; B equals tan β, the slope of the curve at the origin ($x = 0$); and C represents the height at the origin, see Figure 5.9(a). If the origin is located at a point where the tangent to the curve is horizontal, i.e. at the lowest or the highest point, Figure 5.9(b), then the term B becomes zero and the equation reduces to a more convenient form (5.8b), and if the zero tangent point coincides with the origin, Figure 5.9(c), then the equation reduces to its simplest and most convenient form (5.8c).

$$y = Ax^2 + Bx + C \tag{5.8a}$$

$$y = Ax^2 + C \tag{5.8b}$$

$$Y = Ax^2 \tag{5.8c}$$

It is not, however, convenient to express all possible parabolas in a profile in terms of Equation (5.8c). Equations (5.9a) and (5.9b) are used in such cases as more convenient replacements for Equation (5.8b) and (5.8c) respectively.

$$y = A(x - x_0)^2 + C_0 \tag{5.9a}$$

$$y = A(x - x_0)^2 \tag{5.9b}$$

where x_0 = distance to the lowest point
C_0 = tendon height at the lowest point (see Figure 5.10).

A parabola may be required to pass through three known points x_1, y_1; x_2, y_2 and x_3, y_3. For the general Equation (5.8a), values of A, B and C are given by:

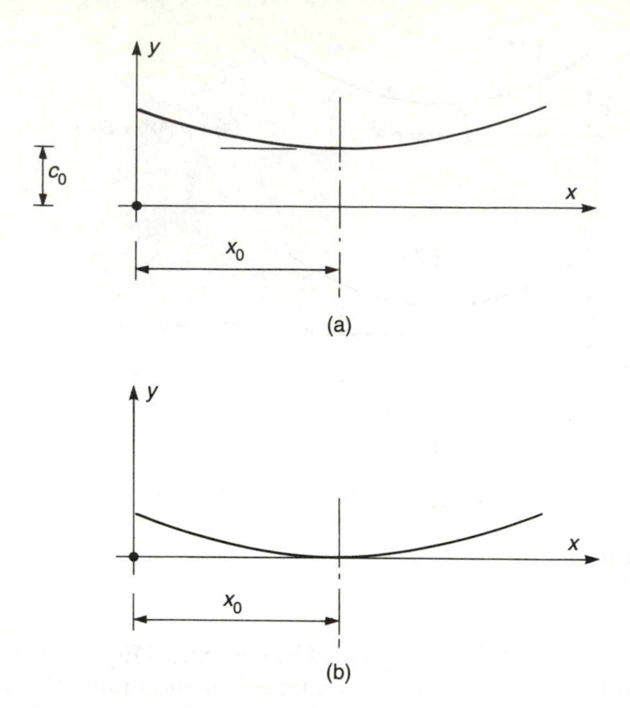

Figure 5.10 *Alternatives to Figures 5.9(b) and 5.9(c)*

$$\begin{aligned} A &= (m_1 - m_2)/(x_1 - x_3) \\ B &= m_1 - A(x_1 + x_2) \\ C &= y_1 - Ax_1{}^2 - Bx_1 \end{aligned} \tag{5.10}$$

where $m_1 = (y_1 - y_2)/(x_1 - x_2)$
$m_2 = (y_2 - y_3)/(x_2 - x_3)$

Parabolas of the form of Equations (5.8) and (5.9) exhibit a few properties which are convenient for working out the geometry of a tendon and its equivalent loads.

As stated earlier in this chapter, the minimum radius of curvature of a parabola is $1/(2A)$.

For Equations (5.8b) and (5.8c) the slope at any point is $2Ax$, which equals $2y/x$. Therefore, the tangent at any distance x_1 bisects the distance x_1 as shown in Figure 5.11. Also, the tangent at a point (x_1, y_1) intersects the y-axis at $-y_1$, which means that the slope of the tangent is twice the slope of the line joining the origin and the point (x_1, y_1).

For a general parabolic profile, Figure 5.9(a) and Equation (5.8a),

$$\begin{aligned} M_x &= \text{moment at distance } x \text{ from left} \\ &= Py_x = P(Ax^2 + Bx + C) \end{aligned}$$

$$\begin{aligned} w_e &= \text{equivalent load} \\ &= -d^2(M_x)/dx_2 \\ &= -2PA \end{aligned} \tag{5.11}$$

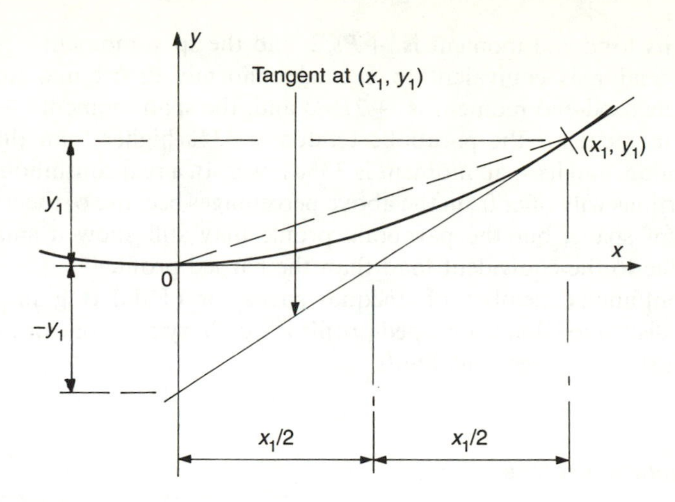

Figure 5.11 *Tangent to a parabola*

Besides the prestressing force P, the value of w_e depends only on the coefficient A; it is independent of B and C in Equations (5.8) and (5.9). Therefore, all parabolas sharing the same value of A are equivalent so far as their equivalent load is concerned. This result can, in fact, be inferred from Section 5.4.1, where it was shown that a straight profile can be superimposed on any other profile without affecting its equivalent load.

This means that *a tendon can be draped to follow any section of a given parabola without changing its equivalent load*. Of course, each section will produce its own secondary moments in an indeterminate structure but the final moment diagram, the sum of the primary and the secondary moments, will be exactly the same for each section.

The relationship between s, the total sag of the parabola, the coefficient A and the equivalent load can be calculated as follows.

$$M_s = \text{midspan moment}$$
$$= -Ps = -w_e.L^2/8 = -2PAL^2/8 \text{ which gives}$$
$$A = +4s/L^2 \tag{5.12}$$

$$w_e = -2PA$$
$$\text{or } w_e = -8Ps/L^2 \tag{5.13}$$

5.5.4 Harped versus parabolic profile

The discussion so far has implied that for a uniformly distributed load the suitable tendon profile is parabolic and for a concentrated load it is harped. In fact, given the right circumstances, a harped profile may be used with advantage for a uniformly loaded member.

Consider the two profiles in a beam element with fixed ends and with the same sag. The harped tendon corresponds to a concentrated equivalent load W_e of

$-4Ps/L$; its fixed-end moment is $+Ps/2$, and the span moment $-Ps/2$. The parabolic tendon is equivalent to a total uniformly distributed load W_e of $-8Ps/L$; its fixed-end moment is $+2Ps/3$ and, the span moment $-Ps/3$. The fixed-end moment for the parabolic tendon is 33% higher than that for the harped tendon, and its span moment is 33% lower. In a real continuous member the proportions will differ from the above percentages because of the influence of the adjacent spans, but the parabolic profile may still show a smaller span moment due to the equivalent load than the harped profile.

In a continuous member of unequal spans, or of differing intensities of uniformly distributed load, a harped profile is worth trying in one or more spans if the span stresses exceed the limits.

5.6 Composite profiles

Apart from ground slabs, straight tendons are rarely used. Simple second-degree parabolas are used in suspended floors only when an anchorage is set in a pocket for stressing from the top of the slab. Normally, the anchorages are positioned with their axes parallel to that of the slab, or the beam, and, therefore, even in the case of a simply supported span, the tendon profile may consist of three parabolas and two straight lines—a large parabola in the middle and two short ones near the anchorages, and two short straight lengths at each anchorage. If the load to be supported is predominantly concentrated then two straight lengths may be interposed between the three parabolas.

This section looks at a few of the composite profiles; their geometry and their equivalent load patterns are discussed where appropriate.

5.6.1 General harped profile

Consider a general harped profile suitable for a predominantly concentrated load, such as shown in Figure 5.12. The profile consists of three short parabolas joined together by two straight lines. The lengths a, b and d, and the heights at the supports are known. The geometry of the profile is calculated below for the distance b; for the remaining distance $(L-b)$ a similar procedure can be followed.

For Figure 5.12, the following three equations represent the three curves, 1-3, 3-4 and 4-2. In the second equation m_1 is the slope of the straight line, tan θ.

Curve 1–3:	for $x < a$	$y = A_1x^2 + y_1$
Length 3–4:	for $a < x < b - d$	$y = m_1x + C$
Curve 4–2:	for $b - d < x < b$	$y = A_2(b - x)^2$

Equating the slopes and ordinates at points 3 and 4, gives

$$A_1 = -y_1/a(2b - a - d)$$
$$A_2 = +y_1/d(2b - a - d) \qquad (5.14)$$

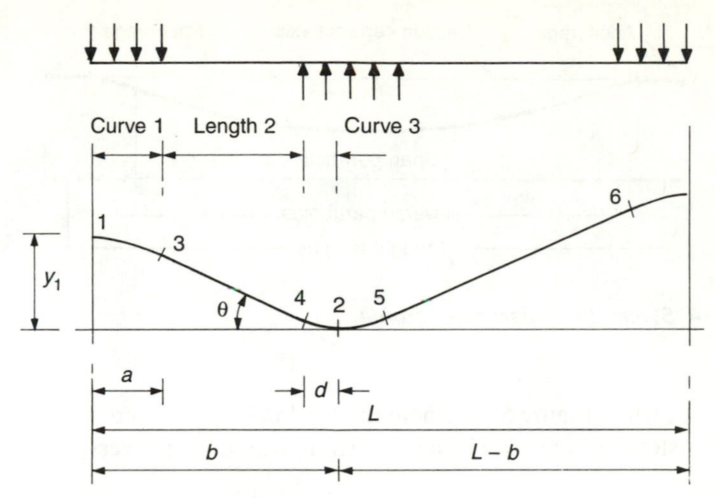

Figure 5.12 *General harped profile*

A part of the concentrated load is picked up by the tendon between points 4 and 2, and deposited at the left support as a uniformly distributed load over length a.

$$V_1 = -2A_1Pa = +Pm_1 \tag{5.15}$$

There is a similar shear at end 2 of magnitude Pm_2, where m_2 is the tendon slope in length 5-6.

In transfer structures where more than one concentrated load is to be carried, the harped tendon can be given as many kinks in it as is practical. Figure 5.13 shows a profile which may suit two concentrated loads.

For Figure 5.13,

$$\begin{aligned} V_a &= P \tan \theta_1 \\ V_b &= P \tan \theta_3 \\ W_1 &= P(\tan \theta_1 - \tan \theta_2) \\ W_2 &= P(\tan \theta_3 + \tan \theta_2) \end{aligned} \tag{5.16}$$

Note that the minus and the plus signs in equations for W_1 and W_2 represent the

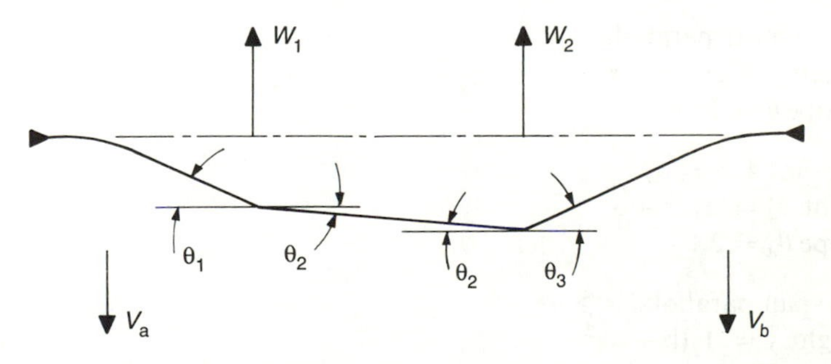

Figure 5.13 *Harped profile for two loads*

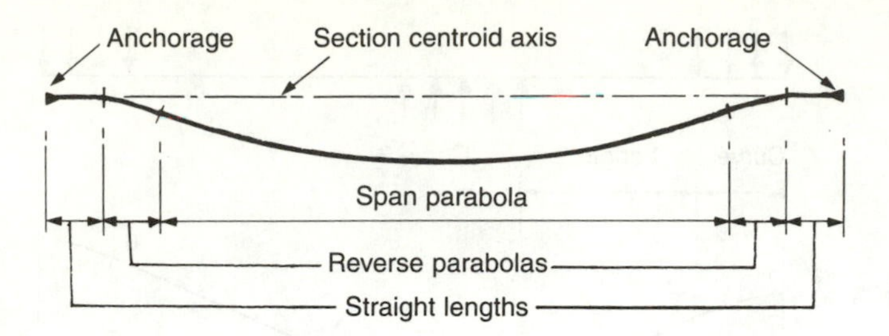

Figure 5.14 *Symmetrical parabolic profile*

profile as shown in Figure 5.13, where the tendon is higher under W_1 than under W_2. If this slope was reversed then the signs would also reverse.

5.6.2 Symmetrical parabolic profile

Figure 5.14 shows a profile for a simply supported span carrying a uniformly distributed load. In floors, as against bridges, the longitudinal axis of the anchorage is normally placed parallel to the plane of the slab. This allows a simple shape for the pocket formers, and during stressing the jacks are at right angles to the slab edge. A short length of tendon in and immediately behind an anchorage avoids a sharp bend in the tendon which would otherwise affect the strength of the strand and the efficiency of the anchorage. The length of the straight section depends on the number and configuration of strands in a tendon and varies slightly from system to system. The main part of the profile being a parabola, the transition from a horizontal to the parabolic shape in the span requires short reverse parabolic curves near the supports.

A symmetrical parabola, of course, is also used in continuous members, possibly in interior spans; in this case the tendon at the supports is usually not at the section centroid, see Figure 5.1 span 2–3. Figure 5.15 shows half of such a profile, drawn to an exaggerated scale. The profile consists of two parabolas tangentially meeting at point 4. The total sag of the profile s and lengths a and b are known.

For the support parabola, $x \leqslant a$

$$\text{height } y = A_1 x^2 + s \qquad (5.17a)$$
$$\text{slope } \theta = 2Ax$$

and at point 4, $x = a$,

$$\text{height } y_4 = A_1 a^2 + s$$
$$\text{slope } \theta_4 = 2A_1 a$$

For the span parabola, $x \geqslant a$

$$\text{height } y = A_2(\text{b} - x)^2 \qquad (5.17b)$$
$$\text{slope } \theta = -2A_2(b - x)$$

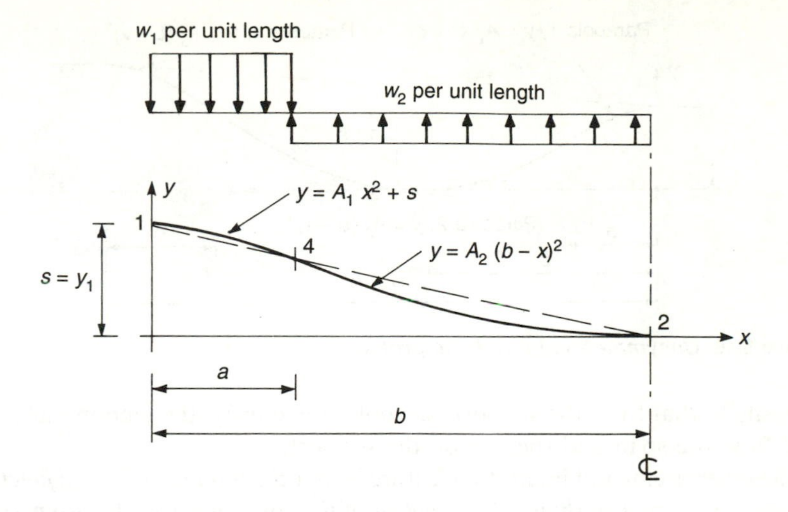

Figure 5.15 *Half of a symmetrical profile*

and at point 4, $x = a$

height $y_4 = A_2(b - a)^2$

slope $\theta_4 = -2A_2(b - a)$

Equating the two sets of values of y_4 and θ_4 gives

Support parabola: $x < a, \quad A_1 = -y_1/(ab)$ (5.18)

Span parabola: $x > a, \quad A_2 = +y_1/(b^2 - ab)$ (5.19)

and at point 4: $x = a, \quad y_4 = y_1.(b - a)/b$ (5.20)

$\theta_4 = -2y_1/b$

Equations (5.18), (5.19) and (5.20) can be used where the tendon profile is symmetrical in a span, such as in a single span member or in the symmetrical interior spans of a continuous member.

A useful geometrical property can be deduced from the above equations, that point 4, where the two parabolas meet, lies on the straight line joining points 1 and 2. It follows that the *parabola 1–4 is a scaled down version of the parabola 4–2.*

Now consider the equivalent load for the tendon profile in Figure 5.14. The support parabola exerts a downward acting total force V_1 distributed uniformly over length a, and the span parabola exerts an upward acting total force $-W_e$ uniformly distributed over length $(b - a)$. For a tendon force P, the intensities of the two loads can be calculated from Equations (5.13). For the half span shown in Figure 5.15,

$$V_1 = w_1 a = +2Py_1/b$$
$$W_e = w_2(b - a) = -2Py_1/b \qquad (5.21)$$

It is not surprising that the total load from the span half parabola is equal and

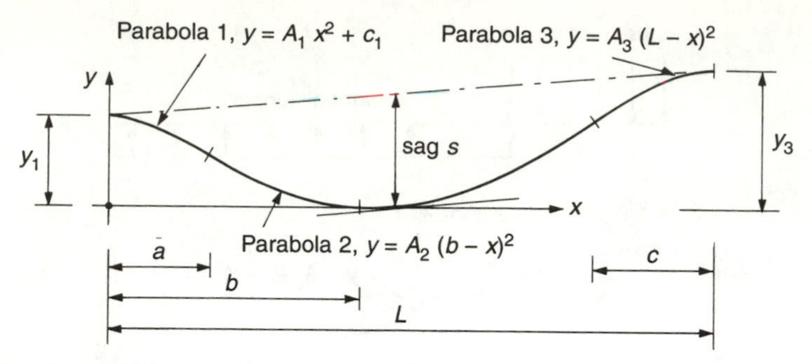

Figure 5.16 *Unsymmetrical parabolic profile*

opposite to that from the support parabola. Effectively, the tendon picks up its load from the span and sheds it on the support.

The other useful and interesting feature is that *the total load W_e is a function of the sag s and span length; it is independent of the position where the two parabolas meet*. This load is distributed uniformly over length a acting downwards, and over length $(b - a)$ acting upwards.

5.6.3 Unsymmetrical parabolic profile

Now consider an unsymmetrical profile, such as that in span 3-4 of Figure 5.1, where the tendon profile is different over the two supports. This is the most often employed profile and, therefore, it is dealt with in some detail. Equations are given for its geometry and certain parameters of the geometry are tabulated in non-dimensional form; these can be used to calculate the tendon profile heights without having to solve the equations.

In this case, the length b is not known, see Figure 5.16. Equations for parabolas 1 and 2 are the same as those for the symmetrical profile shown in Figure 5.15. Let parabola 3, over the right-hand support in Figure 5.16, be represented by:

$$y = A_3(L - x)^2 + y_3$$

Equating the deflection and the slope of parabolas 2 and 3 at their common point 5 gives a quadratic equation for b/L, whose solution is:

$$b/L = [-Y - (Y^2 - 4XZ)^{0.5}]/(2X) \tag{5.22}$$

where $X = (1 - y_3/y_1)$
$Y = ay_3/y_1L + c/L - 2$
$Z = 1 - c/L$

The equations of the three parabolas and their equivalent loads are then:

For $0 < x < a$, $\quad y = A_1x^2 + y_1$

$$\theta = 2A_1x \tag{5.23a}$$

For $a < x < L - c$, $y = A_2(b - x)^2$
$$\theta = 2A_2(x - b) \qquad (5.23b)$$

For $L - c < a < L$, $y = A_3(L - x)^2 + y_3$
$$\theta = -2A_3(L - x) \qquad (5.23c)$$

where $A_1 = -y_1/(ab)$
$A_2 = +y_1/[b(b - a)]$
$A_3 = -y_3/[c(L - b)]$

In the above equations values of A_1 and A_3 are expected to be negative for reverse parabolas over the supports. Note that:

If $a = 0$ then $A_1 = 0$
If $c = 0$ then $A_3 = 0$
If $y_1 = y_3$ then $b = (L - c)L/(2L - a - c)$
If $y_1 = y_3$ and $a = c$ then $b = L/2$ and $A_1 = A_3$

The load on parabola 1 is equal and opposite to the load from the span parabola between points 4 and 2, and similarly the load on parabola 3 is equal and opposite to the load from the span parabola between points 2 and 5. Points 4 and 5, where the span parabola meets the support parabolas, are normally arranged so that the support parabolas shed their loads inside the critical shear zones. A common rule of thumb is that distances a and c are about 10% of the span length. Remembering that $1/(2A)$ represents the minimum radius of curvature, to limit the minimum radius over a support to 2.5 m (8 ft), the values for A_1 and A_3 should not exceed 0.2 m (0.0625 ft). Note that the minimum radius of a bonded tendon is often governed by the capacity of the sheath to bend without damage.

For convenience, non-dimensional values of b/L and other parameters are shown in Table 5.1 for $a = c = 0.1L$ and $a = c = 0.05L$, which can be used for calculating heights of a profile at various points, given its position at supports. The notations are defined in Figure 5.17.

In Table 5.1, b is measured from the point where the tendon height is y_1. For each value of y_3/y_1, the ratios y_0/y_1 and y_4/y_1 are the inverses of each other, and similarly the ratios y_5/y_3 and y_6/y_3 are inverse pairs. Figure 5.18 is a graphical representation of Table 5.1 for $a/L = c/L = 0.1$ only. The curves give a better indication of the range and the sensitivity of the various ratios than the table.

Equation (5.22) for b/L simplifies to Equation (5.24) if a/b is assumed to equal $c/(L - b)$, i.e., the lengths of the two support parabolas are assumed in proportion to the lengths of the two portions of the span parabola either side of its lowest point. The values of b/L from Equation (5.24) are within 5% of those from Equation (5.22) for y_3/y_1 in the range 0.5 to 1.0.

$$b/L = 1/[1 + (y_3(y_3/y_1)^{0.5}] \qquad (5.24)$$

5.6.3.1 *Approximating an unsymmetrical profile*

For preliminary calculations, before the tendon profile has been finalized, it is rather cumbersome to have to evaluate the unknowns from the set of Equations

Table 5.1 *Ordinate ratios for Figure 5.17*

$a/L = c/L$	y_3/y_1	b/L	y_0/y_1	y_4/y_1	y_5/y_3	y_6/y_3
	0.1	0.730	1.159	0.863	0.630	1.588
	0.2	0.669	1.176	0.851	0.698	1.434
	0.3	0.630	1.189	0.841	0.730	1.370
	0.4	0.600	1.200	0.833	0.750	1.333
0.1	0.5	0.576	1.210	0.826	0.764	1.309
	0.6	0.556	1.219	0.820	0.775	1.291
	0.7	0.540	1.228	0.815	0.783	1.277
	0.8	0.525	1.235	0.809	0.790	1.267
	0.9	0.512	1.243	0.805	0.795	1.258
	1.0	0.500	1.250	0.800	0.800	1.250
	0.1	0.746	1.072	0.933	0.803	1.245
	0.2	0.681	1.079	0.927	0.843	1.186
	0.3	0.638	1.085	0.922	0.862	1.160
	0.4	0.607	1.090	0.918	0.873	1.146
0.05	0.5	0.581	1.094	0.914	0.881	1.136
	0.6	0.560	1.098	0.911	0.886	1.128
	0.7	0.542	1.102	0.908	0.891	1.123
	0.8	0.526	1.105	0.905	0.894	1.118
	0.9	0.512	1.108	0.902	0.897	1.114
	1.0	0.500	1.111	0.900	0.900	1.111

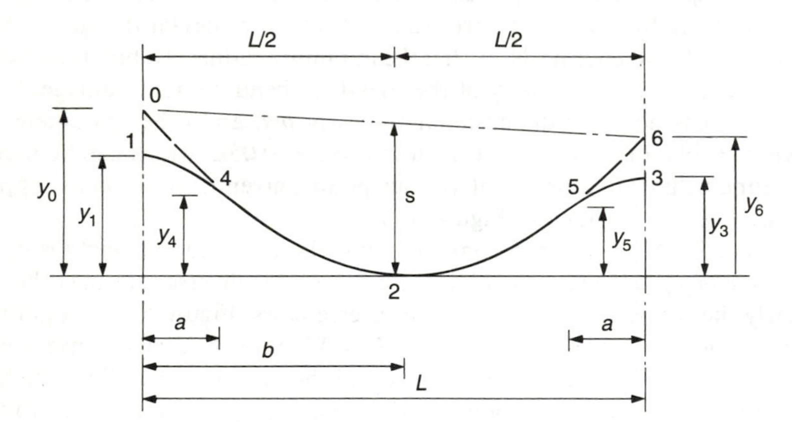

Figure 5.17 *Notations for Table 5.1*

(5.22) and (5.23) for each trial profile. At this stage the designer is interested only in an approximate value of the equivalent load, and this can be calculated from the total sag *s*, Figure 5.16.

In an unsymmetrical profile, the maximum sag should be measured at the point where the tangent to the parabola is parallel to the line joining points 1 and 3 in Figure 5.16. However, it is sufficiently accurate for preliminary purposes to measure the sag at midspan.

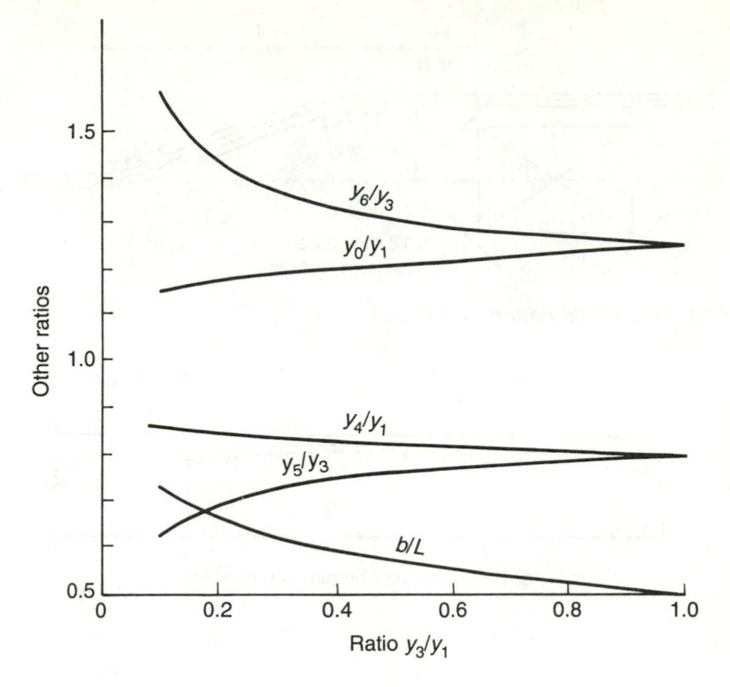

Figure 5.18 *Graphical representation of Table 5.1 for a/L = c/L = 0.1*

$$s \approx (y_1 + y_3)/2 \qquad (5.25)$$
$$W_e = 8Ps/L$$

The total equivalent load W_e is assumed uniformly distributed over the span length L. A slightly better accuracy is obtained if W_e is distributed over the length $(L - a - c)$, the length of the span parabola.

5.7 Tendon deviation in plan

Slab tendons are often moved out of line to allow holes to be formed, as shown in Figure 5.19. Such a deviation is not strictly related to its profile in the sense that it does not affect its equivalent load, although it may affect calculation of losses. The curvature, assumed circular of radius R, produces radial forces of magnitude P/R per unit length to be resisted by the concrete in contact with the tendon, as indicated. A small radius gives a high radial force per unit length, and the tendon may be more difficult to hold in position than in the case of a larger radius. Sufficient reinforcement should be provided to prevent the tendon bursting out of the concrete; a low stress in the reinforcement, of 200 N/mm^2 (30 000 psi), would ensure that the steel is well within the elastic range. It is, of course, preferable to have the curves away from the hole. Recommendations of ACI 318 are shown in Figure 5.19 with regard to the preferred clearances between hole and the tendons.

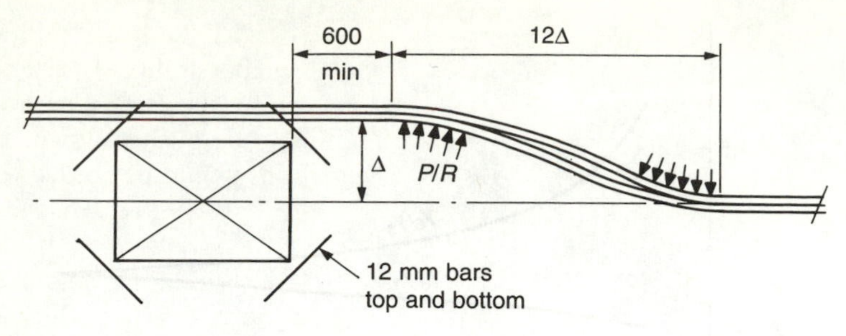

Figure 5.19 *Tendon deviation at a hole*

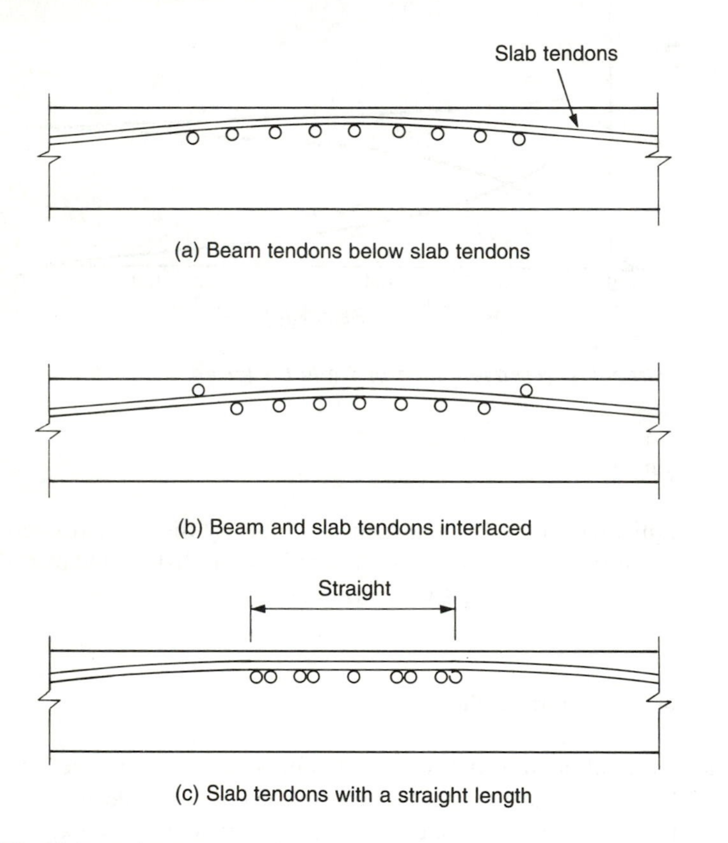

Figure 5.20 *Slab tendons over strip beam*

5.8 Clash of beam and slab tendons

It is desirable to place the tendons at the maximum possible eccentricity in a section, in the slabs as well as in the beams. At the column in a continuous slab spanning over a continuous beam, tendons from both the members are at the top. Both of them cannot run at their maximum eccentricities, and to avoid the clash, one set of tendons must run at a smaller eccentricity.

The slab tendons are often arranged to run on top of the beam cage at the support; this simplifies fixing and there is, of course, no possibility of a clash in the span zone of the beam. Over the column support, either the reverse parabola of the slab tendons would require the beam tendons to be pushed down, Figure 5.20(a), or a certain amount of weaving of the tendons would be needed, Figure 5.20(b); both arrangements result in an effective loss of eccentricity of the beam prestress.

The problem can be avoided by bunching the beam tendons together, and providing a short straight horizontal length in the slab tendons over the width of the beam, as shown in Figure 5.20(c). Of course, the shear force carried by the slab tendons is shed at the beam edges in this case and this must be allowed for in the design of the beam strip.

6 FLEXURE IN THE SERVICEABILITY STATE

The serviceability state covers the flexural stresses in a member, and its deflection and vibration under service loads. This chapter deals with the flexural aspects only; deflection and vibration are discussed in Chapter 9.

6.1 The design process

Figure 6.1 illustrates the sequence of steps in the design of a post-tensioned member. For completeness, it includes the calculations relating to deflection, shear strength and ultimate flexure, though these are discussed in other chapters.

Analysis and design of a concrete member is a longer process in post-tensioned than in reinforced concrete, because the post-tensioning design requires several additional steps not needed in reinforced concrete, and a wider choice is available for optimizing the design. The process tends to be iterative, in that certain assumptions are made about the section and the prestress, which are adjusted in the light of successive calculation results.

In reinforced concrete, serviceability calculations are normally not needed; the usual procedure is to choose a concrete section and calculate the reinforcement required to carry the factored loads corresponding to the ultimate state; deflection criteria are normally deemed to be satisfied by adopting prescribed span-to-depth ratios. If calculations find the initial section to be inadequate, then it is increased, there being no real alternative in the matter. If it is found that the initially chosen section is larger than it needs to be, then it is normally accepted; it is rather rare to repeat calculations for a smaller section.

In post-tensioning, until calculations are carried out, it is not obvious whether the serviceability or the ultimate strength state, or both, are critical. Therefore, the serviceability checks for stresses and deflection are an essential part of the design. At the serviceability state, the stresses in the concrete are required to be within specified limits in compression and in tension, both at the time of prestressing with a minimum of load (*initial stage*) and in the long term under full applied load (*final stage*). If the serviceability state is satisfactory then the calculations proceed to considerations of the ultimate strength.

Serviceability calculations follow the classical elastic theory, where stress is

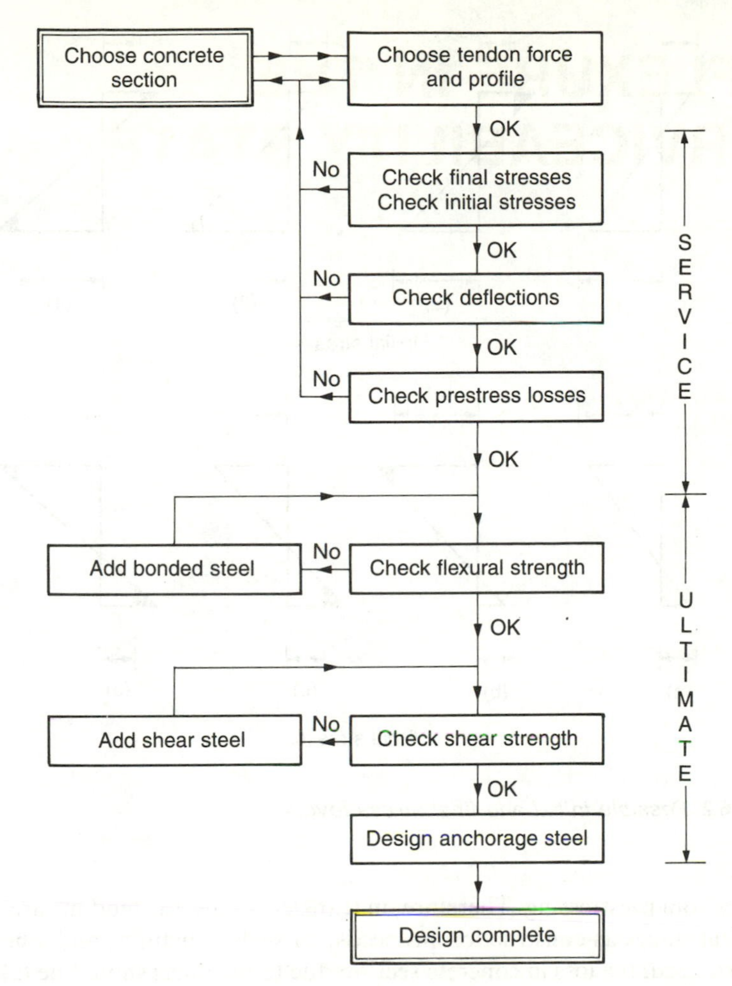

Figure 6.1 *The design process*

proportional to strain and the compression block is triangular. Bonded steel can be allowed for in calculating the moment of inertia of the section, by replacing it with an equivalent area of concrete at an appropriate value of the modular ratio. In designing post-tensioned floors, however, the normal practice is to take the gross concrete section and ignore bonded steel in calculating the section moduli. The area of tendon ducts is not normally deducted in the design of floors, though ACI 318 requires the effect of loss of area due to open ducts to be considered. Ducts may be open where access to a tendon is required after concreting, such as at couplers or where tendons are to be stressed from open pockets at the top of a floor.

In transfer beams, the amount of prestress required is high; usually, the member cannot be stressed in one operation because the dead load from the self-weight alone is insufficient to contain the initial tensile stresses that would

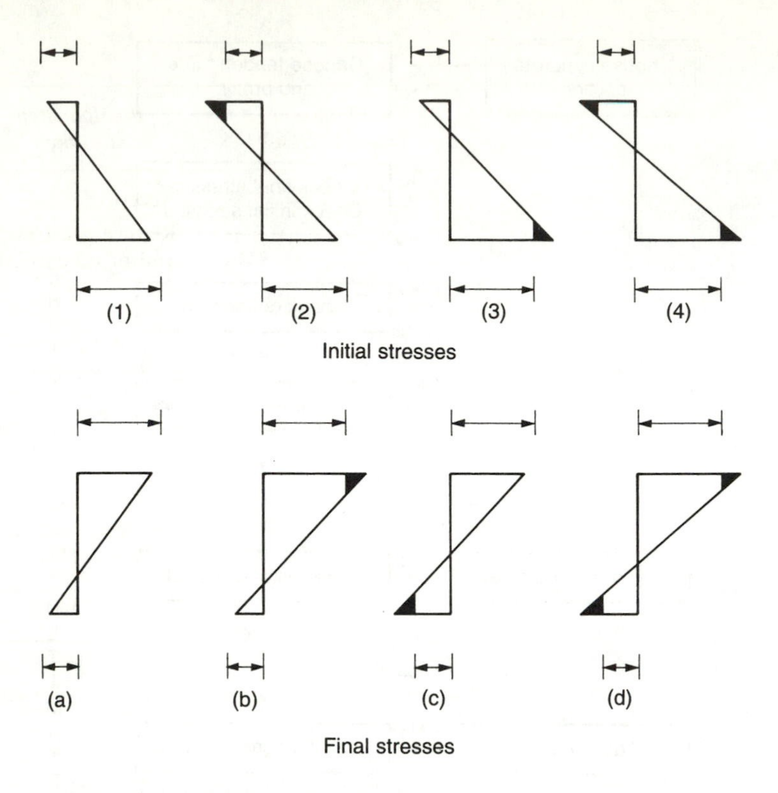

Figure 6.2 *Possible initial and final stress levels*

develop from prestressing. Therefore, in transfer beams, the tendons are stressed in several stages as construction proceeds. In such structures, and where open ducts are used, the loss in concrete section due to the ducts should be taken into account.

6.2 Options in a design

An understanding of the design sequence, and the choices available at various stages, will be useful in deciding on the step to be taken if the initial and/or the final stress(es) exceed the specified limits. The possible stress levels, being acceptable or not, are shown in Figure 6.2. The arrows indicate the acceptable levels; the diagrams are shaded where the stress exceeds the limit.

Table 6.1 lists the sixteen possible combinations of initial and final stresses at the top and bottom of a section. The second column refers to the initial and final stress combination from Figure 6.2. It has been assumed that the moment is sagging. A zero (0) denotes a case where the stress exceeds the limit and some action is required on the part of the designer.

Table 6.1 *Design options*

Case	*Stress diagram*	*Initial*		*Final*		*Possible step to take*
		Top	*Btm*	*Top*	*Btm*	
1	1a	·	·	·	·	Accept
2	4a	0	0	·	·	Reduce prestress
3	2a	0	·	·	·	and/or eccentricity
4	3a	·	0	·	·	
5	1d	·	·	0	0	
6	1b	·	·	0	·	Increase prestress
7	1c	·	·	·	0	and/or eccentricity
8	2c	0	·	·	0	
9	3b	·	0	0	·	
10	3c	·	0	·	0	
11	4b	0	0	0	·	
12	2b	0	·	0	·	
13	4c	0	0	·	0	Increase section
14	2d	0	·	0	0	
15	3d	·	0	0	0	
16	4d	0	0	0	0	

In the first case, when all stresses are within the limits, the design can either be accepted or a smaller section may be considered if so desired.

When permissible stresses are exceeded in the same position at initial and at final stages then the section must be increased. In Case 5, if the section is a Tee then increasing the rib width is another option. This option is also available whenever the bottom is overstressed in a T-section.

The options in Table 6.1 are directly applicable to a simply supported member. For continuous members, it is easier to think in terms of the equivalent loads. An increase of eccentricity and/or prestressing force in a span increases the equivalent load, and it is tantamount to a reduction of the applied load.

Consider two adjacent spans of a continuous string, shown in Figure 6.3. An increase of prestressing force, or eccentricity, in span AB increases the upward equivalent load, which results in a reduction in the moment at support B, but it increases the moments in span BC and at support C. The effect at support C is, of

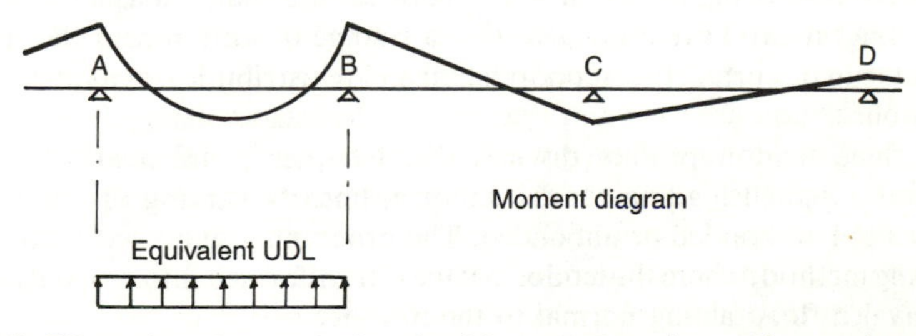

Figure 6.3 *Effect of tendon equivalent load in one span of a continuous beam*

course, smaller than that at support B—in the case of a prismatic member, half at C and a quarter in span BC. Therefore, to reduce the moment at support B, the equivalent load can be increased in either of the two spans AB and BC. To reduce the moment in span BC, the equivalent load must be either reduced in span AB, or increased in span BC.

Most of the floors are designed for uniformly distributed loads and have parabolic profiles. In a continuous member of unequal spans, or where the load intensities differ in adjacent spans, a harped profile may be worth trying, as discussed in 4.5.4.

6.3 Computer programs

The methods of analysis used for a post-tensioned floor are the same as for reinforced concrete. Manual design can be carried out using moment distribution or similar methods. However, the design procedure for a post-tensioned floor, except in the simplest cases, requires iterative calculations, often with a change in section, prestress or eccentricity, to obtain a satisfactory design. A computer program, though not essential, allows the designer to try a number of solutions quickly, and possibly more accurately, to arrive at the optimum design.

A post-tensioned floor can be analysed with any of the normal analysis programs, such as plane frame, flexibility matrix, stiffness matrix, or grid programs. More complicated geometries may be analysed more easily with a finite element program. After the analysis of loads and tendon forces, calculation of stresses may need the use of another program, or it can be carried out manually.

Special programs have been developed for the design of post-tensioned floors, which carry out the analysis of the structure and then design each member at critical sections. Most of these are written in the United States and they primarily comply with the American regulations; versions are available for use in other countries, with the local regulations incorporated into the program. However, compliance with the local regulations and practices is often offered as a choice for modifying the design parameters, and the basic feel of the program remains US biased. Nevertheless, they are very useful, particularly for multispan floors where manual calculations can be tedious and lengthy.

The programs are capable of handling a wide variety of geometrical shapes: the section can be rectangular or ribbed; a floor can be analysed as a frame or as a beam string; it can have drop panels or a change of section near the supports; moments can be curtailed at support faces; and redistribution of moments can be carried out.

The three tendon profiles discussed in Chapter 5 are available in these programs. Applied loading can be uniform, linearly varying or concentrated. Tendons can be bonded or unbonded. The programs usually work on the load balancing method, where the tendon profile is transformed into an axial force and an equivalent load acting normal to the member axis.

The programs are usually interactive, in that they may suggest a prestressing force and a tendon profile for use as a starting point. A more economical arrangement is nearly always arrived at by varying the magnitude of the prestressing force and adjusting tendon drape in each span. The magnitude of prestressing force can be varied from span to span.

The programs calculate prestress losses, serviceability stresses, ultimate flexural strength and shear strength, though not necessarily in this order. The required amounts of bonded rod reinforcement and shear reinforcement are also calculated. Losses are calculated after the adjustments of prestressing force and the tendon profile have been carried out, and if the losses are found to be substantially different from the initial assumptions then the whole process has to be repeated, this time with a better assessment of the prestressing forces.

Generally, the programs are not able to cope with the dispersion of the axial component of prestress which, for example, may occur when a high level of prestress is provided in a beam, with the adjacent slab panels lightly prestressed or in reinforced concrete, see Chapter 4. In such a case the stresses at the critical sections, allowing for the dispersion, may have to be checked manually, or the allowable stresses for a program may be modified to suit the particular case.

Design examples given at the end of this chapter use hand calculation methods in order to give a better understanding of the procedures involved.

6.4 Partial prestressing

In reinforced concrete, the tensile strains are of such a magnitude that the concrete on the tension face cracks at serviceability loadings; without this cracking, the reinforcement cannot develop sufficient tension to make any significant contribution to the strength of the member. In the early period of development of prestressing, the practice was to apply sufficient prestress to eliminate flexural tension from the concrete section under serviceability loadings. With experience, it was realized that an intermediate strategy may be more economical while still being technically satisfactory. The level of prestress where tensile stresses are not allowed to develop is now termed *full prestressing*. There is no agreed definition to distinguish between full and partial prestressing. In this book, the term *partial prestressing* refers to the system in which tension is allowed to develop in the concrete under full service loading.

In current practice, the allowable tensile stress in concrete corresponds to a strain of the order of 2.5×10^{-4} in post-tensioning; in reinforced concrete the average strain in concrete adjacent to tension reinforcement may be 12.5×10^{-4}. It follows that a concrete member may have a strain of either under 2.5×10^{-4} as in post-tensioning, or a strain approaching 12.5×10^{-4} as in reinforced concrete. The range of tensile strain in concrete between 2.5×10^{-4} and 12.5×10^{-4} is at present not used.

It should be appreciated that full prestressing does not necessarily eliminate tension or micro-cracks in concrete; towards the support the principal stress due to the combination of prestress and shear is still tensile. Also, most floors are stressed in one direction only and are cracked in the non-prestressed direction. Observations of the loading history of normal floors show that the actual load on a floor rarely approaches its full design value, so that in practice even the concrete in a partially prestressed floor is not usually subjected to any tension.

The distinction between full and partial prestressing refers to the serviceability state only; the design procedure for ultimate strength follows the same lines for both. Therefore, the ultimate strength of a member is not affected by this classification.

A fully prestressed member may develop unacceptable camber if it normally carries only a small proportion of its full design load. Partial prestressing is, therefore, a better and more economical solution for floors designed for occasional heavy loads or where the normal load intensity is small compared with the design load.

In the current practice, except for specialized structures such as reservoirs, almost all floors are partially prestressed. The question arises as to how much tension should be allowed in concrete. Various national standards specify limiting stresses which correspond to much lower strains than those in reinforced concrete. There are two approaches to specifying the limit: the tensile stress itself may be limited to a safe value so that the concrete does not crack, or the concrete may be allowed to crack but the crack width may be controlled. The cracked section of the latter case will have a larger deflection.

The British Standard 8110 recognizes three classes for the level of prestress in structures at the serviceability stage.

1. Where tension is never allowed in concrete. This class is meant for specialized uses such as liquid retaining structures, which are not covered in this book.
2. A limited tension is allowed but with no visible cracking. However, in serviceability calculations the concrete is assumed to remain uncracked.
3. Tension is allowed to a higher level and the concrete section is assumed to be cracked. Surface cracks must not exceed 0.1 mm for members in very severe environments and 0.2 mm for all other members. The 0.2 mm limit is applicable to post-tensioned floors in normal buildings.

   In Class 3, though the section is cracked, BS 8110 requires notional tensile stresses to be computed, assuming an uncracked section, for complying with the specified limits. Some bonded rod reinforcement is required to take the theoretical tensile stresses in a cracked section.

In the UK, more of the floors in buildings are designed to comply with the Class 2 requirements than Class 3. The latter, however, has the advantage that the extra rod reinforcement on the tension face is useful in controlling early shrinkage and crack distribution.

6.5 Permissible stresses in concrete

In the UK there are some differences in the permissible stresses specified in BS 8110 for Class 2 and Class 3 structures. No such distinction is made in ACI 318,

Table 6.2 *BS 8110 Class 2. Permissible stresses* (N/mm^2)

Initial stresses	
tension	$0.36 \sqrt{f_{ci}}$
compression	$0.50f_{ci}$
Final stresses	
tension	$0.36 \sqrt{f_{cu}}$
compression: span	$0.33f_{cu}$
compression: support	$0.40f_{cu}$

Note: The $0.4f_{cu}$ allowable compression stress at support does not apply to cantilevers.

which does not use the classification system.

The tensile stresses given in Table 6.2 are for normal prestressed concrete, without any enhancing ingredient such as steel fibres. Where such an ingredient is used in the concrete, its tensile properties should be determined by tests.

6.5.1 BS 8110

BS 8110 specifies the following limits for Class 2 structures at the serviceability limit state. At the initial stage, if the stress diagram is near-rectangular, then the compressive stress is limited to $0.4f_{ci}$.

At the final stage, compression in continuous members in the support region can be a maximum of $0.4f_{cu}$; in spans it is limited to $0.33f_{cu}$. If the imposed load is of a temporary nature and is exceptionally high in comparison with the normal load then the allowable tensile stress may be increased by 1.7 N/mm^2 provided that the stress is normally compressive.

These stress limits are used for both bonded and unbonded construction. The allowable tensile stresses shown in the Table 6.2 include a partial safety factor of 1.3.

In Class 3 members the allowable initial tensile and compressive stresses are the same as those for Class 2. For the final tensile stress, however, although cracking is allowed, it is assumed that the concrete section is uncracked and that design hypothetical stresses exist at the limiting crack widths. The allowable final stress is related to the concrete strength, crack width and to the section depth, as shown in Table 6.3.

BS 8110 allows the tensile values given in Tables 6.2 (modified for exceptional loading where applicable) and 6.3 to be exceeded under two conditions.

Firstly, if additional reinforcement is contained within the tension zone, and is positioned close to the tension faces of the concrete, these modified design stresses may be increased by an amount that is in proportion to the cross-sectional area of the additional reinforcement (expressed as a percentage of the cross-sectional area of the concrete in the tension zone). For 1% of additional reinforcement, the stresses may be increased by 4.0 N/mm^2. For other percentages of additional reinforcement, the stresses may be increased in proportion up to a limit of $0.25f_{cu}$.

Secondly, when a significant proportion of the design service load is transitory so that the whole section is in compression under the permanent (dead plus

Table 6.3 *BS 8110 Class 3, bonded tendons. Permissible final tensile stress*

Crack width mm	*Member depth mm*	*Design stress for concrete grade* 30	40	50 and over
0.2	200	4.18	5.50	6.38 N/mm^2
	400	3.80	5.00	5.80
	600	3.42	4.50	5.22
	800	3.04	4.00	4.64
	≥1000	2.66	3.50	4.06
0.1	200	3.52	4.51	5.28 N/mm^2
	400	3.20	4.10	4.80
	600	2.88	3.69	4.32
	800	2.56	3.28	3.84
	≥1000	2.24	2.87	3.36

frequently occurring imposed) load, the hypothetical tensile stresses may be exceeded under the full service load.

BS 8110 does not give any limits for tensile stresses in Class 3 structures where unbonded tendons are used; in practice, it is assumed that the limits shown in Table 6.3 apply. A Class 3 member with unbonded tendons is prone to more cracking under full design load if adequate bonded rod reinforcement has not been provided.

6.5.2 Concrete Society

The Concrete Society (1994) allows the 0.1 mm crack width Class 3 permissible stresses to be used for unbonded tendons provided that the tension is carried on bonded rod reinforcement.

For two-way spanning flat slabs, Concrete Society recommends the permissible stresses given in Table 6.4. The allowable tensile stresses are significantly lower in the support regions, compared with Table 6.3, due to the peaking of the moments.

In Table 6.4, the support zone is assumed to extend for a distance of $0.2L$ from the support; any section beyond this point is considered to be in the span zone. Bonded reinforcement may consist of either rod reinforcement or the tendons themselves.

6.5.3 ACI 318

In the USA two authorities specify allowable stresses in prestressed concrete: The American Association of State Highway and Transportation Officials (AASHTO) and The American Concrete Institute (ACI). The AASHTO specification is the more conservative as it applies to bridges, where the exposure conditions are much more severe than in buildings. ACI 318: 1989 gives the limits for serviceability stresses shown in Table 6.5.

Table 6.4 *Permissible stresses in two-way flat slabs* (*Concrete Society, 1994*)

Location	*Compression*	*Tension with bonded reinforcement*	*Tension with unbonded reinforcement*
Support	$0.24 f_{cu}$	$0.45 \sqrt{f_{cu}}$	0
Span	$0.33 f_{cu}$	$0.45 \sqrt{f_{cu}}$	$0.15 \sqrt{f_{cu}}$

Table 6.5 *ACI 318 Permissible stresses in concrete*

Stage	*Mode*	*Metric N/mm²*	*Imperial psi*
Initial	tension	$0.25 \sqrt{f'_{ci}}$	$3.0 \sqrt{f'_{ci}}$
	compression	$0.60 f'_{ci}$	$0.60 f'_{ci}$
Final	tension	$0.50 \sqrt{f'_{c}}$	$6.0 \sqrt{f'_{c}}$
	compression	$0.45 f'_{c}$	$0.45 f'_{c}$

ACI 318 allows the permissible initial tensile stress at the ends of simply supported members to be increased to double the tabulated value, i.e. to $0.5\sqrt{(f_{ci})}$ N/mm² ($6\sqrt{f_{ci}}$ psi). An increase in the final permissible tensile stress to $\sqrt{f_c'}$ N/mm² ($12\sqrt{f_c'}$ psi) is allowed in one-way spanning members subject to compliance with the deflection and cover requirements.

Where the tensile stresses exceed the permissible values, the total force in the tensile stress zone may be calculated assuming an uncracked section and reinforcement provided on the basis of this force at a stress of $0.6f_y$ but not more than 200 N/mm² (30 000 psi).

Stresses in the concrete are calculated on the basis of an uncracked section. A minimum average prestress of 1.0 N/mm² (150 psi) is required on the gross concrete section, after allowing for all losses in the prestressing force.

6.6 Permissible stresses in strand

BS 8110 specifies that:

> The jacking force should not normally exceed 75% of the characteristic strength of the tendon but may be increased to 80% provided that additional consideration is given to safety and to the load/extension characteristic of the tendon. At transfer, the initial prestress should not normally exceed 70% of the characteristic strength of the tendon, and in no case should it exceed 75%.

ACI 318 specifies the following maximum values of stress in low relaxation strand:

During stressing: $0.94f_{py}$ but not more than $0.8f_{pu}$ and the maximum recommended by the manufacturer.
Immediately after stressing: $0.70f_{pu}$.

A strand carries the highest force during stressing; thereafter the force reduces when the anchorage is locked, and as creep and shrinkage take effect. The initial strand force is normally limited to a maximum of 80% of its breaking load. In practice a force of 70% to 75% of the breaking strength is often aimed for when the strand is stressed; this leaves a useful small reserve for contingencies.

6.7 Analysis

For analysis, a floor is normally divided into a series of strips which are then analysed, generally in the same manner as for reinforced concrete. Each of the critical sections is then checked and designed for adequacy at the serviceability and ultimate states.

In the case of simply supported spans, the analysis consists merely of calculating the midspan moments and end shears for the maximum load. The required prestress and rod reinforcement are then calculated if the section is adequate.

Continuous beams, being one-way elements, can be analysed either as strings, or as frames integral with the columns. In the latter case, the columns are assumed to be fixed at their far ends. The design of continuous members is affected by the longitudinal shortening of the member much more in post-tensioned than in reinforced concrete. In reinforced concrete design, the effect of creep is to increase the deflection, and shrinkage is normally ignored. In post-tensioned members, elastic shortening due to axial prestress, creep and shrinkage cause a shortening in the length of the member, which results in lateral deflection of the columns. The forces developed by the shortening of the member should be considered in the design of post-tensioned members and the vertical elements of the frame.

In strip beams, transverse moments develop across the width of the beam, and are significant if the slab is continuous. Such moments are normally treated as part of the slab design. They may be higher along the column lines, where a short length of the beam collects and transfers the loads to the column. These moments should be taken into account in the design of post-tensioning or reinforcement along the column lines.

One-way slabs are normally analysed in unit widths, supported on beam lines. A slab section in line with the columns and supported on a strip beam may need to be designed for the additional negative moment from the beam as mentioned above. A floor slab and its supporting columns may also be analysed as a frame of width equal to the panel width, or half the panel width for an outer frame; this method, however, leads to the problem of apportioning the total moment between the column and middle strips.

Two-way floors may be analysed in several different ways. The methods usually mentioned in the context of post-tensioned floors are listed below.

- As a two-way slab spanning on to a grid of beams. In each direction the slab panel is considered in strips of unit width, with a proportion of the net load acting in each direction, the proportion along span L_1 being $L_2^4/(L_1^4 + L_2^4)$ as discussed in Section 3.3.2. The sum of all loads on the slab panels and the beams in each direction should not be less than the total load.
- As a grid. This is a useful method with complicated loading patterns, or concentrated loads. It requires the use of a computer because of the large number of variables involved.

  The moments are expected to peak around the columns, and in order to achieve an accurate value of the peak moment, the strips in the vicinity of the columns should be as narrow as practically possible. This, however, also gives an unnecessarily large number of points along the column lines.
- Using finite element programs. The extra effort and expense involved in using finite elements may be justified if the plan shape of the floor and/or the loading pattern cannot be accommodated by any of the other methods.
- As frames in two directions, taking the full load in each direction. The width of a frame is chosen so that its edges coincide with the lines of zero shear in the other direction.

  This method gives the total moment for the whole width of the frame; the moment per unit width, in fact, varies over the frame width, being higher on the column lines.
- As a flat slab, divided into middle and column strips, following the practice for reinforced concrete. This is an empirical method for the distribution of loads on column and middle strips in a panel. There are limitations on the span ratios, number of panels on each direction, and the loads.

  The rules for the distribution of the load specified in the national standards yield only the ultimate moments; no guidance is given for moments at serviceability state. It, therefore, becomes difficult to check the design for compliance with the serviceability requirements.

The first mentioned is the most commonly used method and the last is used only in very special circumstances. Other methods of analysis which give only the ultimate moments, such as the yield line method, are not suitable for designing post-tensioned floors, because adequacy of the structure at the serviceability state cannot be verified; the yield line method can be used in the calculation of strength, provided that the serviceability state requirements are checked using a method of elastic analysis.

6.7.1 Load combinations

As indicated earlier, the serviceability check comprises two stages—initial and final. The initial stage represents the state of the member immediately after the application of the prestress, before any of the long-term prestress losses have taken place. The prestressing force is at its highest, and the worst condition at this stage is that of minimum applied load. No external loads are, therefore, imposed at this stage, and the loading consists of only the dead load and the effect of the initial prestress. The final stage represents the long-term state, when all losses in prestress have taken place. The self-weight of the member and the final prestress are combined with other applied loads in the most adverse manner for each of the critical sections of the member.

In the case of a simply supported member, the load to be considered in the final stage consists of all imposed loads acting simultaneously. In the analysis of continuous members, the normal service loads are combined to yield the maximum value of the total moment at each support and in each span. Support moments may be taken at the support faces rather than at their centrelines.

Normally, only the dead and live loads are involved, in which case the moment at a support is maximum when full load is applied on adjacent spans either side of the support and on alternate spans thereafter. The moment in a span is a maximum when full load is applied on alternate spans. The moments in a floor may also be affected by the imposition of live load on the floors above and below, but this is normally ignored.

The above combinations are, however, generally considered unrealistic and too severe, and national standards allow design to be based on less onerous configurations. BS 8110 and ACI 318 allow the arrangement of live load to be limited to:

- dead load on all spans with live load on two adjacent spans, and
- dead load on all spans with live load on alternate spans.

6.8 Simply supported span

There being no redundancies, calculations for a simple span are easy to carry out manually. Four stress conditions need to be checked in a simply supported member, viz., initial and final stresses at the top and bottom fibres.

Let M_o = moment due to own weight of concrete section
M_s = moment due to other applied loads

Then the stress at the top due to the initial prestress is

$$\begin{aligned}\sigma_{pti} &= P_i/A_c - P_i e_p/Z_t \\ &= (P_i/A_c).(1 - e_p A_c/Z_t) = S_t P_i/A_c \qquad (6.1a)\end{aligned}$$

where $S_t = 1 - e_p A_c/Z_t$

and similarly at the bottom,

$$\sigma_{pbi} = (P_i/A_c).(1 + e_p A_c/Z_t) = S_b P_i/A_c \qquad (6.1b)$$

where $S_b = 1 + e_p A_c/Z_b$

Similarly the stresses due to the final value of the prestressing force are:

$$\sigma_{ptf} = (P_f/A_c)S_t \tag{6.1c}$$

$$\sigma_{pbf} = (P_f/A_c)S_b \tag{6.1d}$$

The four limiting stresses can now be formulated. The initial stresses in the top and bottom fibres must satisfy:

$$\sigma_{ti} = \sigma_{pti} + M_o/Z_t > p_{ti} \tag{6.1a}$$

$$\sigma_{bi} = \sigma_{pbi} - M_o/Z_b < p_{ci} \tag{6.2b}$$

and the final stresses must satisfy:

$$\sigma_{tf} = \sigma_{ptf} + (M_o + M_s)/Z_t < p_{cf} \tag{6.2c}$$

$$\sigma_{bf} = \sigma_{pbf} - (M_o + M_s)/Z_b > p_{tf} \tag{6.2d}$$

where p_{ti} = permissible initial tensile stress (−ve)
p_{ci} = permissible initial compressive stress (+ve)
p_{tf} = permissible final tensile stress (−ve)
p_{cf} = permissible final compressive stress (+ve)
See Section 6.5

Normally, the final tensile stress condition, Equation (6.2d), is expected to govern. Rearranging gives the minimum required value of P_f.

$$P_{fmin} = [p_{tf} + (M_o + M_s)/Z_b]A_c/S_b \tag{6.3a}$$

The maximum prestressing force which can be imposed at a section is usually governed by the initial compressive stress and can be calculated from equation (6.2b).

$$P_{imax} = (p_{ci} + M_o/Z_b)A_c/S_b \tag{6.3b}$$

If R_p is the ratio of final to initial prestress, P_f/P_i, then the maximum final force is given by

$$P_{fmax} = R_p(p_{ci} + M_o/Z_b)A_c/S_b \tag{6.3c}$$

Obviously, for a design to be feasible the value of P_{fmax} must be larger than the P_{fmin} value. If the value of P_{fmax} is found to be less than P_{fmin} then the concrete section is inadequate and it must be increased.

6.8.1 Shape factors

The non-dimensional terms S_t and S_b, Equation (6.1), depend on the shape of the section and on the tendon eccentricity. For a tendon placed below the section centroid (positive eccentricity), S_t is expected to be negative, because the prestress would tend to induce tension at the top fibre, and S_b to be positive. S_t and S_b are ratios of extreme fibre stresses to the average stress and, therefore, are representative of the efficiency of the section.

$$S_t = 1 - e_pA_c/Z_t \tag{6.1a}$$

$$S_b = 1 + e_pA_c/Z_b \tag{6.1b}$$

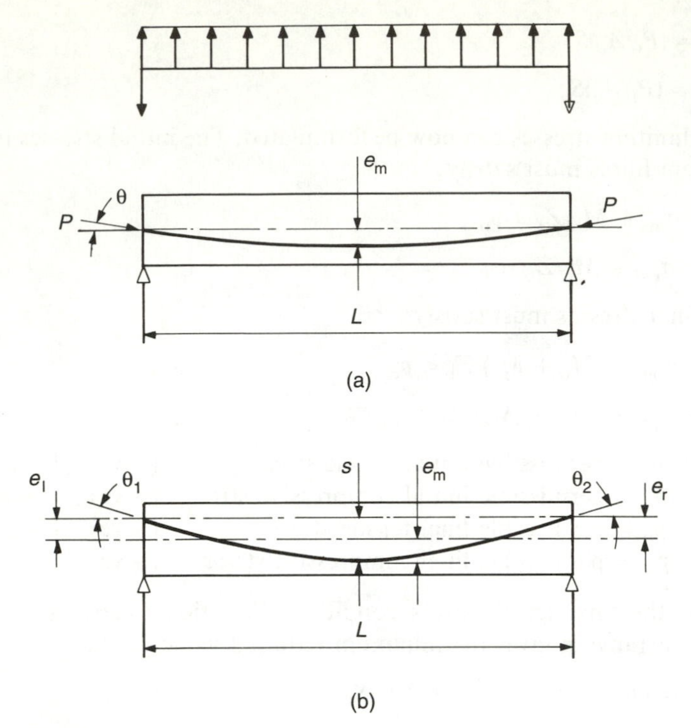

Figure 6.4 *Simply supported beam*

From Equation (6.1a), it can be seen that S_t becomes zero at an eccentricity of Z_t/A_c. At this eccentricity the stress in the top fibre remains zero for all magnitudes of the prestressing force. If the eccentricity exceeds this value then the top stress is negative (tensile); the top fibre stress is positive (compressive) if the eccentricity is less than this value. For a rectangular section, $Z_t/A_c = d/6$.

An eccentricity of $-Z_b/A_c$ has a similar effect on the stress in the bottom fibre. A negative eccentricity is, however, unlikely to be used.

6.8.2 Equivalent load method

The sets of Equations (6.2) and (6.3) take direct account of the eccentricity of prestress rather than using the load balancing method; in the case of simply supported members the former approach is easier. For comparison, the load-balancing approach is shown below for the initial condition at the top, Equation (6.2a). It demonstrates the equivalence of the load-balancing method and the method used in deriving the sets of Equations (6.2) and (6.3), and may be skipped if the demonstration is not of interest.

Assume that the tendon profile is parabolic, e_m is the eccentricity at midspan and the tendons are anchored at each end at the section centroid, Figure 6.4(a). Then from Equation (5.13),

$$w_{ei} = \text{initial equivalent load}$$
$$= -8P_i e_m/L^2 \quad (6.4)$$

$$M_i = \text{moment due to this load}$$
$$= w_{ei}L^2/8 = -P_i e_m$$

$$\sigma_{ti} = P_i/A_c + (M_o + M_i)/Z_t > p_{ti} \quad (6.5)$$

The term $P_i e_m$ is used here as an applied moment, whereas in Equation (6.1) the equivalent term $P_i e$ is used in calculating the stress due to prestress.

The equivalent load w_{ei} is statically balanced by the vertical components of the tendon forces at the end, each of magnitude $P_i.\tan\theta$, so that

for Figure 6.4(a), $w_{ei}L = 2P_i \tan\theta$
for Figure 6.4(b), $w_{ei}L = P_i(\tan\theta_1 + \tan\theta_2)$.

The equivalence of equations (6.2a) and (6.5) is easily demonstrated.

$$\sigma_{ti} = P_i/A_c + (M_o + M_i)/Z_t \quad (6.5)$$
$$= P_i/A_c + M_o/Z_t - P_i e_m/Z_t$$
$$= (P_i/A_c)(1 - e_m.A_c/Z_t) + M_o/Z_t$$
$$= \sigma_{pti} + M_o/Z_t \quad (6.2a)$$

If the tendon was not anchored at the section centroid but the eccentricities at left and right were e_l and e_r as shown in Figure 6.4(b), then the sag of the parabola will be larger than e_m, the midspan eccentricity, but the end eccentricities will need to be considered to arrive at the correct equivalent load and moment.

$$s = \text{sag of parabola} = e_m - (e_1 + e_r)/2$$
$$w_e = \text{equivalent load} = -8P_i s/L^2$$

$$M_e = \text{Moment at midspan due to end eccentricities}$$
$$= -P_i(e_1 + e_r)/2$$

$$M = \text{Net midspan moment}$$
$$= w_e L^2/8 + M_e$$
$$= -P_i s - P_i(e_1 + e_r)/2$$
$$= -P_i[e_m - (e_1 + e_r)/2] - P_i(e_1 + e_r)/2$$
$$= -P_i e_m$$

It can be seen that the two approaches lead to exactly the same result and that for a simply supported span the load balancing method is unnecessarily cumbersome.

6.9 Continuous spans

The design of a post-tensioned member can be considered to comprise two distinct operations—calculation of moments, and calculation of stresses. The calculation of moments is much more involved in the case of continuous spans;

once the moments have been obtained, the stresses at critical sections are obtained from the simple equation

$$\sigma = P/A_c \pm M/Z \qquad (6.6)$$

where the moment M includes the tendon eccentricity, the moment due to self-weight and applied loads, and the secondary moment.

The required prestressing force cannot be directly calculated for a string of spans in the manner of a simple span. Instead, an initial guess must be made, allowing for losses along the tendon length, and the stresses can then be checked for compliance with the permissible values. An adjustment to the initially chosen prestress, followed by re-calculation of stresses, may be necessary.

6.9.1 Equivalent load method

The tendon eccentricity represents a moment of magnitude $P_x e_x$ at a point x along the tendon, where P_x is the prestressing force and e_x the eccentricity. If the curvature of a tendon is treated as a moment diagram applied to a member, then the calculations will involve integration of moment diagrams to set up and solve a number of simultaneous equations. Calculation of indeterminate forces can then be carried out using any of the established methods of elastic analysis, such as influence coefficients or finite elements. This process, however, is unnecessarily complex, especially for manual use.

The equivalent load, or load balancing, method provides a much simpler approach. Using the equations developed in Chapter 5, the tendon profile is transformed into an axial load and an equivalent load on each of the members. These equivalent loads are then treated in the same manner as any other applied load in the analysis and calculation of moments and shears.

Primary and secondary moments have been introduced in Chapter 5. Using the equivalent load method, the secondary moments are automatically included in the serviceability calculations. The secondary moments are, however, needed for the ultimate strength and must be extracted from the serviceability calculations.

Secondary moments are caused by the redundancies present in continuous structures and so vary linearly between supports. Some methods of analysis, such as moment distribution, directly give the final moments and the redundant shears are derived from the moments. If the final moment at a point x resulting from the imposition of the equivalent loading is M_x, then the secondary moment M_{sx} is given by

$$M_{sx} = M_x - P_x.e_x \qquad (6.7)$$

The equations developed in Chapter 5 may result in two different half parabolas on either side of an interior support if the adjacent spans are not equal, so that the intensities of the uniformly distributed reactions from the half parabolas differ (Figure 6.5). This is of no consequence as long as the correct equivalent loads are taken in the calculations.

It is important to appreciate that the concentrated end shears due to the

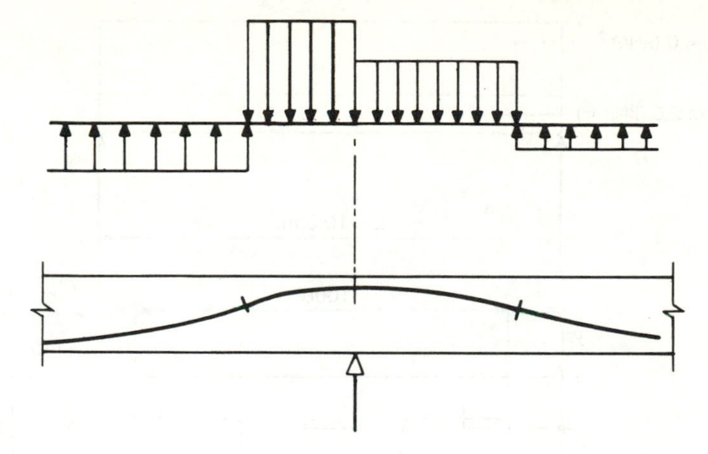

Figure 6.5 *Tendon profile over an internal support with unequal spans*

equivalent loads are internal to the indeterminate system and do not change the overall applied load. The sum of the redundant reactions acting on all the supports due to the equivalent load must be zero. Reactions can only be transferred from one support to another without affecting the sum total of the self-weight and the applied load.

The point can perhaps be better understood by an example. Consider a simply supported beam carrying a load w, and post-tensioned with a parabolic tendon whose equivalent load is w_e. At the support, the shear force in the beam is $(w - w_e)L/2$ whereas the support reaction is $wL/2$. The difference, $w_eL/2$, is in fact provided by the slope of the tendon at the anchorage and equals $P\tan\theta$ in value.

Example 6.1

Calculate the required prestress for the simply supported span shown in Figure 6.6. Assume that the ratio $P_f/P_i = 0.85$, that the tendon centroid at midspan is 40 mm above soffit and that the permissible stresses and design loads are:

$p_{ti} = -1.8$ N/mm^2 $\qquad$ $p_{cf} = +13.2$

$p_{ci} = +12.5$ $\qquad$ $p_{tf} = -2.3$

Imposed loads: dead 2.0, live 4.0 kN/m^2

Solution

Area and section centroid y_t from top:

width × depth	= area (mm^2)	× y	= Ay (mm^3)
1000 × 110	= 110.0 × 10^3	× 55	= 6.05 × 10^6
400 × 170	= 680	× 195	= 13.26
(60/2) × 170	= 5.1	× 167	= 0.85
A_c	= 183.1 × 10^3	A_cy_t	= 20.16 × 10^6

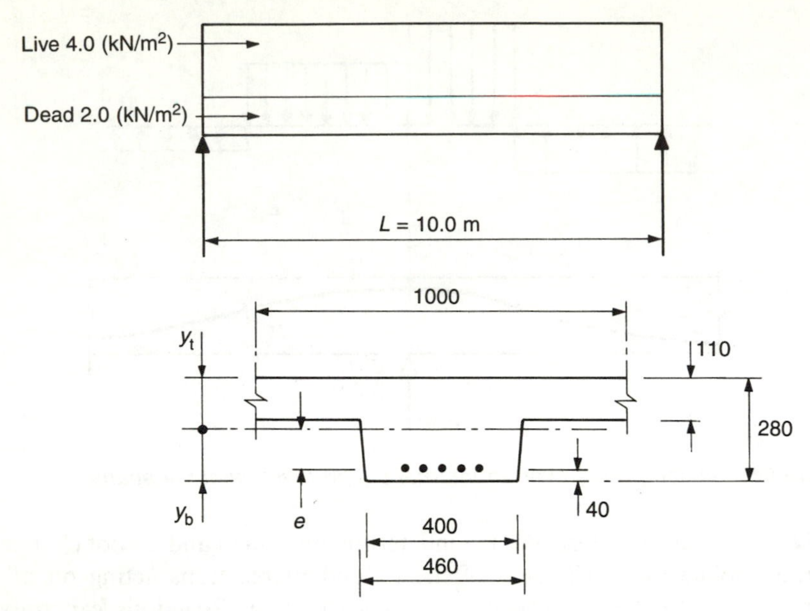

Figure 6.6 *Example 6.1*

$$y_t = A_c y_t / A_c = 110\,\text{mm}$$
$$y_b = D - y_t = 170$$
$$e_m = y_b - 40 = 130$$

Moment of inertia and section moduli:

$$\begin{array}{l} \quad A \qquad\qquad\quad D \qquad\quad y_t \qquad y \\ 110.0 \times 10^3 \times [110^2/12 + (110 - \;\; 55)^2] = 443.67 \times 10^6\,\text{mm}^4 \\ 68.0 \times 10^3 \times [170^2/12 + (110 - 195)^2] = 655.07 \\ 5.1 \times 10^3 \times [170^2/18 + (110 - 167)^2] = \;\; 24.76 \end{array}$$

Moment of inertia $I_c = 1123.40 \times 10^6$

$$Z_t = I_c/y_t \qquad = 10.213 \times 10^6\,\text{mm}^3$$
$$Z_b = I_c/y_b \qquad = 6.608 \times 10^6$$

$$S_t = 1 - e_m A_c/Z_t = -1.331$$
$$S_b = 1 + e_m A_c/Z_b = +4.602$$

Section weight = $24 \times 183.1/1000 = 4.39$ kN/m^2

Moments:

$$M_o = 4.39 \times 10^2/8 = 54.9\,\text{kN.m/m} \quad \text{self-weight}$$
$$M_s = 6.00 \times 10^2/8 = 75.0 \quad \text{other loads}$$
$$M_o + M_s \qquad = 129.9 \quad \text{total}$$

Prestress (equations 6.3a and 6.3b):

$$P_{fmin} = [-2.3 + 129.9/6.608] \times 183.1/4.602 = 690.6\,\text{kN}$$
$$P_{fmax} = 0.85(12.5 + 54.9/6.608) \times 183.1/4.602 = 703.7$$

$P_{fmin} < P_{fmax}$ OK

Use $P_f = 690.6\,\text{kN}$
$P_i = P_f/0.85 = 812\,\text{kN}$

Assuming 160 kN final force per 15.7 mm strand, the number of strands per rib = 690.6/160 = 4.3 ≈ 5

Stresses:

$$P_i/A_c = 812.5/183.1 = 4.44\,\text{N/mm}^2$$
$$\sigma_{pti} = S_t P_i/A_c = -5.91 \text{ (from 6.1a)}$$
$$\sigma_{pbi} = S_b P_i/A_c = 20.43 \text{ (from 6.1b)}$$

$$P_f/A_c = 690.6/183.1 = 3.77\,\text{N/mm}^2$$
$$\sigma_{ptf} = S_t P_f/A_c = -5.02 \text{ (from 6.1c)}$$
$$\sigma_{pbf} = S_b P_f/A_c = 17.36 \text{ (from 6.1d)}$$

Initial stresses (Equations 6.2a and 6.2b)

$$\sigma_{ti} = -5.91 + 54.9/10.213 = -0.53 > p_{ti} \quad \text{OK}$$
$$\sigma_{bi} = +20.43 - 53.4/6.608 = +12.35 < p_{ci} \quad \text{OK}$$

Final stresses (Equations 6.2c and 6.2d)

$$\sigma_{tf} = -5.02 + 129.9/10.213 = +7.70 < p_{cf} \quad \text{OK}$$
$$\sigma_{bf} = +17.36 - 129.9/6.608 = -2.23 > p_{tf} \quad \text{OK}$$

Comment:

A nominal amount of rod reinforcement may be required to comply with the particular national standard. The quantity is subject to ultimate strength calculations. Deflection and shear strength may be calculated as discussed in the relevant chapters.

Example 6.2

In example 6.1 no rod reinforcement is required to carry any forces at the serviceability state, because the concrete section is sufficiently large. Try the smaller section shown in Figure 6.7. Allowable stresses and applied loads remain as before.

Solution

Section properties:

$A_c = 162.8 \times 10^3\,\text{mm}^2$ $\quad I_c = 850.290 \times 10^6\,\text{mm}^4$

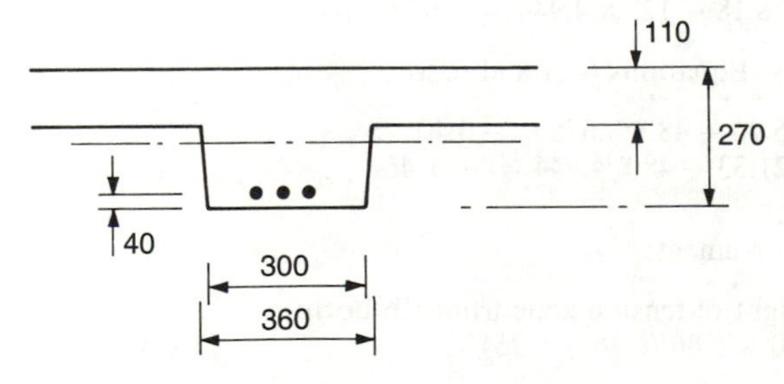

Figure 6.7 *Example 6.2*

$y_t = 98$ mm $\quad Z_t = 8.676 \times 10^6$ mm^3
$y_b = 172$ mm $\quad Z_b = 4.944 \times 10^6$ mm^3
$e_m = 132$ mm $\quad S_t = -1.477$
$S_b = +5.347$

Section weight = 24 × 162.8/1000 = 3.91 kN/m^2

Moments:

$$\begin{aligned} M_o &= 3.91 \times 10^2/8 = 48.8 \text{ kN.m/m} && \text{self-weight} \\ M_s &= 6.00 \times 10^2/8 = 75.0 && \text{other loads} \\ M_o + M_s &= 123.8 && \text{total} \end{aligned}$$

Prestress (Equation 6.3a)

$$P_{fmin} = [-2.3 + 123.8/4.944] \times 162.8/5.347 = 692.4 \text{ kN}$$

Calculations, not shown here, show that the prestressing force of 692.4 kN produces an initial compressive stress at the bottom which exceeds the permissible value. Therefore, the economical solution is to choose a lower value of prestressing force, such that the initial stress is not exceeded, and to provide some rod reinforcement at the bottom of the rib.

From Equation (6.2b), $(P_i/A_c)S_b - M_o/Z_b = p_{ci}$

$$\begin{aligned} P_i &= (p_{ci} + M_o/Z_b)A_c/S_b \\ &= (12.5 + 48.8/4.944) \times 162.8/5.347 = 681.1 \text{ kN} \\ &\qquad \text{say } P_i = 650 \text{ kN} \\ P_f &= 0.85 \times P_i = 553 \text{ kN} \end{aligned}$$

Assuming 160 kN final force per 15.7 mm strand, number of strands per rib = 553/160 = 3.5 ≈ 4

Stresses:

$$\begin{aligned} P_f/A_c &= 553/162.8 = 3.40 \text{ N/mm}^2 \\ \sigma_{ptf} &= S_t P_f/A_c = -5.02 \\ \sigma_{pbf} &= S_b P_f/A_c = 18.18 \end{aligned}$$

$$\begin{aligned} P_i/A_c &= 650/162.8 = 3.99 \text{ N/mm}^2 \\ \sigma_{pti} &= S_t P_i/A_c = -5.90 \\ \sigma_{pbi} &= S_b P_i/A_c = 21.33 \end{aligned}$$

Final stresses (Equations 6.2c and 6.2d)

$\sigma_{tf} = -5.02 + 123.8/8.676 = +9.25 < p_{cf}$ OK
$\sigma_{bf} = +18.18 - 123.8/4.944 = -6.86 < p_{tf}$ A_{st} needed

Initial stresses (Equations 6.2a and 6.2b)

$\sigma_{ti} = -5.90 + 48.8/8.675 = -0.40 > p_{ti}$ OK
$\sigma_{bi} = +21.33 - 48.8/4.944 = +11.46 < p_{ci}$ OK

Tension reinforcement:

h_t = height of tension zone from rib bottom
= 270 × 6.86/(6.86 + 9.25) = 115 mm

b_t = rib width at height h_t

$= 300 + 60 \times 115/(270 - 110) \qquad = 343\ \text{mm}$

$T =$ Total tension
$= [115 \times (300 + 343)/2] \times 6.86/(2 \times 1000) = 126.8\ \text{kN}$

$A_{st} =$ Area of bonded reinforcement per rib
$= 126.8 \times 1000/200 \qquad = 634\ \text{mm}^2$
$= 4 \times 16\ \text{mm dia.}\ (804\ \text{mm}^2)$

The amount of bonded steel (804 mm^2) is 26% more than the required 634 mm^2 and the number of strands provided (4 No.) is 14% more than the required 3.5. An adjustment to the section would be required to arrive at a more efficient solution.

A more economical solution may be obtained if the rib spacing and width are increased by 10%. The section properties and the self-weight per unit width remain unchanged and, therefore, all of the above calculations remain valid; though a revision to indicate the change is required. The reader is invited to try such a change.

Example 6.3

Design the prestressing for a ribbed post-tensioned floor, continuous over three spans of 10, 8 and 11 m lengths, carrying imposed dead and live loads of 2.0 and 4.0 kN/m^2 respectively. The section, spans and loads are shown in Figure 6.8; assume an initial / final prestress ratio of 1.15. The permissible stresses are:

$p_{ti} = -1.8\ \text{N/mm}^2 \qquad p_{cf} = +13.2\ \text{N/mm}^2$
$p_{ci} = +12.5 \qquad p_{tf} = -2.3$

Solution

The solution presented in this example is based on the use of approximate values of equivalent loads. The method is suitable for a preliminary design; the final design should be based on more accurate calculations, whether carried out manually or using a computer program.

The tendon profile, chosen after a previous trial, is also shown in Figure 6.8. The diagram indicates the true sag in each span for the parabola; however, the value taken in the following calculation is an approximation, assuming that the lowest point of the parabola is at midspan, which is not strictly correct but is simpler.

The final prestressing force is assumed a constant 650 kN in all spans. This implies alternate tendons being stressed from opposite ends. The tendon centroid in the centre span is above the section centroid; obviously, the tendon force is too high for this span.

Section properties:

$A_c = 179.6 \times 10^3\ \text{mm}^2 \qquad I_c = 1011.5 \times 10^6\ \text{mm}^4$
$y_t = 106.5\ \text{mm} \qquad Z_t = 9.497 \times 10^6\ \text{mm}^3$
$y_b = 163.5\ \text{mm} \qquad Z_b = 6.186 \times 10^6\ \text{mm}^3$

Self-weight $= 24 \times 0.1796 = 4.3\ \text{kN/m}$

Equivalent loads and their fixed-end moments:

Span 1–2

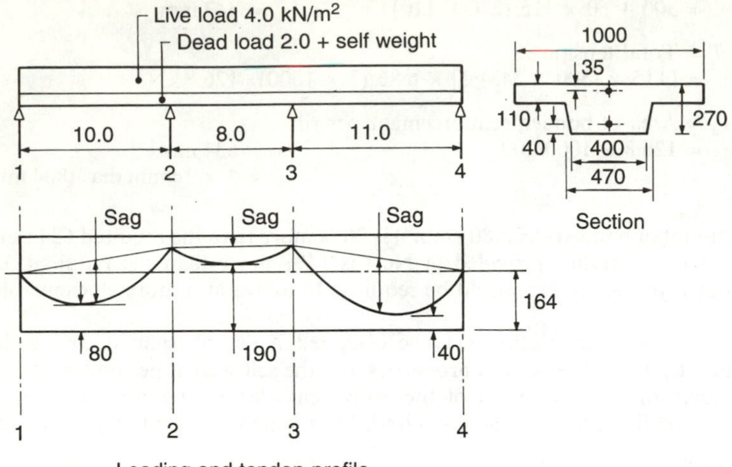

Loading and tendon profile

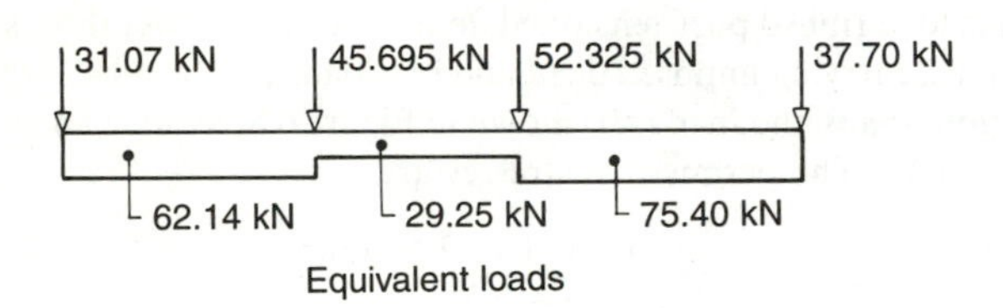

Equivalent loads

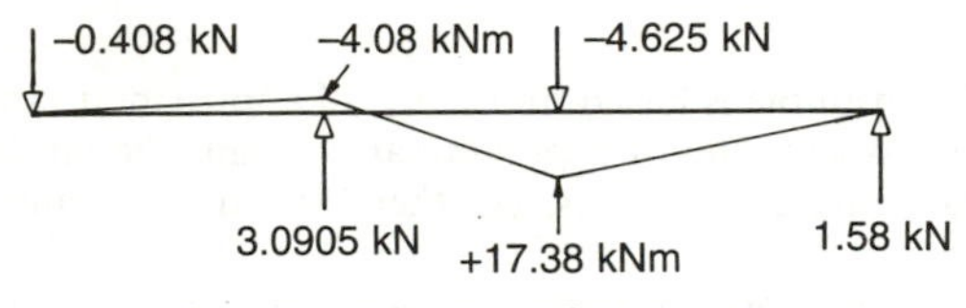

Secondary forces and moments

Figure 6.8 *Example 6.3*

$$s \approx (235 + 164)/2 - 80 \quad = 119.5\,\text{mm}$$
$$w_e = -8 \times 650 \times 0.1195/10^2 = -6.21\,\text{kN/m}$$
$$FEM = 6.21 \times 10^2/8 \quad = 77.63\,\text{kNm}$$

Span 2–3

$$s = 235 - 190 \quad = 45.0\,\text{mm}$$
$$w_e = -8 \times 650 \times 0.045/8^2 \quad = -3.66\,\text{kN/m}$$
$$FEM = 3.66 \times 8^2/12 \quad = 19.50\,\text{kNm}$$

Span 3–4

$$s \approx (235 + 164)/2 - 40 \quad = 159.5\,\text{mm}$$
$$w_e = -8 \times 650 \times 0.1595/11^2 = 6.85\,\text{kN/m}$$
$$FEM = 6.85 \times 11^2/8 \quad = 103.68\,\text{kNm}$$

Analysis results:

Moments resulting from the analysis for dead, equivalent and three alternative live loads are given in the table below.

Support	1	2	3	4
Moments kNm				
1 Eq. load	0	+42.07	+63.53	0
2 Self-weight	0	−35.21	−44.72	0
3 Applied dead	0	−16.38	−20.80	0
4 Live, 1–4	0	−32.76	−41.60	0
5 Live, 1–2, 3–4	0	−21.11	−30.62	0
6 Live, 2–3	0	−11.75	−10.98	0

Final flexural stresses:

Span 1–2

Live load Case 5 is critical

$M_1 = 0$

$M_2 = +42.07 - 35.21 - 16.38 - 21.11 \quad = -30.63$ kNm

$M_s =$ Span moment

$\approx (M_1 + M_2)/2 + (\Sigma w)L^2/8$

$\approx -30.63/2 + (4.3 + 2.0 + 4.0 - 6.21) \times 10^2/8 = 35.76$ kNm

$\sigma_{tf} = 650/179.6 + 35.76/9.497 \quad = +7.38$ N/mm² OK

$\sigma_{bf} = 650/179.6 - 35.76/6.186 \quad = -2.16$ OK

Support 2

Live load Case 4 is critical

$M_2 = +42.07 - 35.21 - 16.38 - 32.76 \quad = -42.28$ kNm

$\sigma_{tf} = 650/179.6 - 42.28/9.497 = -0.83$ N/mm² OK

$\sigma_{bf} = 650/179.6 + 42.28/6.186 = +10.45$ OK

The section would normally be solid over the support, so the section properties for the solid section would be more realistic than those of the T-section used.

The final stresses at other critical points can be checked in the same manner.

Initial flexural stresses:

Initial prestressing force = $1.15 \times 650 = 747.5$ kN

Average stress = $747.5/179.6 = +4.16$ N/mm²

Span 1–2

$M_1 = 0$

$M_2 = +1.15 \times 42.07 - 35.21 \quad = +13.17$ kNm

$M_s =$ span moment

$\approx 13.17/2 + (4.3 - 1.15 \times 6.21) \times 10^2/8 \quad = -28.93$ kNm

$\sigma_{tf} = +4.16 - 28.93/9.497 \quad = +1.11$ N/mm² OK

$\sigma_{bf} = +4.16 + 28.93/6.186 \quad = +8.84$ OK

Initial stresses at other critical points can be checked in the same manner.

Secondary moments:

$M_{sx} = M_x - Pe$
For M_x see Case 1 of Analysis Results

At 2, $M_{sx} = +42.07 - 650 \times (235 - 164)/1000 = -4.08$ kNm
At 3, $M_{ax} = +63.53 - 650 \times (235 - 164)/1000 = +17.38$

Example 6.4

Design the prestressing for a ribbed post-tensioned floor, continuous over three spans of 10, 8 and 11 m length, carrying imposed dead and live loads of 2.0 and 4.0 kN/m^2 respectively. The floor section is shown in Figure 6.9; assume an initial / final prestress ratio of 1.15.

The problem is the same as that in Exercise 6.3, except that this time the calculations are to be based on the actual tendon profiles rather than the approximations, and the tendons are to be curtailed to save on the number of strands.

Solution

Figure 6.9 shows the proposed tendon profile and the final prestressing forces in the three spans. The full prestressing force of the uncurtailed tendons is available at each of the penultimate supports. The influence of the short lengths of the curtailed tendons in the middle span is normally ignored in the calculation of equivalent load and secondary forces. The contraflexure points of the profile are placed at $0.1L$ in each span. The profile and the prestressing forces are the result of an initial attempt, not shown here.

Section properties:

$A_c = 179.6 \times 10^3$ mm^2 $I_c = 1011.5 \times 10^6$ mm^4
$y_t = 106.5$ mm $Z_t = 9.497 \times 10^6$ mm^3
$y_b = 163.5$ mm $Z_b = 6.186 \times 10^6$ mm^3

Self-weight $= 24 \times 0.1796 = 4.3$ kN/m

Equivalent loads: [See Table 5.1 and Equations (5.11) and (5.19)].

Span 1–2 $P_f = 560$ kN

The tendon is higher at the right-hand side; therefore, work from support 2.

$a = 0.1 \times 10 = 1.0$ m
$y_3 = 165 - 40 = 125$ mm
$y_1 = 235 - 40 = 195$ $y_3/y_1 = 0.641$

Interpolating from Table 5.1, $b/L = 0.549$
$b = 5.49$ m measured from support 2

$w_e = -2P_f y_1/(b^2 - ab)$
$= -2 \times 560 \times 0.195/(5.49^2 - 1.00 \times 5.49) = -8.86$ kN/m

W_{12} = shear at support 1 $= w_e.(L - b - a) = 31.10$ kN
W_{21} = shear at support 2 $= w_e.(b - a) = 39.78$ kN

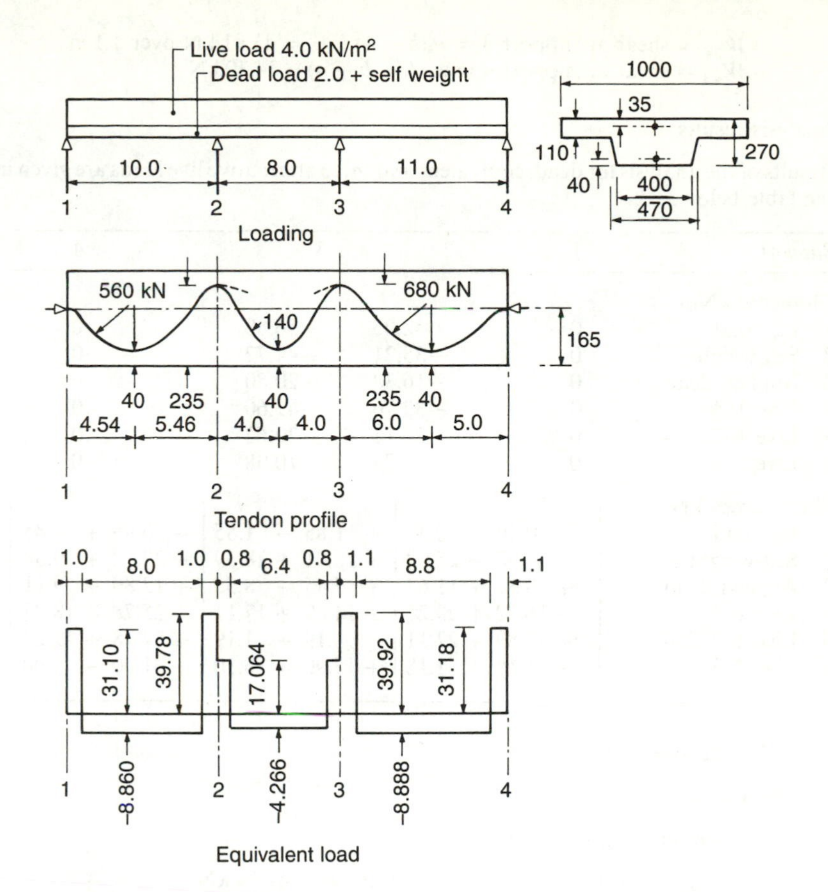

Figure 6.9 *Example 6.4*

W_{12} and W_{21} are uniformly distributed over length $a = 1.0$ m.

Span 2–3 $P_f = 140$ kN

$a = 0.1 \times 8 = 0.8$ m, and by symmetry, $b = 4.0$ m
total sag $s = (235 - 40)/1000 = 0.195$ m

$$w_e = -2 \times 140 \times 0.195/(4.00^2 - 0.80 \times 4.00) = -4.27 \text{ kN/m}$$
$$W_{23} = W_{32} = 4.27 \times (4 - 0.8) = 13.65 \text{ kN over } 0.8 \text{ m}$$

Span 3–4 $P_f = 680$ kN

$a = 0.1 \times 11 = 1.1$ m
$y_1 = 235 - 40 = 195$ mm
$y_3 = 165 - 40 = 125$ $\quad y_3/y_1 = 0.641$

Interpolating from Table 4.1, $b/L = 0.549$
$b = 6.04$ m measured from support 3

$$w_e = -2P_f y_3/(b^2 - ab)$$
$$= -2 \times 680 \times 0.195/(6.04^2 - 1.10 \times 6.04) = -8.89 \text{ kN/m}$$

W_{34} = shear at support 3 = $w_s(b - a)$ = 43.91 kN over 1.1 m
W_{43} = shear at support 4 = $w_s(L - b - a)$ = 34.30 kN

Analysis results:

Results of the analysis for dead, equivalent and three alternative live loads are given in the table below.

Support	1	2	3	4
Moments kNm				
1 Eq. load	0	+42.05	+56.88	0
2 Self-weight	0	−35.21	−44.72	0
3 Applied dead	0	−16.38	−20.80	0
4 Live 1–4	0	−32.76	−41.60	0
5 Live 1–2, 3–4	0	−21.11	−30.62	0
6 Live 2–3	0	−11.75	−10.98	0
Shear forces kN				
1 Eq. load		+ 0.30 − 0.30	+ 1.85 − 1.85	− 0.85 + 0.85
2 Self-weight		+ 17.98 + 25.02	+ 16.01 + 18.39	+ 27.72 + 19.58
3 Applied dead		+ 8.37 + 11.63	+ 7.45 + 8.55	+ 12.89 + 9.11
4 Live 1–4		+ 16.72 + 23.27	+ 14.89 + 17.11	+ 25.78 + 18.22
5 Live 1–2, 3–4		+ 17.89 + 22.11	− 1.19 + 1.19	+ 24.78 + 19.22
6 Live 2–3		− 1.18 + 1.18	+ 15.90 + 16.10	+ 1.00 − 1.00

Final flexural stresses:

Span 1–2:

Live load Case 5 is critical

$V_{12} = +0.30 + 17.98 + 8.37 + 17.89 = +44.54$ kN
$V_{21} = -0.30 + 25.02 + 11.63 + 22.11 = +58.46$ kN
w = net span UDL = $+6.30 - 8.86 + 4.0 = +1.44$ kN/m

Check total load = $(6.3 + 4.0) \times 10.0 = 103.00$ kN
Reactions = $44.54 + 58.46 = 103.00$ OK

X = Zero shear distance from 1
$= (V_{12} - W_{12} + aw_e)/w$
$= (44.54 - 31.10 - 1.0 \times 8.860)/1.44 = 3.18$ m

M_{s1} = Moment in span 1
$= V_{12}X - (W_{12} - aw_e).(X - a/2) - wX^2/2 + M_1$
$= 44.54 \times 3.18 - (31.1 + 1.0 \times 8.860) \times 2.68 - 1.44 \times 3.18^2/2$
$= 27.26$ kNm

Prestress = 560 kN

$\sigma_{tf} = 560/179.6 + 27.26/9.497 = +6.00$ N/mm^2 OK
$\sigma_{bf} = 560/179.6 - 27.26/6.186 = -1.29$ OK

Support 2:

At the support the section will be solid

$A_c = 270 \times 10^{-3}$ mm^2 $\quad Z_b = Z_t = 12.15 \times 10^{-6}$ mm^3

$M_2 = +42.05 - 51.59 - 32.76 = -42.30\,\text{kNm}$
$\sigma_{tf} = 560/270 - 42.38/12.15 \quad = -1.41\,\text{N/mm}^2$ OK
$\sigma_{bf} = 560/270 + 42.38/12.15 \quad = +5.56$ OK

Final stresses in other spans and at support 3, and initial stresses, can be calculated in a similar manner.

Tendon radius over support 2:

From Equation 4.18, $A_1 = -y_1/ab = -0.195/(1 \times 5.49) = 0.0355/\text{m}$
Radius $= 1/2A \quad = 14.1$ m OK

In this example, the negative moment stress is checked at the support centreline; the check may be made at the support face where the moment would be somewhat smaller.

The following further calculations must be carried out to complete the design.

- Check on deflections.
- Prestress losses and, if necessary, calculation of flexural stresses corresponding to the new prestressing forces.
- Ultimate flexural and shear strength calculations. Bonded steel requirement and the point where the section is to change from ribbed to solid.
- Anchorage zone reinforcement

7 PRESTRESS LOSSES

Some of the applied prestressing force is lost during and after the stressing operation, and, therefore, the magnitude of compressive force induced in the concrete is somewhat less than the force applied by the jacking equipment. In the past, the practice was to assume a certain arbitrary value for the overall loss. This is now considered to be too inaccurate and such assumed losses are now accepted to have been under-estimated.

For initial design and approximate quantities, an average final force of $0.60f_{pu}$ to $0.65f_{pu}$ is often assumed in the current practice. Assuming that the strands are initially stressed to $0.75f_{pu}$, these values represent a long-term loss of about 20% to 15% respectively. For the final design the losses should be calculated.

In this chapter the causes of the loss in prestress are considered in roughly the order in which they occur.

7.1 General

There are several factors which contribute to the loss of prestress. Some of these are immediate, as they occur during the stressing operation, while the others are time related and take place over a long period. Therefore, the losses can conveniently be considered in two groups:

a. Immediate, which occur during the stressing operation. These are of interest in assessing the prestressing force at the *initial stage*, i.e. immediately after stressing, before any of the imposed load is applied. This group comprises:

   - friction loss due to tendon curvature and wobble
   - elastic shortening of concrete under the induced compression
   - and anchorage draw-in.

b. Long term, which occur at a gradually decreasing rate over the life of the member. These are considered only at the *final stage* when imposed loads have been applied. This group consists of:

   - shrinkage of concrete
   - creep of concrete
   - and relaxation of steel.

In order to calculate the losses, values must be assumed for the initial stresses at critical sections and the number of strands must be guessed. The initial assumptions will, almost certainly, need to be modified as a result of calculation of losses, and a revised level of stress and a different number of strands will ensue. Therefore, the loss calculations carried out have also to be revised. The losses include the effect of the cable profile whose adequacy, or inadequacy, will not be judged until after most of the stress calculations have been carried out. A further complication is that the losses due to the various causes are interdependent. The calculation process, therefore, is iterative and cumbersome, particularly if an accurate force profile is to be obtained for the length of the member.

Calculation of stresses requires fairly accurate values of the prestressing force at the critical sections in a member. An accuracy of about 5% is considered adequate; endeavouring for a greater accuracy in calculating losses is unlikely to result in a greater degree of accuracy in the stresses, because of the uncertainties and unpredictabilities of material characteristics and workmanship. For example, the modulus of elasticity of concrete may vary from the assumed value, accuracy of installation of tendons is unknown, and the allowable dimensional tolerances permit a variation in section properties.

7.1.1 Simplifying assumptions

The following simplifying assumptions are made in the calculation of losses.

- The angle between the tendon profile and the member axis is small so that $\sin \theta \approx \tan \theta \approx \theta$.
- The increase in tendon length because of its curvature can be ignored and the tendon length in a span equals the span length.

  For an external span the tendon length is measured from the anchorage face, and not the support centreline.
- The losses vary linearly along the tendon length.

In a given span, the force varies from one support to midspan to the other support, but it is considered to be sufficiently accurate to assess the magnitude of the net prestressing force at midspan and use these values for the whole of that span. If the losses are judged to be critical then it is necessary to calculate the losses at the supports of a continuous member as well. This may be necessary if the calculated stresses are very close to the allowable values, or when tendons are curtailed near a support.

7.1.2 High losses

The loss in prestressing force increases with the number of tendons. If the total loss is unacceptable then after stressing all the tendons, each of them can be re-stressed. Alternatively, each tendon can be stressed to a different load—the first tendon to the highest load and the last to the lowest—so that after stressing each of them will have approximately equal force.

Either of these procedures will substantially reduce the loss; the former will be the more efficient. However, they will also add to the work load of the site staff and put an undue extra responsibility on them for what is really a design problem. Therefore, it may be preferable to accept the higher losses if possible, even at the expense of a few additional tendons.

7.1.3 Long tendons

Accurate assessment of the loss of prestressing force is seldom carried out because the values of E_c, μ, K, shrinkage strain, creep coefficient, etc., are all based on previous experience, and the values actually effective for the particular tendon being stressed may well be quite different. Another difficulty in the accurate determination of the loss stems from the interdependence of the various factors contributing to the loss. For example, the elastic shortening loss at a section depends on the stress at the section, which should be the value after elastic loss; creep loss should take into account the elastic loss and the creep loss itself.

In long tendons, running over several continuous spans, the loss at the far end is over-estimated if calculations are based on forces at the jacking point. In such tendons, better accuracy results if the overall member length is divided into several sections and each section is considered in succession. For immediate losses in the second section, the jacking force is taken as the force after friction and elastic losses at the end of the first section, and the force for long-term losses in the second section is taken as the force at the end of the first section, after all losses.

For a long tendon stressed from both ends, the maximum loss occurs near its mid-length. If the losses are calculated in sections, then the loss calculations must also be carried out from each end up to the mid-length.

Where the loss of prestress is unacceptably high, the following options are available.

- Reduce friction loss by making the tendon profile as flat as possible, particularly in the interior spans.
- Provide additional short lengths of tendons in the far spans.
- Apply a higher initial jacking force, and then reduce the jacking force to the nominal design value before locking the anchorage.

  The procedure is, however, unreliable and gives only a marginal advantage. Most national standards limit the maximum jacking force to 80% of the tendon strength so that the procedure can only be used where the design jacking force is lower than this limiting value.
- Re-stress the tendons after some of the shrinkage and creep losses have taken place.

  The procedure also reduces the elastic shortening loss, but requires access to the anchorages for a longer period, thereby slowing down the construction and risking damage to the stressed tendons.
- Stress from both ends.

  Prestressing forces in outer bays are higher than those obtainable from alternate end stressing.

- Stress alternate tendons from the two ends.
  This method results in a more uniform prestressing force over the tendon length than both end stressing, because at each end, alternate tendons have the maximum and the minimum force.
- Stress each bay as it is cast, using intermediate anchors.
  This procedure allows good control over the prestressing force and minimizes shrinkage cracking of concrete. However, extra hardware is required and each tendon requires several stressing visits.

7.2 Friction losses

A proportion of the jacking force is lost during stressing through contact friction between the strand and its surrounding duct—greased plastic extrusion for unbonded tendons and preformed sheath for bonded tendons. The contact may be intentional due to tendon curvature, as when a tendon is draped in the shape of a parabola, or unintentional, caused by minor deviations of the tendon from its intended profile. The loss due to the intentional tendon contact is referred to as *curvature friction*, and that due to the unintentional contact as *wobble* or *parasitic friction*.

In the first case, the contact force, and hence the loss, between the tendon and the duct is proportional to the angle through which the tendon turns starting at the jacking end; where the profile consists of a series of curves the loss is proportional to the sum of all angles. Loss at a point in the tendon length due to unintentional deviation is statistically shown to be proportional to the length of the tendon from the jacking end. In both cases, the loss increases from zero at the live anchorage to a maximum at the dead anchorage, or at the mid-point of the tendon length if it is stressed from both ends.

Some loss also occurs due to friction in stressing jacks and anchorages, particularly in multistrand tendons where the strands comprising a tendon spread out to allow sufficient space for the jack to grip them. The loss from these factors varies from one system to another and is usually accounted for in the calibration of the equipment.

The two losses, due to friction and wobble, are grouped in the following equation:

$$P_x = P_o e^{-(\mu\theta + Kx)} \tag{7.1}$$

where e = Naperian logarithm base
θ = Angle of change of tendon direction in radians
x = distance from anchorage

Note that K has the dimension 1/length and therefore, its value depends on the unit of length being used.

The percentage loss of prestress is given by the expression $100(1 - e^{-(\mu\theta + Kx)})$. For small values of $(\mu\theta + Kx)$, say less than 0.3, the expression $e^{-(\mu\theta + Kx)}$ approximates to $(1 - \mu\theta - Kx)$, and, therefore the expressions for P_x and loss

Table 7.1 *Values of* μ and *K*

Duct material	*BS 8110* μ	*BS 8110* *K/m*	*ACI 318 Commentary* μ	*ACI 318 Commentary* *K/metre*	*ACI 318 Commentary* *K/ft*
Rigid galvanized	0.25	0.0017	0.15–0.25	0.0005–0.0020	$0.15–0.61 \times 10^{-3}$
Greased plastic	0.12	0.0025	0.05–0.15	0.0003–0.0020	$0.09–0.61 \times 10^{-3}$
Non-rigid		0.0033			

percentage can be simplified to:

$$P_x = P_o(1 - \mu\theta - Kx) \tag{7.2}$$
$$\text{Percentage loss} = 100(\mu\theta + Kx)$$

Table 7.1 gives the values of μ and K as specified in BS 8110 and ACI 318. Although the coefficient of friction, μ, is lower for a greased tendon in an extruded plastic sheath, such a tendon is not as rigid as a metal duct and, therefore, it is liable to have a larger deviation of profile. Recognizing this, the wobble factor, K, is higher for the greased strand in plastic extrusion. The 0.0033 value of K, given for a non-rigid duct is meant for sheaths which are not sufficiently rigid to resist being displaced during concreting.

Drawn strand has a smoother surface and, in greased plastic extrusion, its coefficient of friction is much lower than that for the normal shaped strand. The value of μ for drawn strand can be as low as 0.03 for this product.

Suppliers of post-tensioning hardware give values of μ and K for their products which are generally lower than those specified in BS 8110—typically, $\mu = 0.19$ and $K = 0.0008$/m for rigid metal sheath, and $\mu = 0.06$ and $K = 0.0005$/m for greased strand. The supplier's values may be used if known. If not, it may be prudent to use the coefficients specified in the relevant standard.

7.3 Anchorage draw-in

In most of the post-tensioning systems presently available for use in floors, strand is gripped by the conical wedges as they slide into a tapered hole in the barrel or the bearing plate. In the process, the length of the stressed tendon is reduced by the amount of *draw-in* needed to lock the anchorage, which is usually a minimum of 6 mm (0.25in). Hardware manufacturers provide figures for the draw-in for their systems but these do not make any allowance for the strand slipping in the wedges before the grip becomes effective. The possibility of strand slippage is much less with jacks where the wedges are driven home hydraulically. However, an overall allowance of 8 mm (0.3in) draw-in is reasonable at the design stage.

The slip may be only a few millimetres but it is unpredictable. Being unknown and variable, it is difficult to calculate accurately the loss of prestress at the design stage. The normal practice is to measure the actual total draw-in during stressing and make allowance for any such slip at that stage by adjusting the extensions for the tendons still to be stressed. If the measured draw-ins are found to be outside the range specified by the designer, then the tendon forces are calculated from the draw-ins and the serviceability stresses are checked. Some of the tendons may have to be re-stressed if the stresses are found to be unacceptable.

Loss of prestress due to anchorage draw-in becomes increasingly significant as the tendon length reduces; in short tendons this loss may be very high. A short length of tendon needs a relatively small total elongation of the strand for the desired prestressing force and an 8 mm draw-in loss is more significant in this case than it would be in the case of a long tendon where the initial elongation is higher. For example, a 5 m long tendon, with a jacking extension of 30 mm, would lose 26% of the force if the slip was 8 mm; a longer tendon would lose a proportionately smaller amount of force.

The high loss of prestress due to draw-in, coupled with the uncertainty over the amount of strand slip, makes it difficult to post-tension short members. A useful procedure to follow in the case of short tendons is described in Chapter 13.

When the jack is pulling the tendon, friction between the tendon and the duct opposes the jacking force, so that the tendon force reduces gradually away from the jack. During the anchorage locking operation the tendon moves in the opposite direction—into the duct—and friction now opposes this movement. After locking, the net stress in the tendon, therefore, increases away from the anchorage for a distance and then it reduces just as it did when jacking. At this stage, the loss is confined to a definite length of the tendon. This is shown diagrammatically in Figure 7.1; Figures (a) and (b) show the forces immediately before and after lock-in.

Calculation of the length Z, within which this loss takes place, is necessary to check if any of the critical sections fall in this zone, and if any does, then the tendon force is calculated at that point. In this calculation the loss is assumed to be linear along the length of the member.

7.3.1 Unbonded tendons

Unbonded tendons, not being grouted, remain free to move in their greased plastic sheathing, allowing relative movement to occur between the strand and the concrete. The movement may possibly be initiated by vibration, temperature changes, shrinkage or creep. Such a movement tends to spread the draw-in loss beyond the initial distance Z, and gradually to the whole length of the tendon. Therefore, the long-term prestress distribution in an unbonded tendon differs from that shown in Figure 7.1(b). The draw-in loss for unbonded tendons can be calculated on the basis of the simple assumption that the whole tendon length is affected.

The slip movement in fact starts within a few minutes after stressing and the tendon force may reach a uniform level within a few hours thereafter. If the initial loss in a tendon is judged to be excessive then it may be re-stressed within a few hours of the first stressing.

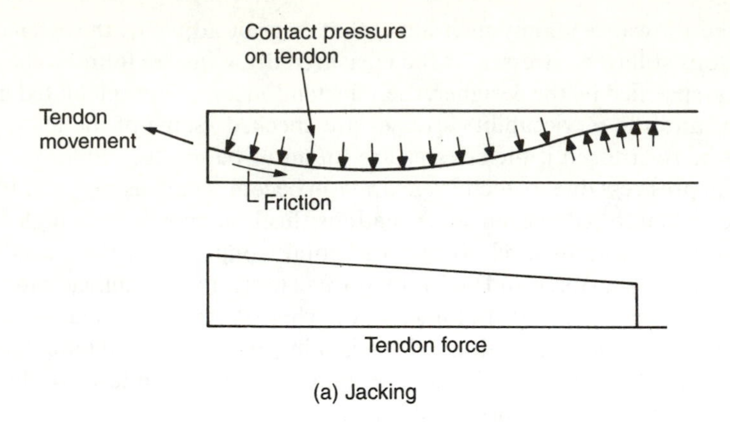

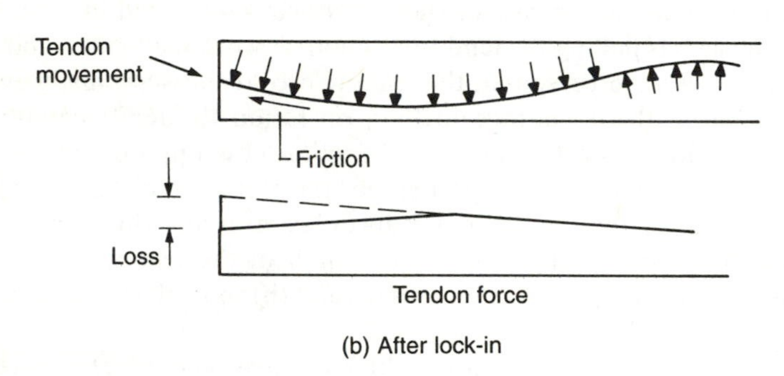

Figure 7.1 *Draw-in loss of prestress*

$$\text{Loss in force} = (\delta/L_t)A_pE_s \text{ for unbonded tendon} \tag{7.3}$$

where δ = Anchorage draw-in

7.3.2 Bonded tendons

In the case of bonded tendons, the duct is grouted soon after stressing and the draw-in loss is, therefore, locked within the distance Z. Figure 7.2(a) shows the case of a long bonded tendon, where the length Z affected by the draw-in is less than the total tendon length L_t. Line ABC represents the tendon force at jacking, before the tendon is locked in, and line CDE represents the force after lock-in, when the loss due to draw-in has taken place. Points B and D are at the mid-lengths of AC and CE respectively. The loss of prestressing force at any point within the length Z is represented by the vertical distance between lines ABC and CDE.

Let β represent the slope of line ABC, its units being loss of force per unit length. The value of β can be determined by calculating friction loss at a convenient point away from the anchorage, such as at the next support or at

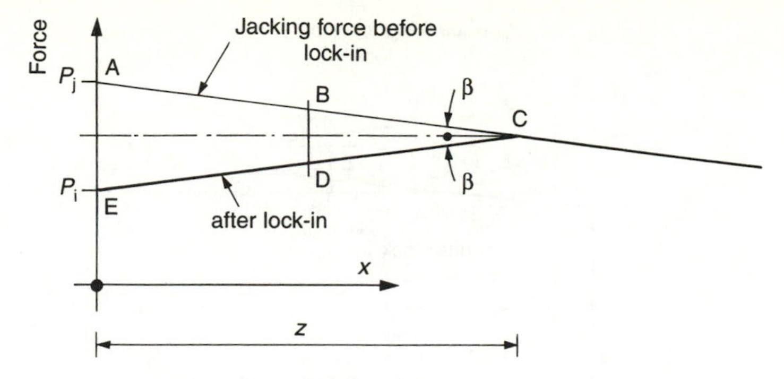

Figure 7.2 *Loss of prestress in a bonded tendon*

midspan, and then dividing this loss by the distance. It is reasonably accurate to assume the slope of line CDE as $-\beta$, because the values of μ and K remain the same but the direction of friction has changed. This is not strictly accurate because the slope of line CDE should correspond to the tendon force at C, which is lower than the original jacking force at A, but it is a reasonable assumption.

For the value of β, calculate the loss over the span length L. Let θ represent the angular deviation of the tendon over length L. Then

$$\begin{aligned} P_L &= \text{Force at distance } L \\ &= P_j e^{-(\mu\theta + KL)} \approx P_j(1 - \mu\theta - KL) \end{aligned} \tag{7.4}$$

$$\beta = (P_j - P_L)/L \approx P_j(\mu\theta/L + K) \tag{7.5}$$

The loss of force between A and C is βZ. The loss of force at B, represented by the line BD, is also βZ. This is the average loss over length Z and the corresponding change in tendon length must equal the draw-in δ. Therefore,

$$\begin{aligned} \delta &= (\beta Z/A_p E_s)Z \quad \text{and so} \\ Z &= (\delta A_p E_s/\beta)^{0.5} \end{aligned} \tag{7.6}$$

Knowing P_j, and β and Z from Equations (7.4) and (7.6), the tendon force P_i and the force at any distance x from the anchorage can now be calculated.

$$\begin{aligned} &P_i = P_j - 2(P_j - P_Z) = P_j - 2\beta Z \\ &\text{For } x \leqslant Z, \qquad P_x = P_i + \beta x \\ &\text{For } x \geqslant Z, \qquad P_x = P_j - \beta x \end{aligned} \tag{7.7}$$

The above discussion has assumed that the bonded tendon is longer than the distance Z, so that the tendon length near the dead end is not affected by the draw-in loss. In a short tendon this may not be the case and the whole of its length will then suffer a loss. This case is shown in Figure 7.3. Line AF represents the distribution of the force before lock-in, and line GE after lock-in. L_t is the length from the anchorage A to the dead end at F, points M and N being at the middle of lengths AF and GE respectively.

The value of β can be calculated from Equation (7.5), substituting L_t for L. Line

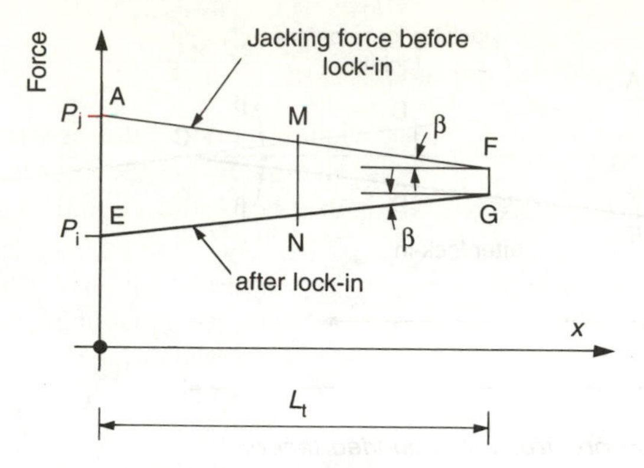

Figure 7.3 *Draw-in loss in short bonded tendons*

MN represents the average loss of force over the tendon length L_t; let this loss be represented by P_{MN}. Then, the change in tendon length under the influence of P_{MN} must equal the draw-in δ.

$(P_{MN}/A_p E_s)L_t = \delta$, and so

$$P_{MN} = \delta L_t / A_p E_s$$
$$P_i = P_j - \beta L_t - P_{MN} \quad (7.8)$$
$$P_x = P_i + \beta x$$

x is less than or equal to L_t.

7.4 Elastic shortening

As a tendon is stressed, it induces compression in the concrete, which causes a slight reduction in the length of the member. The shortening in length and the application of prestress being simultaneous, there is no loss of force in the particular tendon being stressed and, after stressing, the tensile force in the tendon is equal to the compression in concrete.

When the second tendon is stressed then it suffers no loss from its own stressing but the elastic shortening caused by its stressing reduces the force in the first tendon. Similarly, when the third tendon is stressed, the forces in the previously stressed two tendons reduce.

Consider a number of tendons in close proximity with each other, which are stressed sequentially.

P_L = Loss of force in a tendon when another tendon is stressed

$$= p_c(E_s/E_{ci})A_1 \quad (7.9)$$

where

P_L = Loss of force in a tendon when another is stressed
p_c = stress in concrete at tendon level due to P_1

A_1 = the area of one tendon
P_1 = force in one tendon after lock-in

As explained above, there is no loss when the first tendon is stressed. When the second tendon is stressed, it carries its full force P_1 but the force in the first tendon is reduced by P_L. The average loss for the two tendons equals half the total loss, i.e. $0.5P_L$.

When the third tendon is stressed then it carries its full force, the second tendon now loses P_L and the first tendon has now sustained a loss of $2P_L$. The total loss in the three tendons is $3P_L$. With four tendons stressed sequentially the total loss of force is $6P_L$.

For N tendons the total loss of force is $0.5N(N-1)P_L$, and the average loss per tendon is $0.5(N-1)P_L$, which is equal to half the loss in the first tendon due to the stressing of the remaining $(N-1)$ tendons.

Using Equation (7.9), the loss for N tendons, of total area A_p, is given by

$$\begin{aligned} P_{LN} &= \text{Total loss in } N \text{ tendons} \\ &= 0.5(E_s/E_{ci})[(N-1)p_c].[NA_1] \\ &= 0.5[(N-1)p_c].(E_s/E_{ci})A_p \qquad (7.10) \end{aligned}$$

What exactly is p_c? As stated above, it is the stress in the concrete at the level of the tendons. While the slab is supported on formwork, the stress induced by a tendon is a uniform compression on the concrete section and $p_c = P_1/A_c$. As more tendons are stressed, the slab lifts off the formwork and then p_c is the result of the axial force, the moment produced by the eccentricity of the tendons and the self-weight moment, given by

$$\begin{aligned} p_c &= P_1/A_c + P_1 e_p^2/I_e - M_o e_p/(NI_c) \\ &= P_1/A_c + [P_1 e_p - M_o/N]e_p/I_c \qquad (7.11) \end{aligned}$$

Equation (7.11) is not strictly correct, because of the change in the value of p_c as the member becomes self-supporting. Also, in continuous members the value of p_c is modified by the redundant forces. However, the loss of prestress due to elastic shortening is only one of several other losses, none of which can be calculated with exact accuracy; further inaccuracy arises because of the uncertainty in the value of E_{ci}. Therefore, an exact calculation of the elastic shortening loss is unnecessary, and Equations (7.10) and (7.11) are considered valid for use in the design of post-tensioned floors using bonded tendons.

Equations (7.10) and (7.11) can be simplified by assuming $N \approx N-1$, so that

$$p_{cn} = P_i/A_c + [P_i e_p - M_o]e_p/I_c \qquad (7.12)$$

$$P_{LN} = 0.5.p_{cn}.(E_s/E_{ci})A_p \qquad (7.13)$$

where p_{cn} = the stress in the concrete at tendon level due to N tendons and
P_i = the force in N tendons after all immediate losses, including that due to elastic shortening.

The losses in the tendon force are normally calculated at midspans and supports, where the stresses are high. This is the correct procedure for bonded tendons and Equations (7.12) and (7.13) can be used as they stand. For unbonded tendons, the tendon strain is the average over its length. Assuming a parabolic tendon profile and a parabolic shape of the self-weight moment diagram, the values of the last two terms in Equation (7.12) average out to 0.53. Using 0.5, for unbonded tendons

$$p_{cn} = P_i/A_c + 0.5[P_ie_p - M_o]e_p/I_c \tag{7.14}$$

$$P_{LN} = 0.5p_{cn}(E_s/E_i)A_p \tag{7.13}$$

The term $[P_ie_p - M_o]$ in Equations (7.12) and (7.14) represents the net moment at the section under consideration. For continuous members the terms P_ie_p and M_o are replaced with the moments obtained from the analysis for the tendon equivalent loads and the permanent loads respectively.

In most of the post-tensioned floors, the self-weight is balanced by the tendon equivalent load, so that the term $[P_ie_p - M_o]$ in Equations (7.12) and (7.14) becomes small. Neglecting this term, the two sets of equations are identical; the loss due to elastic shortening can be calculated for both bonded and unbonded tendons from the much simpler Equation (7.15).

$$\begin{aligned} P_L &= \text{the total loss of force for all tendons} \\ &= 0.5[P_i/A_c](E_s/E_{ci})A_p \end{aligned} \tag{7.15}$$

P_i is the total prestressing force in all tendons after all immediate losses, including that due to the elastic shortening. It is normally assumed as 90% of the jacking force P_j.

7.5 Shrinkage of concrete

Shrinkage of concrete has been discussed in Chapter 2 and the relationship between relative humidity, section depth and shrinkage shown in Figure 2.3. In the UK climate, for plain concrete a strain of 300×10^{-6} is taken for indoor exposure and 100×10^{-6} for outdoor exposure. More accurate values may be obtained from Figure 2.3 and modified for water content of the concrete mix, if known, as discussed in Chapter 2. In the UK, the ambient relative humidity is taken as 45% for indoor exposure and 85% for outdoor.

The bonded rod reinforcement normally provided in beams and slabs resists any change in concrete length. Shrinkage strain is, therefore, reduced by the bonded reinforcement. A slab normally has reinforcement on only one face and the quantity is smaller than that in a beam, which has a reinforcement cage with steel on both faces. Consequently, the shrinkage is smaller in a beam than in a slab. It would, however, be impractical to assume different shrinkage rates for a slab and its beams, and the normal practice is either to ignore the reinforcement content, or to assume a constant value for the whole floor. BS 8110 gives the following relationship between reinforcement content and shrinkage strain, assuming a symmetrical placement of bonded reinforcement.

$$\varepsilon_s = \varepsilon_p/(1 + pK_s) \tag{7.16}$$

where ε_p = the ultimate shrinkage strain in plain concrete
p = proportion of steel area to concrete
K_s = a constant, 25 for indoor exposure and 15 for outdoor exposure

Some shrinkage will already have taken place when the concrete is stressed and, therefore, only the residual shrinkage need be considered in calculation of losses. Post-tensioned slabs are normally stressed within a fortnight of casting, and sometimes within three days. It would be sufficiently accurate to assume that 20% of shrinkage will have taken place in two weeks and that within this period the rate of shrinkage is linear.

$$\text{Loss of prestressing force} = \varepsilon_s E_s A_p \tag{7.17}$$

For a residual shrinkage strain of 240×10^{-6} and a 0.2% rod reinforcement content, the loss in prestress in a floor in the UK indoor environment would be $0.045\ \text{kN/mm}^2$.

7.6 Creep of concrete

Creep of concrete has been discussed in Chapter 2 and the relationship between relative humidity, section depth and creep coefficient is shown in Figure 2.4 for loading at different ages. The appropriate value of the creep coefficient, C_c, can be read from these curves and then

$$\begin{aligned}\text{Loss of prestressing force} &= (\sigma\varepsilon_c)E_s A_p \\ &= (\sigma C_c/E_c)E_s A_p \\ &= \sigma A_p C_c E_s/E_c \end{aligned} \tag{7.18}$$

where σ = local stress at tendon level
C_c = creep coefficient
ε_c = creep after stressing

On stressing, the member becomes self-supporting, and creep is a long-term phenomenon. Therefore, the local stress σ in Equation (7.18) should include for prestress and the moments due to self-weight of the member and other permanent dead loads; live load is generally ignored. Therefore, for bonded tendons,

$$\sigma = (P_i/A_c) + (P_i e_p - M_o - M_d).e/I_c \tag{7.19}$$

The change in the strain at any point along a bonded tendon is the same as that in the concrete at the tendon level. Therefore, the creep loss varies along the span in accordance with the variation of concrete stress.

In an unbonded member, tendons can move relative to the adjacent concrete. Therefore, the loss of prestress is due to the average value of σ for the tendon length and not the local value. Assuming parabolic moments and tendon profile,

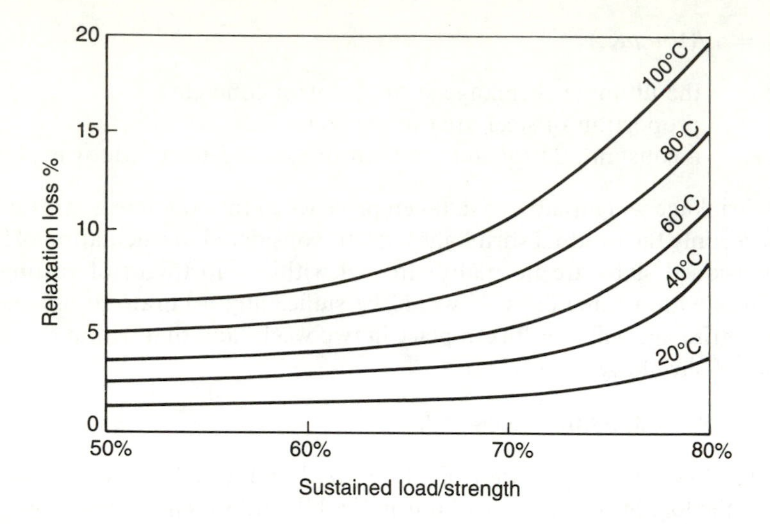

Figure 7.4 *Long-term relaxation of low relaxation strand*

the value of the second term in Equation (7.19) averages out to 0.53. Using 0.5, the average value of σ to be used for unbonded tendons is

$$\sigma = (P_t/A_c) + 0.5(P_i e_p - M_o - M_d)e_p/I_c \tag{7.20}$$

In most post-tensioned floors the permanent loads are approximately balanced by the equivalent tendon load, so that the stress on the concrete section is nearly uniform; in such a case, the second term in Equations (7.19) and (7.20) can be neglected. The expression for σ then reduces to P_i/A_c and the loss of prestressing force can be calculated from Equation (7.21), which is applicable to both bonded and unbonded tendons.

$$\text{Loss of prestressing force} = C_c(P_i/A_c)A_p(E_s/E_c) \tag{7.21}$$

where E_c is the modulus of elasticity of concrete at the time of loading.

7.7 Relaxation of tendons

Relaxation of strand has been discussed in Chapter 2 in some detail. Use of low relaxation strand is now common in post-tensioned floors; however, local circumstances may require normal strand to be used.

Figure 7.4 shows the long-term relaxation for low relaxation strands at different ambient temperatures and various initial stress levels. In the absence of definite data from the manufacturer of strand, the relaxation percentage can be read from these curves.

$$\text{Loss of prestressing force} = p\%\ P_i/100 \tag{7.22}$$

where p% = percentage loss

The tendon force to be taken in the calculation of relaxation loss is the force after the short-term losses have occurred, i.e. the jacking force less the losses due to elastic shortening, friction and draw-in. This force varies along the tendon length, and an average value may be taken, unless the desired degree of accuracy requires the actual forces to be taken at each critical point.

7.8 Tendon elongation

During jacking the tendon force is measured with a hydraulic pressure gauge. A pressure cell may also be used as a confirmatory device. In addition to these direct measuring devices, the normal practice is to use the elastic elongation of the tendon to monitor the force. Elongation is measured at full jacking load before locking the anchorage; the amount of draw-in is also recorded.

The expected elongation, and the margin of acceptable variation, is specified by the designer for each tendon. If the measured elongation falls outside the specified range then immediate steps can be taken to remedy the situation if it is considered necessary. The elongation of a tendon stressed by a force P is:

$$\delta = PL_t/(A_pE_s) \tag{7.23}$$

where P = average force in a tendon

P should be taken as the average value of the prestressing force, i.e., either the force at the midpoint of the tendon length, or the average of the jacking force (before locking) and the force at the dead end.

A tendon, particularly a bonded one, nearly always has a slack in it before it is stressed. During the stressing of bonded tendons, initially the jacking effort goes into taking the slack up, and then it stresses the tendon. Unbonded tendons have virtually no slack. The normal procedure is to partially stress the tendon, to about 10% of the specified jacking force, and take this as the reference zero point for measuring elongation. This procedure automatically allows for any tendon slack and any slip at the dead end. The elongation, measured from the reference point, must be multiplied by a correction factor so that it can be compared with the calculated δ.

$$\begin{aligned}\delta &= \text{corrected value of actual elongation}\\ &= \delta_m P/(P - P_r)\end{aligned} \tag{7.24}$$

where δ_m = measured elongation
P_r = Tendon force at the reference point

7.9 Tendon force from elongation

Sometimes, a tendon may not achieve the elongation specified by the designer, or a tendon may be lost during stressing through failure at the anchorage or the breaking of the tendon. In such circumstances, if the adjacent tendons are perceived to possibly have sufficient reserve to make up for the loss of force, then the forces in the stressed tendons are calculated from the data recorded during stressing. This is also a good discipline for short tendons where the loss due to the tendon draw-in may be significant. The tendon force is given by:

$$P_i \approx P_j - (\delta_d/\delta_m).(P_j - P_r) \qquad (7.25)$$

where P_i = force in tendon, allowing for immediate losses
δ_d = measured draw-in
δ_m = measured elongation from P_r to P_j
P_j = recorded jacking force
P_r = recorded tendon force at the reference point.

Example 7.1

Calculate the losses in span 1-2 of Example 6.3.
$A_c = 179.6 \times 10^3$ mm^2 $\quad$ $I_c = 1011.5 \times 10^6$ mm^4
$y_b = 163.5$ mm

$f_{cu} = 40$ N/mm^2 $\quad$ $E_c = 28.0$ kN/mm^2
$f_{ci} = 25$ N/mm^2 $\quad$ $E_{ci} = 21.7$ kN/mm^2

Solution

Tendons:

$P_f = 650$ kN, assume $P_j = 830$ kN and $P_i = 740$ kN

Try 4 superstrands 15.7 mm dia,
$A_p = 4 \times 150 = 600$ mm^2

The prestressing force is assumed constant for all spans; half of the strands are stressed from each end.

1. Friction loss

Tendon slope
at a = 2 × 0.084/4.66 = 0.0361

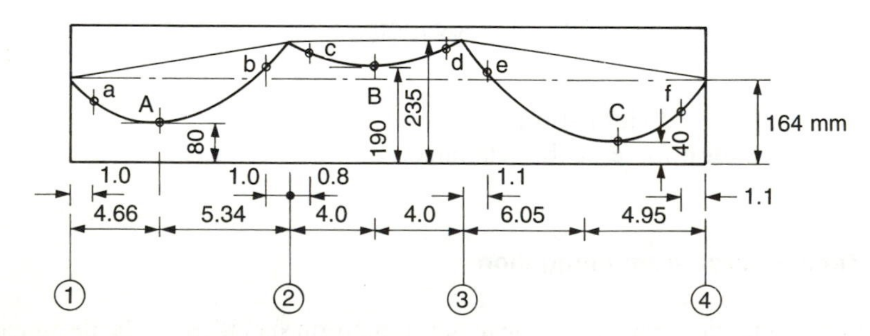

Figure 7.5 *Example 7.1: Tendon profile*

at b = 2 × 0.155/5.34 = 0.0581
at c = 2 × 0.045/4.00 = 0.0225
at d = 2 × 0.045/4.00 = 0.0225
at e = 2 × 0.195/6.05 = 0.0645

Take $\mu = 0.1$ and $K = 0.001$/m for unbonded tendons.

Point	Cumulative θ	$\times (\mu\theta + Kx)$	
A	$2 \times 0.0361 = 0.0722$	× 4.66	= 0.0119
2	$\theta_A + 2 \times 0.0581 = 0.1884$	× 10.00	= 0.0288
B	$\theta_2 + 2 \times 0.0225 = 0.2334$	× 14.00	= 0.0373
3	$\theta_B + 2 \times 0.0225 = 0.2784$	× 18.00	= 0.0458
C	$\theta_3 + 2 \times 0.0645 = 0.4074$	× 24.05	= 0.0648

Assume that for tendons stressed from end 4, $(\mu\theta + Kx)$ at A is 0.0648, the same as at C for tendons stressed from end 1.

Then, loss in force at A = 740 × (0.0119 + 0.0648)/2 = 28 kN

2. Anchorage draw-in

Allow 8 mm draw-in.
Tendon length = 29 m
Loss of force = (8/29000) × 600 × 195 = 32 kN

3. Elastic shortening

Use Equations (7.14) and (7.13)

See Example 6.3, the equivalent load of 650 kN is taken in the tables for moments and shears. In the present example P_i is assumed 740 kN.

$$w_e = -6.21 \times (740/650) = -7.07 \text{ kN/m}$$
$$M_2 = +42.05 \times (740/650) - 35.21 = +12.66 \text{ kNm}$$
$$M_s = 12.66/2 + (4.3 - 7.07) \times 10^2/8 = -21.97 \text{ kNm}$$

$$p_{cn} = P_i/A_c + 0.5[P_i e - M_o]e/I_c$$
$$= 740/179.6 + 0.5[-21.97] \times (164 - 80)/1011.450 = 3.21 \text{ N/mm}^2$$
$$P_{LN} = 0.5 p_{cn}(E_s/E_{ci})A_t$$
$$= 0.5 \times 3.21 \times (195/21.7) \times 600/1000 = 9 \text{ kN}$$

Short-term losses:

Friction	28 kN	Total 69 kN
Draw-in	32	
Elastic	9	

For $P_j = 830$ kN, $P_i = 830 - 69 = 761$ kN
The assumed value of 740 kN is close enough not to need repetition.

4. Shrinkage of concrete:

From Figure 2.03, for a 180 mm deep section and at 45% relative humidity, the 30-year shrinkage is 400×10^{-6}. Assume 80% residual shrinkage, and a bonded reinforcement content of 0.13%. Then, from Equations (7.16) and (7.17),

$$\varepsilon_p = 0.8 \times 400 \times 10^{-6}/(1 + 25 \times 0.0013) = 310 \times 10^{-6}$$

Loss of prestress = $310 \times 10^{-6} \times 195 \times 600 = 36$ kN

5. Creep of concrete:

From Figure 2.04, for a 180 mm deep section, at 45% relative humidity, and final stressing at 14 days age, the 30-year creep coefficient is 3.3. The average stress in concrete may be taken as

$P_i/A_c = 761/179.6 = 4.2\ N/mm^2$

Assuming that 20% creep has taken place before the concrete is loaded, from Equation (7.21),

Loss of force $= 0.8 \times 3.3 \times 4.2 \times 600 \times 195/21.7/1000 = 60$ kN

6. Relaxation of strand

$P_i/P_u = 761/(4 \times 265) = 0.72$
From Figure 7.4, for an ambient temperature of 20°C, relaxation loss is 2%.
Loss of force = 2% of 761 kN = 15 kN

Final prestressing force = 761 − 36 − 60 − 15 = 650kN OK
No adjustment is required in the assumed jacking force of 830 kN.

8 ULTIMATE FLEXURAL STRENGTH

This chapter deals with the ultimate flexural strength of post-tensioned members of a floor and the strength requirements in the anchorage zone; shear strength is discussed in Chapter 10.

The purpose of checking the ultimate strength of a member is to ensure that, if loaded beyond the nominal design value, it will not fail before the load exceeds the design load multiplied by a predetermined factor. Since a number of combinations of different loads (self-weight, other dead loads, live load and perhaps lateral loads) are considered, the obvious basic requirement for flexural strength is that, at any section, the design ultimate moment of resistance M_u be not less than the maximum required moment of resistance M_r for the combinations.

8.1 Failure mechanisms

As the load on a member is increased, the corresponding moments and stresses increase in proportion to the load, until the tensile stress in the steel reaches its yield point, or the compressive stress in the concrete reaches the limit of proportionality at a section. After this, the rotation in the short yielded length is disproportionately large for any further increase in the load. It can be assumed that the section has reached its peak moment of resistance and any attempt to increase the load will only increase the rotation; the section, in fact, behaves as a hinge. Hinges, obviously, form at the peak points of the bending moment diagram.

For a suspended member to collapse, a number of plastic hinges must form at the critical sections, transforming the whole or part of the floor into a mechanism. An understanding of the failure mechanisms is useful for the calculation of the ultimate strength of post-tensioned floors, particularly those with unbonded tendons.

8.1.1 Modes of failure

Failure at a hinge may take place in one of three possible modes, depending on the level of prestress.

1. For lightly loaded members, the magnitude of the required prestressing force may be very low, so that in the serviceability state the tensile strength of the concrete may provide most of the required flexural strength. If overloaded, the concrete will crack, and its contribution to the strength of the member will be lost. In such a case, it is quite possible for the ultimate strength of the section based on the steel content to be lower than its cracking strength. Failure then occurs as soon as the section cracks, with no warning.

   This state is to be avoided. ACI 318 recommends a minimum average prestress of 1.0 N/mm^2 on the gross section after all losses, and requires sufficient tendon area to be provided to develop a factored load at least 1.2 times the cracking load.
2. At a higher level of prestress than that of Case 1, with increasing load, the tendon reaches its limit of proportionate strain. Beyond this point, the stress-strain relationship for the strand becomes increasingly non-linear and plastic hinges form at the critical sections. Noticeable rotation takes place at each hinge, the neutral axis rises, and the depth of the compression block reduces. The section eventually fails either when the tendon breaks or when the concrete crushes.

   This mode of failure is always accompanied by a large deflection and ample warning. Most post-tensioned floors would fail in this mode if loaded to failure.
3. At a very high level of prestress, the concrete crushes before the tendon reaches its limit of proportionality. Failure in this case is without warning, and explosive.

   Such a high level of stress would exist if the concrete section were reduced to the minimum. High levels of prestress are best avoided. However, if unavoidable, then adequate binding reinforcement should be provided around the concrete in the compression block.

8.1.2 One-way spans

Formation of the first hinge in a one-way spanning continuous member does not lead to its collapse. Figure 8.1 shows the hinges required in one-way spanning members to transform them into mechanisms. A cantilever needs one hinge at its root, as does a simply supported span; the end span of a continuous string needs two hinges if it is free to rotate at the outer support; and the inner span needs three hinges, as does the outer span if it is not free to rotate at the end support.

The ultimate load carrying capacity of a member can be calculated if its moments of resistance are known at the critical points for the loading pattern. Consider the inner span in the ultimate state, carrying a uniformly distributed load. For static equilibrium,

$$w_u L^2/8 = M_{ub} - (M_{ua} + M_{uc})/2 \qquad (8.1)$$

where w_u = ultimate load per unit length

M_{ua}, M_{ub} and M_{uc} are the values of the moments of resistance at a, b and c; normally, the support moments M_{ua} and M_{uc} are expected negative.

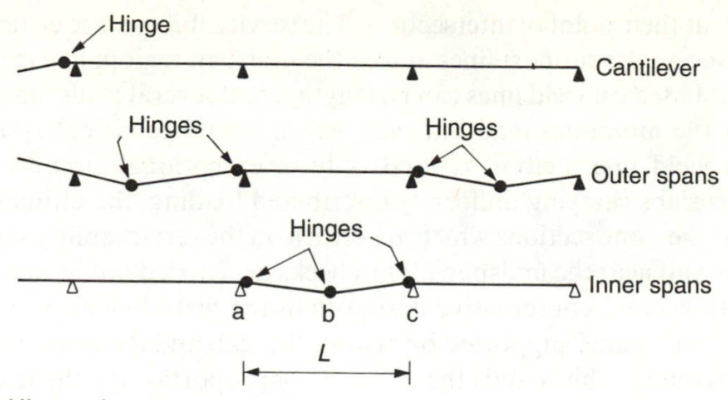

Figure 8.1 *Hinges in one-way spans*

In the case of an external span, M_{ra} is zero, and for a simply supported span both M_{ra} and M_{rc} are zero.

8.1.3 Two-way spans

In a floor containing a number of two-way spanning panels in each direction, it is extremely unlikely that all panels in a line will get overloaded at the same time, and the possibility of collapse occurring in the manner shown in Figure 8.1 is virtually nil, especially if only vertical loads are imposed on the floor. More probable is the failure of one panel initially, which may lead to the collapse of adjacent panels. A single panel out of a two-way continuous floor can fail only when yield lines have developed so that the panel is divided into several zones, see Figure 8.2.

The yield line mechanism is a failure pattern and it does not reflect the serviceability stresses. Serviceability checks are based on the proportioning of loads in two orthogonal directions, so as to equate the deflection of two strips, at

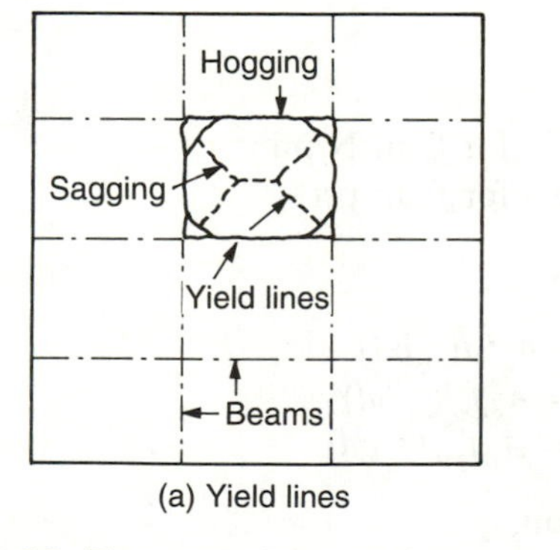

(a) Yield lines

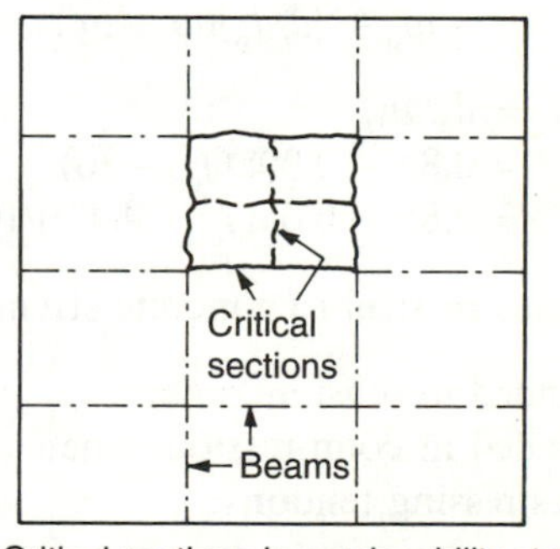

(b) Critical sections in serviceability state

Figure 8.2 *Two-way slab panel*

right angles, at their point of intersection. The serviceability state critical sections are located along the support lines and in the midspan region, in each direction. Calculations based on yield lines can be lengthy, and several trials may be needed to arrive at the minimum load intensity which would cause collapse. For this reason, the yield line method is used only in exceptional circumstances; for normal floor slabs carrying uniformly distributed loading, the ultimate strength is checked at the same sections which are critical in the serviceability state—along the supports and near the midspans. The checks are carried out in each direction. This leads to a more conservative design than the yield line approach.

In the case of a panel supported by beams, the slab and the beam elements are checked separately. This avoids the difficulty of proportioning the load between the column and middle strips if the whole panel width were to be checked as one unit. The sum of the loads carried by the slab and the beams in each direction must equal the total load.

8.2 Level of prestress

Using the tensile strength of concrete, it is theoretically possible to satisfy the serviceability requirements of a floor spanning a short distance with zero prestress—the concrete tensile strength being sufficient to take the flexural tension. At the other extreme, a very thin section can be prestressed to such an extent that, when overloaded, failure occurs by crushing of the concrete before the tendons fail.

Neither of the two is desirable in a practical design. The first case does not meet the ultimate strength requirement, and for both cases failure will be sudden and without warning.

ACI 318 limits the maximum area of steel—tendons plus rod reinforcement—so that

$$\begin{aligned} 0.36\beta_1 &> \omega_p \\ &> \omega_p + (d/d_p)(\omega - \omega') \end{aligned} \tag{8.2}$$

where $\beta_1 = d_c/dn$
$= 0.85 - 0.008(f_c' - 30)$ for f_c' in N/mm^2
$= 0.85 - 0.05(f_c' - 4000)/1000$ for f_c' in psi

Ratios of steel to concrete strengths:

Bonded in tension zone: $\omega = A_s f_y/(f_c' bd)$
Bonded in compression zone: $\omega' = A_s' f_y/(f_c' bd)$
Prestressing tendons: $\omega_p = A_p f_{ps}/(f_c' bd)$

b = width of concrete in compression,
= web width in the case of a T-beam
f_{ps} = tendon stress at design load

8.2.1 Cracking load

In the absence of self-weight and applied loads, a post-tensioned member carries only the prestressing stresses and the whole of the section is in compression. When a small load is applied, the compression reduces on one face. Under an increasing load, the pre-compression reduces to zero on the tension face, and then the stress becomes tensile. As long as the tensile stress is less than the modulus of rupture of the concrete, the section remains uncracked and the deflection corresponds to the moment of inertia of the gross section.

When the tensile stress reaches the modulus of rupture, the section cracks, and the tensile force, which was up to now carried by the uncracked concrete, is suddenly transferred to the tendon. The action is accompanied by an immediate increase of deflection. The rate of increase of deflection is now higher because of the smaller moment of inertia of the cracked section.

The moment at which the flexural tensile stress on the section equals the modulus of rupture is called the *cracking moment*. In calculating the cracking moment, the final prestressing force is used, after allowing for all losses.

$$M_{cr} = f_t Z + P/A_c(Z + e_p A_c) \tag{8.3}$$

where M_{cr} = cracking moment
f_t = modulus of rupture
Z = section modulus of the extreme tension fibre

In BS 8110, the cracking moment and the principal tensile stress are used to determine if the cracked or the uncracked section should be used in the calculation of shear strength. The limiting value of the principal tensile stress recommended for use in this case is $0.24(f_{cu})^{0.5}$ in N-mm units. This topic is dealt with in detail in Chapter 10.

8.3 Applied loads

The minimum value of the required moment of resistance M_r at a section is calculated by imposing serviceability loads multiplied by the load factors. Where a member is required to resist a lateral force, the lateral force multiplied by the appropriate load factor is included in the calculation of the required moment of resistance.

A number of load combinations are normally considered in order to arrive at the most adverse condition at each critical position. The value of a load factor depends on the importance of the load combination being considered, and, for a particular combination, the load factor may be different for each load type. The dead load and the secondary moments are, of course present in all cases; however, the corresponding load factor varies between 1.0 and 1.4 for dead loads depending on the load combination and remains constant at 1.0 for the secondary moments.

The loading patterns for strength calculation are the same as for the serviceability state, viz:

Dead load on all spans with live load on two adjacent spans, and
Dead load on all spans with live load on alternate spans.

BS 8110 requires the following load combinations to be considered; the numerical values of the load factors are given in the expressions.

$$
\begin{aligned}
&\text{(a)}\quad 1.4G + 1.6Q \\
&\text{(b)}\quad 1.4G + 1.4W \\
&\text{(c)}\quad 1.2G + 1.2Q + 1.2W \\
&\text{(d)}\quad 1.05G + 0.35Q + 0.35W \\
&\text{(e)}\quad 1.05G + 1.05Q + 0.35W
\end{aligned}
\tag{8.4}
$$

where G = dead load
Q = nominal live load, including dynamic load
W = wind load

Cases (d) and (e) above represent abnormal loading cases and are used if it is necessary to consider the probable effects of excessive loads caused by misuse or accident, or when considering the continued stability of a structure after it has sustained localized damage. Only those loads likely to be acting simultaneously, or those likely to occur before measures are taken to repair or offset the effect of the damage need be included. For buildings used predominantly for storage or industrial purposes or where the imposed loads are permanent, the whole of the imposed load should be taken and expression (e) applied; for other cases expression (d) should be used.

The load combinations required by ACI 318 and the corresponding values for the load factors are:

$$
\begin{aligned}
&\text{(a)}\quad 1.4G + 1.7Q \\
&\text{(b)}\quad 0.75(1.4G + 1.7Q + 1.7W) \\
&\text{(c)}\quad 0.9G + 1.3W \\
&\text{(d)}\quad 0.75(1.4G + 1.7Q + 1.87E) \\
&\text{(e)}\quad 0.9G + 1.43E \\
&\text{(f)}\quad 0.75(1.4G + 1.7Q + 1.4T) \\
&\text{(g)}\quad 1.4G + 1.4T
\end{aligned}
\tag{8.5}
$$

where E = load effects of earthquake
T = effect of temperature, settlement, shrinkage or creep

Expressions (d) and (e) of the ACI 318 requirements are applicable only if the effect of earthquake is to be considered, and (f) and (g) are to be used where effects of differential settlement, creep, shrinkage, or temperature change are significant in design.

8.4 Procedure for calculating strength

The procedure for calculating the flexural strength of a post-tensioned member is similar to that for a reinforced concrete member, the only difference being the calculation of stress in the tendons at ultimate load. The basic assumption in both cases is that plane sections remain plane.

The applied loads are increased by specified *load factors* and the stresses in the materials are reduced by applying *partial safety factors* to their strengths. The load factors take account of the possible inaccuracies in the assessments of loads, and the tolerances. The partial safety factors for materials allow for the differences between laboratory and actual strengths, local weaknesses and inaccuracies of the properties of materials. The load factors and the partial safety factors are specified by the various national standards; BS 8110 and ACI 318 requirements are given in this chapter.

The required ultimate moment is calculated, at each of the critical sections, by combining the moments for the factored loads in the most adverse manner. Although the tendon force is higher at the ultimate state than at serviceability state, the effect of this increase on the secondary moments is ignored and the secondary moments are added unfactored to each of the load combinations.

The level of prestress in a post-tensioned floor is usually too low for an explosive failure; failure is likely to be initiated by the stress in the steel—tendons and bonded reinforcement—reaching the limit of linearity of stress and strain. This, of course, does not apply to highly stressed members, such as transfer beams.

The design ultimate moment capacity of a section is calculated by considering static equilibrium, with the tendon in the non-linear state. The total tensile force in all steels, tendons and rod reinforcement, must equal the total compressive force in the concrete. The fact that some of the force in the tendons is applied as prestress is irrelevant.

Concrete is assumed to have reached its failure strain at the extreme fibre. Tests indicate that in the ultimate state the shape of the compression block is nearly parabolic for post-tensioned concrete, as for reinforced concrete. The compressive force and its centroid can be obtained from the stress-strain curve for the concrete being used, and the partial safety factors applied. The procedure, however, is complex, and it does not necessarily lead to a more accurate assessment of the ultimate strength capacity of a member, because of the possible variations in the material properties. The shape of the compression stress block is, therefore, normally simplified to a rectangle. In order for the rectangular block to be equivalent to that obtained from tests, or the stress-strain diagram, it is necessary to apply certain factors to the intensity of the stress and to the depth of the compression block. The factors defining the equivalent rectangular compression block differ in BS 8110 and ACI 318.

The tendon stress in the ultimate state is usually between the yield point and the breaking stress, and can be determined by strain compatibility at the section being considered. Strain compatibility calculations are, however, rather complicated, particularly for unbonded tendons, and the various national standards give simple rules for determining the stress in bonded and unbonded tendons. BS 8110 and ACI 318 recommendations are given in this chapter.

Figure 8.3 shows a typical stress diagram for a post-tensioned member.

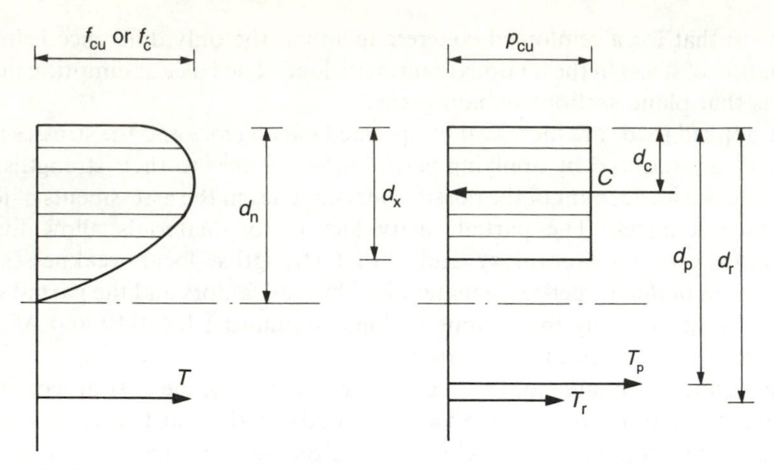

Figure 8.3 *Ultimate state stress diagrams*

$$C = \text{Compression} = p_{cu}bd_x \tag{8.6a}$$

$$T = \text{Tension} = f_{pb}A_p + f_{sb}A_s = T_p + T_r \tag{8.6b}$$

where p_{cu} = compressive stress in concrete at failure, as given in Section 8.5.1
$T_p = f_{pb}A_p$ = tension in prestressing tendons
$T_r = f_{sb}A_s$ = tension in bonded rod reinforcement
f_{sb} = stress in bonded reinforcement

Taking moments of the tensile forces about the centre of compression C gives the ultimate capacity of the section M_u, which, of course, should not be less than the required ultimate moment of resistance M_r.

$$M_u = f_{pb}A_p(d_p - d_c) + f_{sb}A_s(d_r - d_c) \tag{8.6c}$$

where d_c = the depth from compression face to the centre of the compressive force.

The above equations assume a rectangular section; the equations are valid for a flanged section provided that the neutral axis remains within the depth of the flange. This is normally the case in post-tensioned concrete. A neutral axis located below the flange may be indicative of a potentially brittle failure and should be avoided by increasing the depth of the section or of the flange.

If the tensile force in the tendons alone is not sufficient to provide the required strength, then rod reinforcement may be added; the second term in Equations (8.6) represents rod reinforcement. In this case the calculation process becomes iterative, because the depth of the compression block, and hence the centroid of the compressive force, changes with the amount of bonded rod reinforcement; however, the calculation is not very sensitive. The procedure for the ultimate state check generally consists of the following steps.

1. Calculate the required moment of resistance M_r
2. Calculate T_r and T_p, initially T_r may be assumed zero
3. Calculate d_x and d_c
4. Calculate M_u
5. If $M_u < M_r$ then add bonded reinforcement and repeat from 2.

8.5 Ultimate stresses

The stresses in the ultimate state are assumed to equal the strength of the material modified by the partial safety factor, as discussed above. The factors differ for concrete, bonded rod reinforcement and prestressing tendons.

8.5.1 Concrete stresses

The compressive stress block in a flexure member approaching failure is parabolic, but a rectangular compression block is normally assumed, as discussed above.

If any compression reinforcement is provided, then its stress can be calculated from the strain diagram at the section being considered. The compression reinforcement would require links to prevent buckling. Use of compression reinforcement, however, is very rare in post-tensioned floors, and it is not discussed in any detail in this book.

BS 8110 requires:

- the 28-day concrete strength to be divided by a partial safety factor of 1.5,
- the concrete strain at the compression face to be assumed as 0.0035,
- the rectangular stress block to have an average stress p_{cu} of $0.45f_{cu}$,
- the depth of the rectangular compression block (d_x) to be 0.9 times the depth of the neutral axis (d_n),
- tensile strength of concrete to be ignored.

The above conditions give the total compressive force as $0.405f_{cu}bd_n$, acting $0.45d_n$ from the compression face of the section.

ACI 318 takes a different approach. Rather than specifying partial factors for the individual materials, it requires that the design strength provided by a member be taken as the nominal calculated strength multiplied by a strength reduction factor of 0.9.

For the rectangular compression block, ACI 318 specifies that:

- the concrete strain at the compression face be assumed as 0.003.
- the rectangular stress block be assumed to have an average stress p_{cu} of $0.85f_c'$.
- the depth of the rectangular compression block be taken as β_1 times the depth of the neutral axis. Table 8.1 gives the values of β_1 for different cylinder strengths.
- tensile strength of concrete is ignored.

Table 8.1a *Compression block depth ratio, ACI 318 (Metric units)*

f'_c	30	35	40	45	50	55 N/mm^2
β_1	0.85	0.81	0.77	0.73	0.69	0.65

Table 8.1b *Compression block depth ratio, ACI 318 (Imperial units)*

f'_c	3000	4000	5000	6000	7000	8000 psi
β_1	0.85	0.85	0.80	0.75	0.70	0.65

8.5.2 Stress in rod reinforcement f_{sb}

The stress in the rod reinforcement can be calculated from the stress-strain curve for the steel and the strain diagram at the section being considered. However, in most cases it reaches its yield stress.

BS 8110 specifies an upper limit for the stress in the rod reinforcement equal to its yield stress divided by a partial safety factor of 1.15. The upper limit specified by the ACI 318 is 550 N/mm^2. The strength reduction factor of 0.9 implies a limit of 495 N/mm^2.

8.5.3 Stress in bonded and unbonded tendons

A large rotation takes place over the short length of each plastic hinge in the mechanism, so that the concrete strains at the hinges are much larger than elsewhere, and the strains in the lengths between the hinges can be ignored.

In a bonded tendon, the increase in strain is the same as the local strain in concrete at the tendon level. The tendon stress, therefore, increases at a higher rate locally at a hinge than elsewhere in its length. The behaviour of an unbonded tendon is quite different. It slips relative to the concrete at the hinge, and the local elongation is distributed evenly over the whole length of the tendon; consequently the strain is virtually constant over the tendon length and is much smaller at the hinge than that for a bonded tendon. The increase in stress is also correspondingly smaller. Therefore, a bonded tendon develops a higher flexural strength at the hinge, but an unbonded tendon allows a much larger hinge rotation to take place before it reaches its breaking load.

The average increase in strain, and stress, in an unbonded tendon is the sum total of elongations at all the hinges divided by the tendon length. For convenience, the effective length of an unbonded tendon is taken as the actual tendon length divided by the number of hinges necessary to change the span under consideration into a mechanism. Therefore,

$$L_{te} = L_t/N \quad (8.7)$$

where L_{te} = effective length of an unbonded tendon
N = number of hinges necessary to change the span into a mechanism

In a string of continuous spans, it is theoretically possible for all spans to be overloaded simultaneously, but only one span is considered to approach the ultimate state at a time. Considering the outer span, or the penultimate support of a string of spans with pinned end supports, the increase in tendon strain is due to the rotation at two hinges divided by the tendon length of all spans through which it passes.

8.5.4 Stress in bonded tendons f_{pb}

For bonded tendons, BS 8110 gives the values in Table 8.2 for the tendon stress and the depth of the neutral axis in the ultimate state.

For a given section, values of R_p and (f_{pe}/f_{pu}) are calculated and the corresponding value of $(f_{pb}/0.87f_{pu})$ is obtained from Table 8.2, from which f_{pb} can be deduced.

The ACI 318 requirements for stress in bonded tendons are:

$$f_{pb} = (1 - K_p\gamma_p/\beta_1)f_{pu} \qquad (8.8)$$
$$K_p = (r_p f_{pu}/f_c') + d/d_p)(\omega - \omega')$$

where γ_p = factor for the type of tendon
= 0.55 for f_{py}/f_{pu} not less than 0.80
= 0.40 for f_{py}/f_{pu} not less than 0.85
= 0.28 for f_{py}/f_{pu} not less than 0.90

Table 8.2 *BS 8110 Stress in bonded tendons and depth of neutral axis.*

R_p	$f_{pb}/(0.87f_{pu})$ for f_{pe}/f_{pu} =			d_n/d for f_{pe}/f_{pu} =		
	0.60	*0.50*	*0.40*	*0.60*	*0.50*	*0.40*
0.05	1.00	1.00	1.00	0.11	0.11	0.11
0.10	1.00	1.00	1.00	0.22	0.22	0.22
0.15	0.99	0.97	0.95	0.32	0.32	0.31
0.20	0.92	0.90	0.88	0.40	0.39	0.38
0.25	0.88	0.86	0.84	0.48	0.47	0.46
0.30	0.85	0.83	0.80	0.55	0.54	0.52
0.35	0.83	0.80	0.76	0.63	0.60	0.58
0.40	0.81	0.77	0.72	0.70	0.67	0.62
0.45	0.79	0.74	0.68	0.77	0.72	0.66
0.50	0.77	0.71	0.64	0.83	0.77	0.69

$R_p = A_p f_{pu}/(f_{cu}bd)$
d = depth to steel centroid from compression face
0.87 is the inverse of partial factor 1.15

$\beta_1 = d_c/d_n$ ⎤ as defined
$\omega = A_s f_y/(f_c' bd)$ ⎦ in Section 8.2
$\omega' = A_s' f_y/(f_c' bd)$
$r_p = A_p/bd_p$ = ratio of tendon area

If any compression reinforcement is taken into account then the term K_p shall be taken as not less than 0.17, and d' (distance from extreme compression fibre to centroid of compression reinforcement) shall be no greater than 0.15 d_p.

In the absence of any bonded rod reinforcement, the expression (8.8) reduces to:

$$f_{pb} = f_{pu} - (\gamma_p/\beta_1)[A_p f_{pu}/bd_p f_c']f_{pu} \quad (8.9)$$

Values of f_{pb}/f_{pu} for a range of $(A_p f_{pu}/bd_p f_c')$, calculated from Equation (8.9), are given in Table 8.3.

8.5.5 Stress in unbonded tendons

For unbonded tendons, BS 8110 gives:

$$f_{pb} = f_{pe} + (7000d/L_{te})(1 - 1.7R_p)$$
$$= f_{pe} + \delta_p \quad (8.10a)$$

$$d_n = 2.47dR_p(f_{pb}/f_{pu}) \quad (8.10b)$$

where $R_p = A_p f_{pu}/(f_{eu} bd)$
L_{te} = effective tendon length, as defined in 8.5.3.

The value of f_{pb} should not be taken as greater than $0.7f_{pu}$.

Equation (8.10a) has been derived by taking the length of the zone of inelasticity within the concrete as $10d_n$. The second term, corresponding to δ_p, represents the increase in the tendon stress. Table 8.4 gives the increase of tendon stress in N/mm^2, based on Equation (8.10a), for a range of R_p and L_{te}/d values.

Equation (8.10b), defining the depth of the neutral axis, can be written in a simpler form as:

$$d_u/d = 2.47R_p(f_{pb}/f_{pu}) \quad (8.11)$$

ACI 318 requirement for span-depth ratios of 35 or less is:

$$f_{pb} = f_{pe} + 70 + f_c' bd_p/(100A_p) \quad \text{N.mm}^2 \text{ units}$$
$$= f_{pe} + 10\,000 + f_c' bd_p/(100A_p) \quad \text{psi units} \quad (8.12a)$$

The value of f_{pb} must not exceed f_{py}, nor f_{pe} + 400 in N/mm^2 units (f_{pe} + 60 000 in psi units).

For span-depth ratios higher than 35, ACI 318 specifies

$$f_{pb} = f_{pe} + 70 + f_c' bd_p/(300A_p) \quad \text{N.mm}^2 \text{ units}$$
$$= f_{pe} + 10\,000 + f_c' bd_p/(300A_p) \quad \text{psi units} \quad (8.12b)$$

The value of f_{pb} must not exceed f_{py}, nor f_{pe} + 200 in N/mm^2 units (f_{pe} + 30 000 in psi units).

Table 8.3 *Stress ratio f_{pb}/f_{pu} for bonded tendons, ACI 318*

	$\frac{A_p f_{pu}}{bd_p f'_c}$	f'_c N/mm² (psi)				
γ_p		30 (4350)	40 (5800)	50 (7250)	60 (8700)	70 (10 150)
0.55	0.05	0.968	0.964	0.960	0.955	0.948
	0.10	0.935	0.929	0.920	0.910	0.896
	0.15	0.903	0.893	0.880	0.865	0.844
	0.20	0.871	0.857	0.841	0.820	0.792
	0.25	0.838	0.821	0.801	0.775	0.741
	0.30	0.806	0.786	0.761	0.730	0.689
	0.35	0.774	0.750	0.721	0.684	0.637
	0.40	0.741	0.714	0.681	0.639	0.585
	0.45	0.709	0.679	0.641	0.594	0.533
	0.50	0.676	0.643	0.601	0.549	0.481
0.40	0.05	0.976	0.974	0.971	0.967	0.962
	0.10	0.953	0.948	0.942	0.934	0.925
	0.15	0.929	0.922	0.913	0.902	0.887
	0.20	0.905	0.896	0.884	0.869	0.849
	0.25	0.882	0.870	0.855	0.836	0.811
	0.30	0.859	0.844	0.826	0.803	0.774
	0.35	0.835	0.818	0.797	0.770	0.736
	0.40	0.812	0.792	0.768	0.738	0.698
	0.45	0.788	0.766	0.739	0.705	0.660
	0.50	0.765	0.740	0.710	0.672	0.623
0.28	0.05	0.984	0.982	0.980	0.977	0.974
	0.10	0.967	0.964	0.959	0.954	0.947
	0.15	0.951	0.945	0.939	0.931	0.921
	0.20	0.934	0.927	0.919	0.908	0.894
	0.25	0.918	0.909	0.899	0.885	0.868
	0.30	0.901	0.891	0.878	0.862	0.842
	0.35	0.885	0.873	0.858	0.839	0.815
	0.40	0.868	0.855	0.838	0.816	0.789
	0.45	0.852	0.836	0.817	0.793	0.762
	0.50	0.835	0.818	0.797	0.770	0.736

8.6 Strain compatibility

Figure 8.4 shows the strain history at a section as the load is increased from nil (Line 1) to cracking load (Line 2), and finally to its ultimate capacity (Line 3). The initial strain in the tendon corresponds to the prestress alone and equals

$$\varepsilon_1 = P/(A_p E_s) \tag{8.13a}$$

When the member is loaded, up to the point where the section remains uncracked, the strain line on the diagram swings about the section centroid; the

Table 8.4 *Increase in stress in unbonded tendons* δ_p, *N/mm*^{2} *(Based on BS 8110)*

L_{te}/d	$R_p = f_{pu}A_p/f_{cu}bd$									
	0.05	0.10	0.15	0.20	0.25	0.30	0.35	0.40	0.45	0.50
10	641	581	522	462	403	343	283	224	164	105
20	320	291	261	231	201	172	142	112	82	52
30	214	194	174	154	134	114	94	75	55	35
40	160	145	130	116	101	86	71	56	41	26
50	128	116	104	92	81	69	57	45	33	
60	107	97	87	77	67	57	47	37	27	
70	92	83	75	66	58	49	40	32		
80	80	73	65	58	50	43	35	28		
90	71	65	58	51	45	38	31	25		
100	64	58	52	46	40	34	28			
110	58	53	47	42	37	31				
120	53	48	43	39	31					
130	49	45	40	36						
140	46	42	37							
150	43	39								

Reminder: BS 8110 limits the value of $(f_{pe} + \delta_p)$ to a maximum of $0.7f_{pu}$.

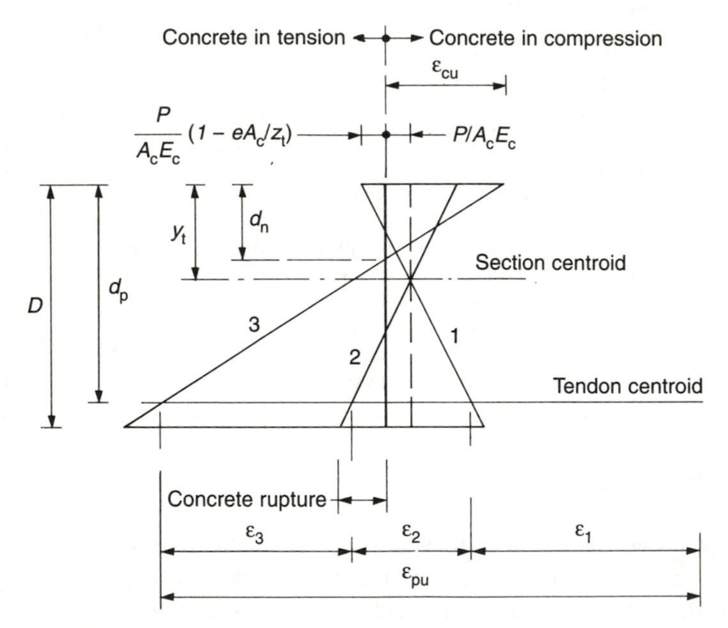

Figure 8.4 *Strain history*

change in strain in the strand ε_2 equals the strain in the concrete at the tendon level.

$$\varepsilon_2 = \{P/(A_c E_c)\}(1 + e_p{}^2 A_c/I_c) \tag{8.13b}$$

When the section cracks and the member is overloaded to failure, the neutral axis rises to d_n below the extreme compression fibre, and the maximum concrete strain reaches its limiting value ε_{cu}, which is 0.0035 according to BS 8110 or 0.003 as given by ACI 318. The increase in tendon strain is

$$\varepsilon_3 = \varepsilon_{cu}(d_p - d_n)/d_n \tag{8.13c}$$

The total strain in the tendon at failure (ε_{pu}) is the sum of the three strains and equals

$$\varepsilon_{pu} = \varepsilon_1 + \varepsilon_2 + \varepsilon_3 \tag{8.14}$$

Knowing the strain and the tendon stress f_{pb}, the depth of the compression block d_x can be calculated by considering the static equilibrium. For static equilibrium, the total compression in concrete equals the total tension in the tendon.

$$p_{cu} b d_x = A_p f_{pb} \tag{8.15}$$

Then the flexural strength (M_u) is found by taking the moment of the tendon force about the line of action of the compressive force in the concrete.

$$M_u = A_p f_{pb}(d_p - d_x/2) \tag{8.16}$$

8.7 Anchorage zone

Prestress is applied to the concrete through an anchorage assembly, which usually consists of a steel casting, though a flat steel plate may sometimes be used. The bearing area of the anchorage is only a small proportion of the concrete area associated with an anchorage. The prestressing force, concentrated on the bearing area, spreads through the concrete, and can be considered to be uniformly distributed on the concrete section at a suitable distance from the anchorage.

Experiments to determine the stresses generated in the anchorage zone have been carried out by applying the force through a thick plate bearing on the surface of a concrete block; the width of the plate was the same as that of the concrete block. In practice, an anchorage is smaller than the concrete section, covering only a proportion of it. In assessing the bursting stresses, the force from the anchorage is assumed to spread within a prism of concrete having the largest available area of concrete placed symmetrically around the anchorage, see Figure 8.6. The combined effect of a group of anchorages is assessed by replacing the group with an equivalent single anchorage, of the same overall area and carrying the same total force as the group, and considering a corresponding symmetrical prism of concrete.

Figure 8.5 shows the spread of the compressive force, applied to a small area on the surface of a concrete block. Along the axis of the force, the stress is

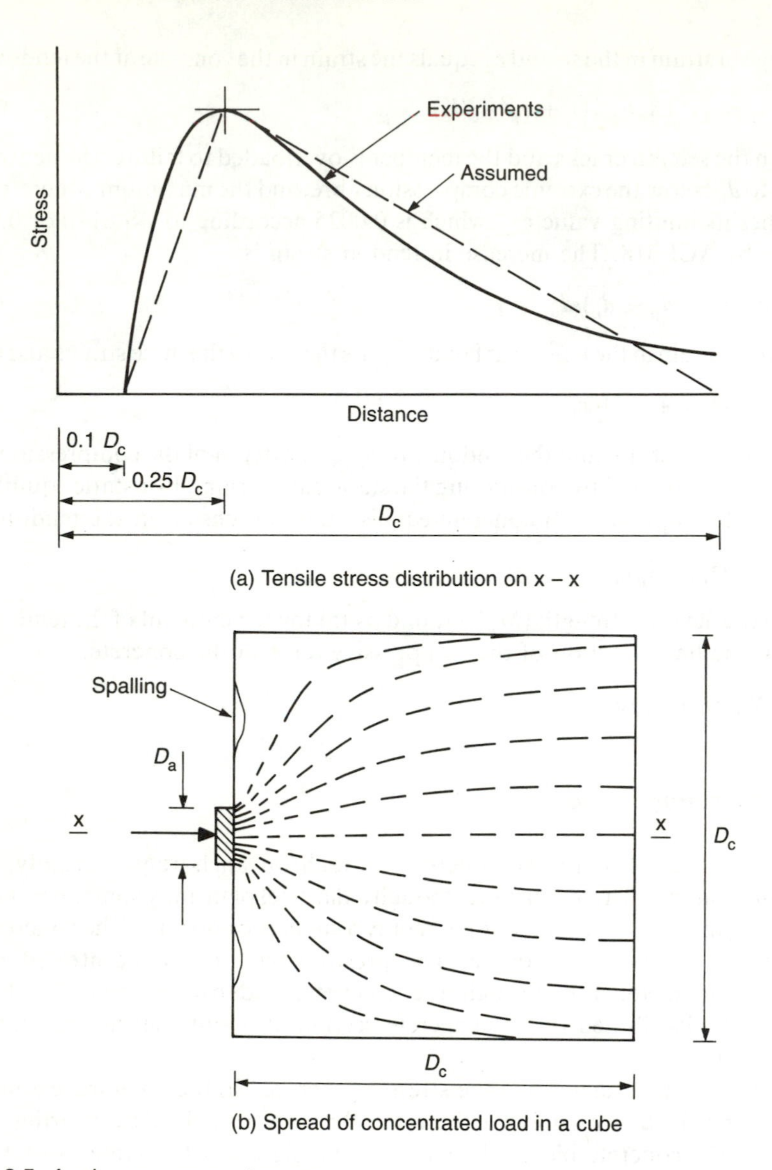

Figure 8.5 *Anchorage zone stresses*

compressive, gradually reducing in intensity away from the loaded face. At a distance equal to the depth of the symmetrical concrete prism the stress can be assumed uniform over the section, the intensity being P/A_c. The stress at right angles to the compression, on a plane through the anchorage axis, is tensile; the tension starts at a distance of about one-tenth of the depth of the symmetrical block from the surface and peaks at about one-quarter of the depth of the block. This tensile force is called the *bursting force*. The distribution of tensile stress along the block axis is usually simplified to a triangular form as shown in Figure

8.5(a). Tension also develops near the surface surrounding the plate area; this force tends to cause spalling of the concrete surface in heavily stressed members. However, in normal post-tensioned floors this effect is negligible.

These experimental studies differ from the true situation in several ways. Firstly, an anchorage casting is approximately conical in shape, with a stiff flange near one end; it transfers the prestressing force to the concrete through direct bearing over the flange area and through the cone action. The proportion of load transferred through the two actions depends, among other factors, on the geometry of the anchorage casting. The wedge action of the conical shape usually increases the bursting force, and affects the peak position of the bursting force. A further deviation from the test conditions is that anchorages are never placed on the surface of concrete; instead, they are located in individual or common pockets, ranging in depth from 75 mm up to 150 mm (3 to 6 inches).

Additionally, the experimental studies have been largely confined to two-dimensional models, whereas the problem is three dimensional. In spite of these limitations, useful data have been obtained and most of the current design methods are based on the results of such studies.

The bursting force is normally contained by rod reinforcement provided as a series of mats or as a series of closed links behind the anchorage; the latter is the more efficient. The spalling tension may be taken care of by a reinforcement mat near the surface, but this stress, being small in post-tensioned floors, is often ignored.

Anchorages are expected to function satisfactorily for tendon forces of around 95% of the tendon strength, though most of the standards limit the jacking force to about 80% of the tendon strength. It may be permissible to take the design force for anchorage zone reinforcement as equal to the specified jacking force, but it is preferable to design for a force of 90 to 100% of the tendon strength as this provides a useful reserve for a deliberate or inadvertent overstressing of a tendon. The stress in the reinforcement is limited to 200 N/mm^2 (30 000 psi).

In designing the anchorage zone, vertical and horizontal sections are considered, first through individual anchorages and then through anchorage groups. The block of concrete, associated with each section, consists of the largest prism concentric with the centroid of the anchorage or group of anchorages. In a concrete floor, however, a vertical crack due to the bursting force is very unlikely to occur (except possibly for anchorages near a corner), because the width of the floor acts as a deep restraint. Design is, therefore, carried out for horizontal cracking at all anchorages and vertical cracking near the slab corners.

Figure 8.6 indicates some of the arrangements of anchorages used in post-tensioned floors and the concrete prism associated with each. The jagged lines on the diagram indicate the positions where tensile stresses should be checked for each arrangement. In the case of a slab anchorage near a corner, bursting stresses should be checked on both horizontal and vertical planes passing through the centre of the bearing plate, while for an interior anchorage of a slab, it is only necessary to check the horizontal plane for bursting stress. The anchorages in the beam are shown positioned in two staggered layers; two checks are necessary in this case: an individual anchorage for bursting on a horizontal

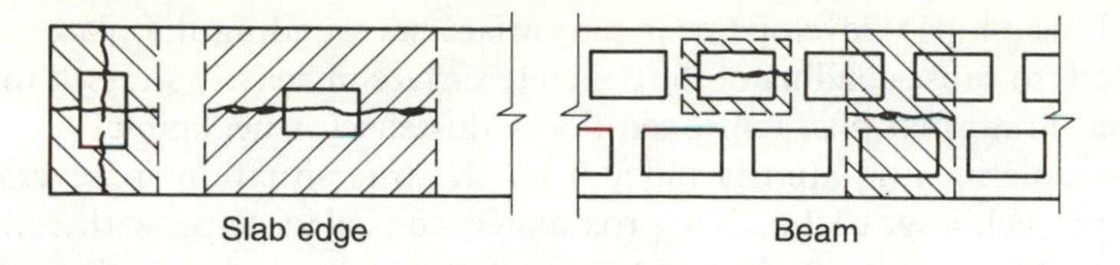

Figure 8.6 *Typical anchorage arrangements*

Table 8.5 *Bursting tensile force*

D_a/D_c	0.20	0.30	0.40	0.50	0.60	0.70
F_b/P_a	0.23	0.23	0.20	0.17	0.14	0.11

D_a = Depth of anchorage F_b = Bursting force
D_c = Depth of concrete prism P_a = Anchorage force

plane through the centre of the bearing plate, and the group bursting on a horizontal plane halfway between the upper and lower anchorages. For the latter, the upper and the lower anchorages are replaced by an hypothetical single anchorage covering the combined area of the two anchorages and the gap between them. For circular anchorages, the loaded surface can be assumed to be a square of an equal area.

Table 8.5, reproduced from BS 8110, gives a simple approach to assessing the bursting force based on a linear relationship with an upper limit of 0.23. A safe rule of thumb for bursting steel around a single tendon is to provide 1.2 mm^2 of link area for each kN of anchorage force.

Example 8.1

Check the strength of the floor in Example 6.4 in span 1-2; its geometry is shown in Figure 6.9. Other data are:

Section:
Section depth = 270 mm
Flange width = 1000 mm depth = 110 mm

Serviceability loads, and shears at support 1:
Dead load = 2.0 + self-weight 4.3 = 6.3 kN/m^2, Shear = 26.35 kN
Live load = 4.0 Shear = 17.89
Shear force from secondary effects = 0.30

Tendons
Final force = 560 kN (4 superstrands 15.7 mm)
Minimum height = 40 mm
Height at 1 = 235 mm
Distance a = 1.0 m
Distance b = 5.49 m (lowest point)

Strand area = 150 mm^2

Breaking load = 265 kN

Note that four superstrands are used in preference to the standard strand because of their superior relaxation characteristics.

Being the end span, support moments are not required.

(i) Solution: BS 8110, Unbonded tendons

Required moment of resistance M_r

Span load = 1.4 × 6.3 + 1.6 × 4 = 15.22 kN/m
Shear at 1 = 1.4 × 26.35 + 1.6 × 17.89 + 0.3 = 65.81 kN
Span moment = 65.81^2 (2 × 15.22) = 142.30 kNm

Moment is maximum at 65.81/15.22 = 4.32 m from support 1

Tendon height

From Equation (5.19),
$A_2 = Y_1/(b^2 - ab) = (235 - 40)/[1000 \times (5.49^2 - 1.0 \times 5.49)] = 0.00791/\text{m}$

From Equation (4.17b),
$y = 0.0079 \times (5.49 - 4.32)^2 \times 1000 = 11\ \text{mm}$

Tendon height at 4.32 m from 1 = 40 + 11 = 51 mm

Tensile force

Tendon area = 4 × 150 = 600 mm^2, Final force = 560 kN
f_{pe} = 560 000/600 = 933 N/mm^2

Three strands are stopped at support 1, one continues for the whole 29 m length of the slab.

Equivalent tendon length = (3 × 10 + 1 × 29)/4 = 14.75 m

Two plastic hinges are required for span 1–2 to become a mechanism, at support and in span

$d = 270 - 51 = 219$ mm,

say 220 mm to allow for reinforcement bars at a larger lever arm

$L_{te}/d = (14.75/2)/0.22 = 33.5$
$R_p = f_{pu}A_p/f_{cu}bd = 265 \times 1000 \times 4/(40 \times 1000 \times 220) = 0.120$

From Equation (8.4),

$f_{pb} = 933 + (7000/33.5)(1 - 1.7 \times 0.12) = 1099\ \text{N/mm}^2$
$0.87f_{pu} = 0.87 \times 265 \times 1000/150 = 1537 > 1099$ OK

Tendon force = 1099 × 600/1000 = 659.4 kN

Provide 2 No. 10 mm diameter bars at bottom, $f_y = 460\ \text{N/mm}^2$
Area = 157 mm^2 Force = (157 × 460/1000)/1.15 = 62.8 kN maximum

Total tension = 62.8 + 659.4 = 722.2 kN

Note that in calculating R_p, the section is taken as rectangular of width equal to the flange width. This is valid if the neutral axis is within the flange depth.

Section strength

Concrete stress in stress block = $0.45 \times 40 = 18$ N/mm^2
Depth of compression block = $699.4 \times 1000/(1000 \times 18) = 38.9$ mm

Lever arm = $270 - 51 - 38.9/2 = 199.6$ mm for tendons
and $270 - 40 - 38.9/2 = 210.6$ mm for reinforcement

Strength = $(659.4 \times 199.6 + 62.8 \times 210.6)/1000 = 144.8$ kNm > 142.30 OK

Check stress in reinforcement bars

$d_n = 38.9/0.9 = 43.2$ mm
distance between bars and neutral axis = $270 - 40 - 43.2 = 186.8$ mm
strain in bars = $0.0035 \times 186.8/43.2 = 0.0151$
$E_s \times$ strain = $200 \times 1000 \times 0.0151 = 3000$ N/mm^2 > 0.87×460 OK

(ii) Solution: BS 8110, Bonded tendons

Assume a flat duct and allow 2 mm rise of tendon in duct

Required moment of resistance = 142.3 kNm as for (i) above
Tendon height = 51 mm
Final tendon stress f_{pe} = 933 N/mm^2

Tensile force

Provide 2 No. 8 mm bars, f_y = 460 N/mm^2
$R_p = 0.12$ as above, f_{pe}/f_{pu} = $933/(265 \times 1000/150)$ = 0.528
From Table 8.1, $f_{pb}/0.87f_{pu}$ = 0.99

Tendon force = $4 \times 0.99 \times 265 \times 0.87$ = 913 kN
Force in 2 bars 8 mm dia. = $2 \times 50 \times 0.87 \times 460/1000$ = 40
Total tensile force = $913 + 40 = 953$ kN

Concrete stress in stress block = 0.45×40 = 18 N/mm^2
Depth of compression block = $953 \times 1000/(1000 \times 18)$ = 53 mm

Lever arm for tendons = $270 - 51 - 53/2$ = 192.5 mm
and for rod reinforcement = $270 - 40 - 53/2$ = 203.5 mm

Strength = $(913 \times 192.5 + 40 \times 203.5)/1000$ = 183.9 kNm.
> 142.30 OK

Stress in reinforcement bars can be checked as in (i).

(iii) Solution: ACI 318, Unbonded tendons

$f'_c = 40$ N/mm^2

Required moment of resistance

Span load = $1.4 \times 6.3 + 1.7 \times 4$ = 15.62 kN/m
Shear at 1 = $1.4 \times 26.35 + 1.7 \times 17.89 + 0.3$ = 67.60 kN
Span moment = $67.60^2/(2 \times 15.62)$ = 146.28 kNm

Moment is maximum at $67.60/15.62 = 4.33$ m from support 1

Tendon height = 51 mm at 4.33 m from support 1

Tensile Force

$f_{pe} = 560\,000/600 = 933$ N/mm^2
$d_p = 270 - 51 = 219$ mm
$f_{pb} = 933 + 70 + 40 \times 1000 \times 219/(100 \times 4 \times 150) = 1149$ N/mm^2

Check: span/depth = 10/0.27 = 37 > 35
Limiting stress = 933 + 200 = 1133, Use 1133 N/mm^2

Tendon force = 1133 × 4 × 150/1000 = 679.8 kN
Force in bars = 157 × 460/1000 = 72.2

Total tension = 679.8 + 72.2 = 752.0

Section strength

Concrete stress in stress block = 0.85 × 40 = 34 N/mm^2
Depth of compression block = 752 × 1000/(34 × 1000) = 22 mm
Lever arm = 270 − 51 − 22/2 = 208 mm for tendons
and 270 − 40 − 22/2 = 219 mm for reinforcement
Strength = 0.9 × (679.8 × 0.208 + 72.2 × 0.219) = 157.2 kNm
> 146.28 OK

Check stress in rod reinforcement

$\beta_1 = 0.85 - 0.008 \times (40 - 30) = 0.77$
$d_n = 22/0.77 = 28.6$ mm
strain in bars $= 0.003 \times (270 - 40 - 28.6)/28.6 = 0.021$
$E_s \times$ strain $= 200 \times 1000 \times 0.021 = 4200$ N/mm^2 > 460 OK

Example 8.2

Design the anchorage zone reinforcement for a 200 mm deep slab with 125 mm × 75 mm anchorages (75 mm is the vertical dimension) spaced at 500 mm centres. The jacking force in each tendon is 200 kN.

Solution

$P_a = 200$ kN $D_a = 75$ mm $D_c = 200$ mm $D_a/D_c = 0.375$

Interpolating from Table 8.5,

$F_t/P_a = 0.2075$ $F_t = 0.2075 \times 200 = 41.5$ kN
$A_{st} = 41.5 \times 1000/200 = 210$ mm^2

Provide 2 links (4 legs) 8 mm dia, at 75 mm and 125 mm, behind the anchorage flange.

9 DEFLECTION AND VIBRATION

Excessive floor deflection or vibration can cause alarm to building occupants and may result in damage to non-structural elements such as partitions or finishes. Sag of roof slabs can also cause ponding of rainwater. The deflection performance of post-tensioned floors is generally better than that of reinforced concrete floors, since the action of the prestressing force causes members to have an initial upwards curvature, or camber, which counteracts the sag due to the self-weight and live loads. However, the greater slenderness of post-tensioned floors makes them more susceptible to vibration problems than reinforced floors. Both deflection and vibration are serviceability problems which should, therefore, be checked using unfactored loads.

9.1 Deflections

Calculation of deflections for a post-tensioned floor is simpler than for a reinforced concrete member, since a post-tensioned section can usually be assumed to be uncracked, so that the properties of the gross concrete section can be used in the calculations. (Strictly speaking, for slabs with bonded tendons the transformed section should be used, while for those with unbonded tendons the net section should be used.) Other assumptions commonly made in deflection analysis are that:

- the force in a tendon is constant along its length
- the slope of the tendons is small, so that the horizontal component of the prestress is constant
- any change in prestress in the tendons caused by the deflections may be neglected

Deflection calculations for beams and one-way spanning slabs are based on elastic beam theory and are relatively straightforward, if somewhat laborious. For two-way spanning slabs, since realistic modelling of the structural behaviour can become extremely complex, it is usual to adopt one of a number of simplified approaches. The resulting loss of accuracy is considered acceptable because of the high degree of uncertainty inherent in such calculations. The designer is unlikely to have an accurate knowledge of the modulus of elasticity and creep properties

Table 9.1 *Deflection limits for floors*

	Member type	*Limit*	*Deflection to consider*
ACI318	Roofs not supporting any non-structural elements	L/180	Immediate under live load
	Floors not supporting any non-structural elements	L/360	
	Floors or roofs supporting brittle non-structural elements	L/480	Deflection occurring after installation of non-structural elements
	Floors or roofs supporting non-brittle non-structural elements	L/240	
BS8110	Any visible structural member	L/250	Total deflection
	Floors or roofs supporting brittle non-structural elements	Smaller of L/500 or 20 mm	Deflection occurring after installation of non-structural elements
	Floors or roofs supporting non-brittle non-structural elements	Smaller of L/350 or 20 mm	

of the concrete, or of the exact loads the structure is likely to carry over a period of several years. Field measurements of floor deflections show extremely wide scatter, even in apparently identical floors within the same building (ACI Committee 435, 1991).

Deflection limits recommended in the British and American concrete codes are shown in Table 9.1, in which L is the shorter span length. The recommended limits for roofs are not intended to safeguard against ponding. Checking against these criteria is likely to require the calculation of both short and long-term deflections, though in most instances the long-term behaviour is the more critical, since it affects the behaviour of finishes and partitions. In order to calculate realistic long-term deflections, it is essential that the loading history and the magnitude of the sustained load are carefully considered.

9.1.1 Influence of loading history

Deflections of post-tensioned slabs have two components; a downwards deflection caused by the self-weight and any imposed loads on the slab, and an upwards camber due to the prestress. Under short-term loading these effects are calculated elastically. For long-term loading, the additional deflection due to creep is calculated by multiplying the short-term deflection by an appropriate creep coefficient C_c. The camber must also be modified to take account of any prestress losses. The total deflection is then the sum of the short-term and creep deflections.

Creep coefficients have been discussed in detail in Chapter 2. In addition to environmental factors, the creep behaviour is dependent on the average load intensity and the age at which loads were applied. The loading history is, therefore, important.

The significance of the loading history can best be illustrated by considering a typical floor. Immediately after stressing the deflection is likely to be negative (i.e. upwards) because the upwards-acting balancing force due to the curved tendons is greater than the self-weight of the floor. If no construction loads or live loads are applied for some months, then the creep deflection will also be negative. It is thus possible that, even after the application of live loads, the resultant deflection will be negative. If, however, construction loads are applied shortly after stressing, and are quickly followed by live loads, then the long-term deflection will be positive.

As can be seen from the deflection limits in Table 9.1, it may be necessary to calculate deflections at several stages within the life of a structure. If detailed information on the construction timetable, including details of back-propping and time of application of live loads, is available, then this can be used in the deflection calculations. Otherwise, it is reasonable to approximate the loading history by the following three-stage sequence:

1. The floor is back-propped, and so effectively carries no load, for between one and four weeks. The exact duration of this phase will depend on the construction programme, but two weeks is a reasonable average.
2. Thereafter, the floor carries its own weight and a construction load which is typically 25% of its self-weight, though this can be subject to large variations. After approximately three months, finishes are added, services are installed and the construction load is removed. However, since the weight of finishes etc. is likely to be of the same order as the construction load, it may be assumed that the load intensity remains unchanged.
3. Once the building is finished, it is occupied and additional live loads are applied. For a large building, this may take over a year, but six months is more typical.

9.1.2 Live load intensity

As well as the time at which they are applied, the magnitude of the loads is important. While the self-weight and other permanent loads can usually be accurately assessed, the value of live load which may be regarded as a sustained long-term load is much harder to determine. Calculating the long-term deflection on the assumption that the maximum live load is applied at all times yields values of deflection considerably larger than the actual, measured values, and is as unrealistic as assuming no live load at all. BS 8110: Part 2 recommends that *expected* values should be used to give a best estimate of the likely behaviour of the structure while, for calculations to satisfy a particular limit state, upper or lower bound values should be used, depending on whether or not the effect is beneficial.

Live load can be considered as comprising a permanent or near-permanent component w_p due to items such as office furniture, and a transient component w_t due to personnel, vehicles etc. In estimating long-term behaviour, it is reasonable to transform these loads into an equivalent permanently sustained load w_s on the basis of the duration for which the transient load is present:

$$w_s = w_p + w_t R_t \tag{9.1}$$

where R_t is the fraction of the total time for which the transient load is present. Note that w_p and w_t are the actual estimated loads, not the recommended values for flexure design. Their sum will not usually equal the nominal live loading specified for a floor.

Applying this formula to public car parks, a typical vehicle weighs approximately 1300 kg and occupies a floor area of around 13 m^2, giving an average transient load intensity w_t of 1.0 kN/m^2 (20 psf). Assuming no finish on the floor, w_p is zero. In shopping areas a car park is usually occupied for around ten hours per day, six days per week (i.e. 35% of the time), but is unlikely to be fully loaded for the whole of this time. Even if full occupancy during these hours is assumed, this gives $R_t = 0.35$ and hence an equivalent sustained live load of 0.35 kN/m^2 (7 psf) is appropriate for the parking bays. The sustained load in the aisles is negligible; a nominal value of, say, 0.1 kN/m^2 (2 psf) may be assumed if desired. Car parks in long-term use, such as at airports, may be occupied for twenty hours per day, seven days per week ($R_t = 0.85$), giving an equivalent sustained load of 0.85 kN/m^2 (17 psf) in the parking bays and, say, 0.25 kN/m^2 (5 psf) in the aisles.

Clearly, these values fall well below the recommended load intensity of 2.5 kN/m^2 (50 psf) for car parks in the United Kingdom. A sustained load of 30% of the recommended design load, applied over the entire floor area including the aisles, would, therefore, be ample for estimating long-term deflections in car parks.

The recommended live load for office floors, excluding storage areas, is 2.5 kN/m^2 (50 psf), but the current trend is towards 4.0 kN/m^2 (80 psf). There is no evidence that a load as large as 4.0 kN/m^2 is ever applied to a normal office floor, with surveys suggesting that a value of 1.25 kN/m^2 (25 psf), excluding partitions and services, is appropriate (Mitchell and Woodgate, 1971; Choi, 1992). Since the transient loads due to personnel in offices are small compared to the weight of furnishings, it is reasonable to take $R_t = 1.0$ in spite of the fact that most offices are fully occupied only about 25% of the time. Therefore, a sustained live load of 1.25 kN/m^2 (25 psf) is suitable for long-term deflection calculations.

Public buildings such as art galleries, theatres and cinemas are likely to be occupied for 25% of the time or less. In this case, the transient load due to crowds of people is the largest component of the live load. A sustained loading of 30% of the design live load is a reasonable upper bound.

Warehouses and storage areas are expected to be fully loaded most of the time and the deflection in such buildings should be calculated for the full design load.

Excepting warehouses, it can be argued that 30% of the design live load represents a sensible upper bound of load intensity for use in long-term deflection calculations. It should be remembered that, since upwards as well as downwards deflections can cause problems, too high a loading is as erroneous as too low.

9.1.3 One-way spanning slabs

Short-term deflections of one-way slabs can be easily calculated using elastic beam theory. The normal procedure is to consider the loads acting on a one metre wide strip of slab. The total downwards load is the sum of the self-weight and any applied loads present, while the equivalent load corresponding to the prestress normally acts upwards. A detailed discussion of equivalent loads due to various tendon profiles is given in Chapter 5. In general, parabolic profiles correspond to uniformly distributed loads and harped profiles correspond to concentrated loads at the point of the reversal in slope. For deflection calculations it is acceptable to ignore the small reverse parabolas near the supports as these will mostly cause compression of the supports rather than bending of the floor span. Similarly, the small parabolas under the load points in harped profiles can be neglected.

Knowing the loads, and treating columns or walls as simple supports, the reactions can be calculated and a bending moment expression set up in terms of the distance x from one end of the slab. The curvature of the member is related to the bending moment by

$$d^2y/dx^2 = M(x)/(E_c I_c) \tag{9.2}$$

and the displacement y is found by integrating (9.2) twice, with the resulting constants of integration determined from the boundary conditions. Under uniform loading the peak deflection of a slab panel, δ, is the value of y at its centre. It may be necessary to calculate deflections in both outer and inner panels in order to determine the worst case.

As an alternative to Equation (9.2), it may be possible to obtain a sufficiently accurate estimate of the deflection using well-known standard results such as those tabulated in the Steel Designers' Manual (SCI, 1992). Some particularly useful results are:

$$\text{simply supported beam: } \delta = 5wL^4/(384EI) \tag{9.3a}$$

$$\text{fixed-end beam: } \delta = wL^4/(384EI) \tag{9.3b}$$

$$\text{cantilever: } \delta = wL^4/(8EI) \tag{9.3c}$$

The peak deflection of an interior panel in a multispan slab can be closely approximated by using the result for a fixed-ended beam.

Having calculated the deflections due to the applied loads and due to the prestress, the net short-term deflection is simply the sum of the two values. The long-term deflection is found by modifying the deflection due to the applied loads to take account of creep, and that due to the prestress to account for both creep and loss of prestress.

9.1.4 Two-way spanning slabs

Accurate calculation of deflections in two-way spanning floors is far more difficult to achieve than for one-way spanning systems, due to the three-dimensional nature of the problem and the difference in prestressing arrangements in the two span directions. A thorough review of the numerous semi-analytical and approximate calculation methods which have been proposed is given by ACI Committee 435 (1974). Only the most widely used methods are discussed here.

Where slabs are supported on beams, the beam deflection often makes a significant contribution to the total slab deflection. Therefore, for the purposes of deflection calculations, two-way slabs should be taken to include slabs supported on beams, even though these are normally treated as one-way spanning in other parts of the design process.

Classical plate theory

For square or rectangular panels whose edges can reasonably be approximated as fixed or pinned, formulae based on classical elastic plate theory can be used. The central deflection of a rectangular plate of plan dimensions $a \times b$ (where $b > a$) under a uniform load w per unit area is given by

$$\delta = \alpha_1 w a^4 / H \qquad (9.4)$$

where α_1 = a factor depending on the aspect ratio
H = the plate bending stiffness
$= (ED^3)/[12(1 - v^2)]$
D = the plate thickness.

For a central point load W, the deflection is

$$\delta = \alpha_1 W a^3 / \mathrm{H} \qquad (9.5)$$

Timoshenko and Woinowsky-Krieger (1959) give tabulated values of α_1 for numerous combinations of edge conditions, of which a few simple cases are reproduced in Table 9.2. Unfortunately, plate solutions are rarely adequate for floor slabs as the boundary conditions are usually considerably more complex than simple pins or fixity. However, these solutions are useful as part of the frame-and-slab method, discussed later.

Timoshenko and Woinowsky-Krieger also give an analysis for an interior panel within a flat slab. Assuming the columns provide point support only, the central deflection under a uniform load is given by

$$\delta = \alpha_2 w a^4 / H \qquad (9.6)$$

where α_2 is taken from Table 9.2.

This formula is suitable for interior panels of large, uniformly loaded floors, but cannot be applied to other cases (outer panels, or non-uniform loads) which are often more critical.

Table 9.2 *Values of α_1 and α_2 for rectangular panels*

b/a	α_1 Fixed edges uniform load	α_1 Pinned edges uniform load	α_1 Pinned edges central point load	α_2 Interior panel uniform load
1.0	0.00126	0.00406	0.01160	0.00581
1.1	0.00150	0.00485	0.01265	0.00713
1.2	0.00172	0.00564	0.01353	0.00888
1.3	0.00191	0.00638	0.01422	0.01105
1.4	0.00207	0.00705	0.01484	0.01375
1.5	0.00220	0.00772	0.01530	0.01706
1.6	0.00230	0.00830	0.01570	0.02150
1.7	0.00238	0.00883	0.01600	0.02664
1.8	0.00245	0.00931	0.01620	0.03254
1.9	0.00249	0.00974	0.01636	0.03623
2.0	0.00254	0.01013	0.01651	0.04672

Crossing beam methods

There are several methods available based on the idealization of a slab as two perpendicular sets of beams, of which the simplest and most widely used is described here (Rice and Kulka, 1960). Strips of the slab in each direction are treated as beams, and the central deflection is then given by the sum of the midspan deflections of a column strip spanning in one direction and a middle strip spanning perpendicular to it, Figure 9.1. Beams running along the column lines can be easily included in the calculation of the equivalent beam properties for the column strips. The deflections of the equivalent beams can be calculated using elastic methods or simple formulae, as discussed in Section 9.1.3.

The correct proportioning of load between the equivalent beam strips can be difficult; usually the column strips of a slab carry rather larger moments than the middle strips. If prior analysis has already yielded values of the moments on column and middle strips, then these may be used directly. If only the average load on the floor is known, then this must be distributed appropriately. For reinforced concrete flat slabs, ACI 318 recommends that column strips be assumed to carry 1.2 times the average load and middle strips 0.8 times the average. If no more accurate information is available, then this rough guide can also be used for post-tensioned slabs. It is sufficiently accurate to assume that the load is uniform along the length of each strip, neglecting any reductions due to sharing of the load where the two strips meet.

This simplified method obviously lacks rigour, but has been found to give reasonably reliable estimates of slab deflections.

Frame-and-slab method

This method is quite similar to the crossing beam method discussed above, but takes rather better account of the two-way spanning behaviour of the central

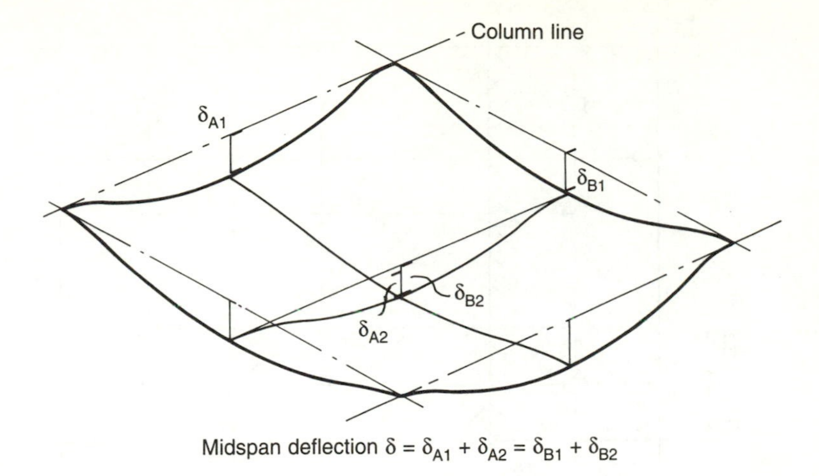

Figure 9.1 *Crossing beam method*

portion of a slab panel. Comparisons with measured values suggest that the method slightly underestimates actual deflections (Nilson and Walters, 1975).

As with the crossing beam method, the slab is first split into strips along the column lines, Figure 9.2(a). Each column strip is assumed to have a width of 0.4 times the relevant span dimension, so that the edges of the strips correspond roughly to lines of contraflexure in the slab. (Of course, the exact lines of contraflexure are hard to determine, and are almost certainly not straight.) The rectangular panel bounded by the column strips is treated as an elastic plate with simply supported edges. Each column strip is assumed to carry uniformly distributed dead and imposed loads, together with line loads along its edges corresponding to the support reactions of the interior panel. The total central deflection of the slab panel is then found as the sum of three components, Figure 9.2(b):

- the centreline deflection of a column strip spanning between columns, δ_1, calculated using elastic beam theory;
- the deflection of the edge of the column strip relative to its centreline, δ_2
- the central deflection of the rectangular panel between column strips, δ_3, calculated using the Equations (9.4) and (9.5).

Finite element method

Numerous sophisticated finite element packages are now available, making it a relatively simple matter to generate and analyse a simple plate or grillage model of a concrete floor. The difficulty of choosing appropriate boundary conditions at the edges of individual panels means that it would be necessary to create a model of an entire floor, or at least of a substantial area, in order to have confidence in the accuracy of the results. The finite element model required would thus be quite large and the cost of the analysis would not be negligible.

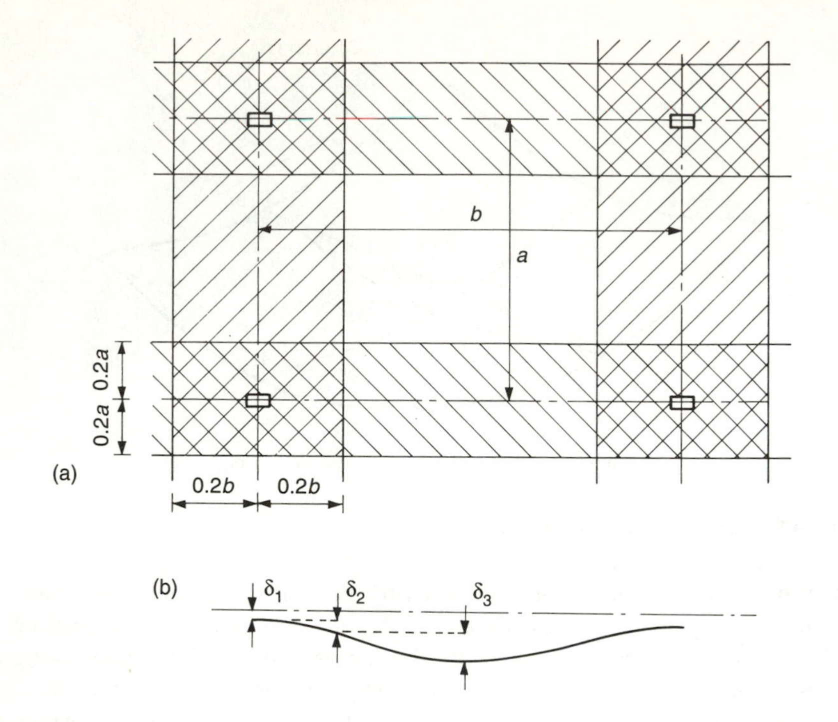

Figure 9.2 *Frame-and-slab method: (a) Division of slab into column strips and central panels (b) Section through slab at midspan, showing component deflections*

Choice of method

The choice of which of the above methods to use in a particular application depends on the structural details and, to an extent, on personal preference. The classical plate solution is only likely to be useful in quite a small number of cases. For most structures both the crossing beam approach and the frame-and-slab method provide reasonable estimates of deflections, with the former usually preferred on the grounds of its simplicity. Given the level of uncertainty surrounding the calculations, the sophistication of the finite element method is only likely to be justified for structures with irregular or unusual geometries. A comparative example of deflection calculations using these methods is given at the end of this chapter.

9.2 Vibration

Concern over vibration problems has traditionally been focused on lightweight floors such as those of timber or composite steel-concrete construction. In recent years, however, it has become increasingly clear that vibrations may also be an important design criterion in concrete floors, particularly those with long spans and high slenderness ratios. The most common problem is annoyance or discomfort to building occupants caused by vibrations set up by human footfall.

It should be emphasized that the overwhelming majority of post-tensioned floors do not give rise to any vibration problems and that, in general, the vibrational performance of such floors is far better than that of other construction types. Nevertheless, as post-tensioned floors of increasing slenderness are constructed the importance of vibration increases. It is, therefore, recommended that some simple order-of-magnitude checks are performed at the design stage, with a more detailed analysis carried out only if the simple checks suggest that there may be a problem.

In assessing the acceptability of a floor vibration, both the frequency and amplitude of the motion are important. Usually, amplitude is expressed in terms of the peak acceleration, though occasionally other measures such as velocity or displacement are used. In general, humans are more responsive to quite low frequency vibrations, so that greater accelerations are acceptable at high frequencies than at lower ones.

A common way of assessing floor vibrations in the majority of buildings is the use of perceptibility scales. These scales are suitable for most conventional forms of building occupancy (offices, apartments, etc.). However, for a very sensitive occupancy such as a microelectronics workshop, a lower vibration threshold may be appropriate. Such cases are likely to require specialist advice, and are not discussed further here.

One of the most useful of the perceptibility scales is the one given in the Canadian steelwork code, CAN3-S16.1-M89, shown in Figure 9.3. This consists of a series of *iso-perceptibility lines* corresponding to different types of excitation and levels of damping. Curves for other damping levels can be found by interpolation. Above 8 Hz, the lines slope upwards, reflecting the greater human tolerance of higher frequency vibrations. For a given floor, a point on the scale is found corresponding to the natural frequency of the floor and the peak acceleration caused by the activity. If the point lies above the relevant iso-perceptibility line then the vibration is excessive, otherwise it is acceptable. The *continuous vibration* line is intended for assessment of vibrations lasting for at least ten cycles, while the *walking vibration* lines are used to assess a floor's acceptability for walking vibrations based on the peak acceleration caused by a heel drop test. The test is performed by a person of average weight rising to the balls of their feet, then dropping onto their heels at the centre of the floor.

The perceptibility lines shown in Figure 9.3 are for relatively quiet occupancies such as offices, where the occupants are likely to notice quite small vibrations. For active occupancies such as car parks and shopping centres, Pernica and Allen (1982) recommend that these acceleration limits be increased by a factor of 3.

In order to assess the floor using such a scale, it is necessary to determine the frequency and damping characteristics of the floor and its response to walking or heel drop excitation. While all these parameters are quite easy to determine by testing, they are difficult to predict accurately at the design stage. The best available methods and suitable numerical values are discussed in the following sections.

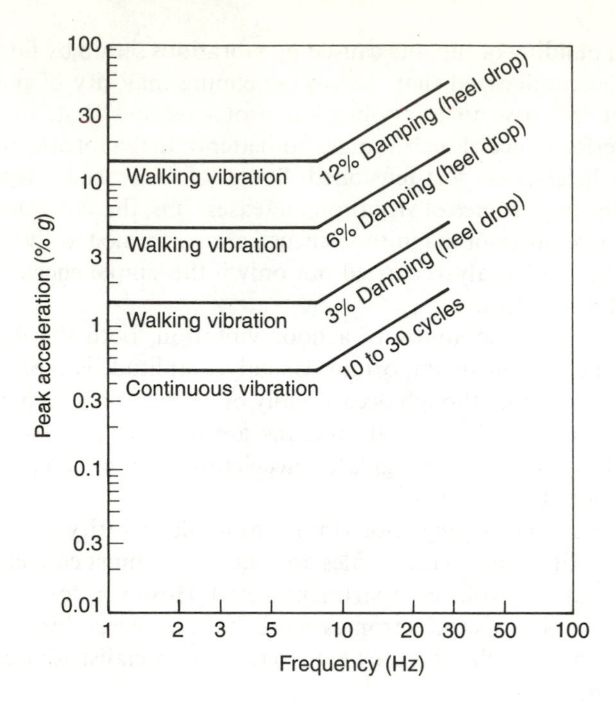

Figure 9.3 *The CSA Vibration Perceptibility Scale*

The relationship between floor natural frequency and vibration amplitude is a complex one. Floors with a high ratio of mass to stiffness have low natural frequencies, so that resonance with walking frequencies is likely. However, such floors also have high inertia, meaning that a given magnitude of loading will produce a smaller response than in a stiffer, lighter floor. Thus, while an estimate of the natural frequency is an essential part of any assessment, it is not alone sufficient. Wyatt (1989), for example, classifies floors into two categories, depending on whether their lowest natural frequency is above or below the third harmonic of the walking frequency (which is typically around 2 Hz). For floors below this cut-off, resonance with walking excitations is likely. For floors above the limiting value, problems may still occur, but the vibrations are caused by the impulsive loading of a single footfall rather than by resonance.

9.2.1 Prediction of natural frequency

Numerous simplified methods exist for the prediction of natural frequencies of floors, but comparison with measured values shows that they err quite considerably on the low side. This error will generally be conservative, so an acceptable procedure would be to use a simplified method for a first estimate and only perform a more sophisticated analysis (e.g. using finite elements) if the simple method indicated that vibration problems were likely.

As with deflections, the behaviour of slabs supported on beams is unlikely to follow simple one-way spanning assumptions. Since the beams used with post-tensioned floors are usually quite wide and shallow, they do not provide sufficient stiffness to act as supports. Therefore, two alternative modes of vibration are possible: local bending of the slab in the shorter span direction, at right angles to the beams; and overall bending of the slab and beams along the length of the beams, i.e. in the longer span direction (Pavic *et al.*, 1994). Without calculation, it is not usually possible to ascertain which of these two modes predominates in a particular floor. Beam-and-slab floors should, therefore, be approached by calculating the natural frequencies of both modes, assuming one-way spanning behaviour in each case, i.e. by considering:

- a one-metre wide strip of slab spanning between the beams, assuming no beam movement, and
- a T-section consisting of the beam together with a tributary width of slab, spanning between supporting columns. The choice of tributary width of slab to use in calculating the T-beam properties should follow code guidelines; for example, BS 8110 recommends using a width of 0.2 times the distance between the points of contraflexure of the slab under uniform loading.

The fundamental mode is then the one having the lower natural frequency.

Equivalent beam method

This approach is widely used for composite floors, and can be applied to concrete floors having a predominantly one-way action. A section of the floor is treated as a simple, one-way spanning beam of length L, and supporting a weight per unit length w. Its frequency f_0 is then determined using the formula

$$f_0 = (\pi/2)[(E_c.I_c.g)/(wL^4)]^{0.5} = 0.18(g/\delta)^{0.5} \quad (9.7)$$

where g = gravitational acceleration (9.81 m/s^2, or 32.2 ft/s^2).
δ = deflection due to dead and imposed loads (not prestress)

For floors consisting of several continuous spans, Equation (9.7) tends to underestimate the natural frequency due to the assumption of simple supports rather than continuity at the member ends. To account for this, it is recommended that the span length L for a continuous floor is reduced by a factor of 0.9 before applying Equation (9.7). This is equivalent to increasing the natural frequency by 24%. In doing so, it should be noted that the effect of end continuity on the dynamic behaviour can vary considerably between floors (Caverson *et al.*, 1994), so that even this reduced value of span length may sometimes give rather low values of natural frequency.

Two-way spanning floors

The U.K. Concrete Society (1994) has proposed a vibration assessment procedure which assumes that a floor vibrates in two independent sets of modes in the two perpendicular span directions. The natural frequency of the lowest mode in each direction is calculated using a beam-type formula, but with modifications to account for the two-way nature of the floor and the continuity at the edges of a panel. It should be noted that even floors which are treated as one-way spanning in other parts of the design process frequently exhibit two-way vibrational behaviour. For this reason, equations are given here for ribbed floors, even though such floors are normally regarded as one-way members. Such floors may reasonably be assessed using either the equivalent beam method or the procedure outlined below.

Formulae are given for a two-way continuous floor. Equations (9.8) to (9.11) and (9.14) relate to vibration of the slab in the x-direction; the characteristics of the y-direction mode are determined by interchanging the x and y subscripts in these equations.

The effective aspect ratio of a slab panel is defined as

$$\lambda_x = (n_x L_x / L_y)[I_y / I_x]^{0.25} \tag{9.8}$$

I_x and I_y are moments of inertia in x and y directions
L_x and L_y are the span lengths in the two directions
n_x and n_y are the numbers of bays in the two directions.

This, in turn, is used to calculate a modification factor k_x.

For solid or waffle slabs: $k_x = 1 + (1/\lambda_x^2)$ (9.9)
For a ribbed slab: $k_x = [1 + (1/\lambda_x^4)]^{0.5}$ (9.10)

For slabs with beams along the column lines, the natural frequency is then

$$\begin{aligned} f_x' &= (k_x \pi/2)[(E_c I_y g)/(w L_y^4)]^{0.5} \\ &= 0.18 k_x (g/\delta_y)^{0.5} \end{aligned} \tag{9.11}$$

where w = the load per unit area.
δ_y = static deflection of a 1 m wide strip spanning in the y-direction.

For slabs without perimeter beams the frequency given by (9.11) is modified by the calculation of a frequency f_b. For solid or waffle slabs:

$$\begin{aligned} f_b &= \frac{(\pi/2)[(E_c I_x g)/(w L_x^4)]^{0.5}}{[1 + (I_x L_y^4)/(I_y L_x^4)]^{0.5}} \\ &= \frac{0.18(g/\delta_x)^{0.5}}{(1 + \delta_y/\delta_x)^{0.5}} \end{aligned} \tag{9.12}$$

where δ_x = static deflection of a 1 m wide strip spanning in the x-direction.

For ribbed slabs:

$$\begin{aligned} f_b &= \frac{(\pi/2)[(E_c I_x g)/(w L_x^4)]^{0.5}}{\{1 + [(I_x L_y^4)/(I_y L_x^4)]^{2/3}\}^{3/4}} \\ &= \frac{0.18(g/\delta_x)^{0.5}}{[1 + (\delta_y/\delta_x)^{2/3}]^{0.75}} \end{aligned} \tag{9.13}$$

The natural frequency is then

$$f_x = f_x' - (f_x' - f_b).[1/n_x + 1/n_y]/2 \qquad (9.14)$$

Comparison with measured values suggests that this method underestimates natural frequencies, sometimes by as much as 50% (Williams and Waldron, 1994). Nevertheless, it remains useful as a quick, order-of-magnitude check. As with deflections, a more accurate assessment is likely to require the creation of a detailed finite element model of the entire floor.

9.2.2 Damping

Damping is a collective term used to cover the numerous mechanisms of energy dissipation within a vibrating structure. A certain amount of damping will be inherent in the material from which the structure is built, but considerable additional damping can arise from non-structural sources such as finishes, services, furnishings and even the building occupants themselves. It is thus not possible to predict analytically the damping which will be present in a floor. Instead, recommended values are averages of numerous measured values.

Damping is usually expressed as a percentage of a notional critical value which would cause vibrations to cease completely in a single half cycle. The following percentages are recommended for post-tensioned floors:

- bare concrete: 2–3%
- with false floors, ceilings, services, furniture: 4–6%
- with non-structural partitions extending for the full height of the floor: 5–8%.

These values are typical for most floors. While it is possible that lower damping values may occasionally occur, designing to such low values would be excessively conservative for the vast majority of floors.

9.2.3 Prediction of floor response

Predicting the peak acceleration of a floor is even more problematic than the estimation of natural frequency. For one-way slabs it is possible to apply the approach recommended by the Canadian Standards Association (CSA) for composite floors, in which the peak acceleration for a heel drop is estimated and used in conjunction with the perceptibility scale shown in Figure 9.3. For two-way floors, the calculation of accelerations is too error-prone, so the magnitude of the response is estimated using an empirical response factor.

The CSA method

This approximate method of calculating the response to a heel drop load for one-way spanning floors assumes that the heel drop provides a sudden impulse H, which results in a peak acceleration a_0 given by

$$a_0 = 2\pi f_0 Hg/W \qquad (9.15)$$

where W is the weight of a simple oscillator having the same dynamic characteristics as the floor. This is dependent on the width of floor which participates in the response and on the mode shape. Typically, it is assumed that the width participating is around 40 times the floor thickness and that the mode shape is sinusoidal; it can then be shown that the weight of the equivalent oscillator is 0.4 times the total distributed load on the floor. Using these values, Equation (9.15) becomes

$$a_0 = (\pi f_0 H g)/(8wLD) \tag{9.16}$$

where D is the thickness of the floor. For ribbed or waffle slabs, D should be taken as the thickness of a solid slab having the same moment of inertia as the ribbed section. The heel drop impulse H is generally taken as 70 Ns (15.7 lb.s).

Knowing the peak acceleration, frequency and damping, the acceptability of a floor can be assessed from Figure 9.3. However, because of the numerous approximations made, it should be remembered that the estimate of a_0 is prone to quite large errors. Therefore, values close to a borderline on the figure should be treated with caution.

Concrete Society method

For two-way floors, the Concrete Society (1994) gives a method of calculating a response factor R which follows on from the natural frequency equations, (9.8) to (9.14). First, two dimensionless response coefficients, N_x and C_x are calculated. For solid or waffle slabs:

$$N_x = 1 + (0.5 + 0.1 \log_e \zeta)\lambda_x \tag{9.17}$$

and for ribbed slabs:

$$N_x = 1 + (0.65 + 0.1 \log_e \zeta)\lambda_x \tag{9.18}$$

where ζ is the fraction of critical damping, as discussed in section 9.2.2.

$$\begin{aligned}
&\text{For } f_x < 3\text{ Hz:} && C_x = 244.8/(f_x^2\zeta)\\
&f_x \text{ between 3 and 4 Hz:} && C_x = 27.2/\zeta\\
&f_x \text{ between 4 and 5 Hz:} && C_x = (83.2 - 14f_x)/\zeta\\
&f_x \text{ between 5 and 20 Hz:} && C_x = 0.88(20 - f_x)/\zeta + 2(f_x - 5)\\
&f_x \text{ greater than 20 Hz:} && C_x = 30
\end{aligned} \tag{9.19}$$

Equation (9.19) is illustrated graphically in Figure 9.4 for the ranges of frequency and damping values likely to be encountered in practice. The response factor in the x-direction is then

$$R_x = 1000C_xN_x g/(wn_x n_y L_x L_y) \tag{9.20}$$

After repeating the frequency and response calculations for the slab spanning in the y-direction, the overall response factor is

$$R = R_x + R_y \tag{9.21}$$

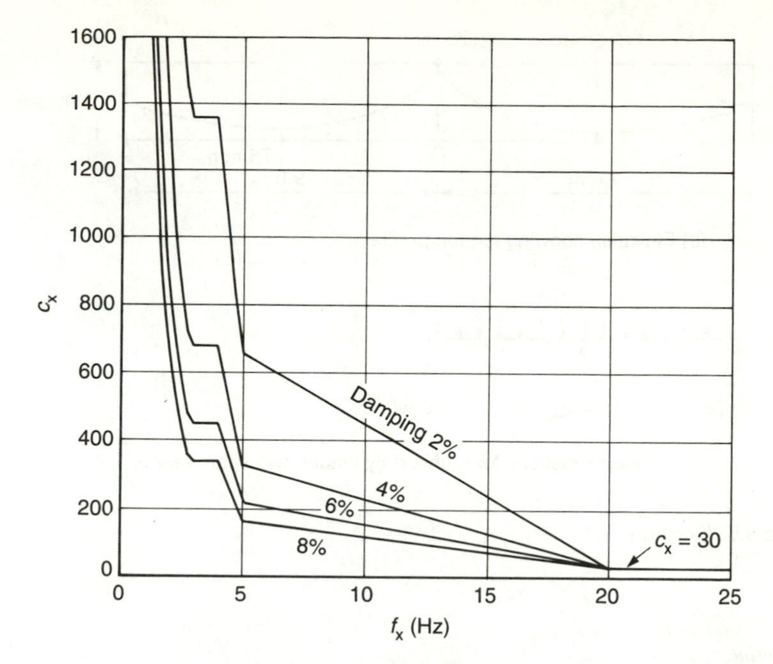

Figure 9.4 *Variation of response coefficient C_x with natural frequency (Equation 9.19)*

For the vibration behaviour to be acceptable, R must not exceed 8 in a normal office or 12 in a busy office where there are frequent visual and audible distractions. In an environment where technical tasks requiring prolonged concentration are performed, R should be limited to 4.

Comparisons with field measurements of heel drop tests indicate that the response factor is a reasonable guide to floor acceptability, even though the calculation method is based on rather conservative predictions of natural frequency (Williams and Waldron, 1994).

Example 9.1 Deflection calculation for a one-way slab.

Calculate the deflection of the two-span car park slab shown in Figure 9.5(a). The applied loads on the slab are:

Construction load $= 1.5\,\text{kN/m}^2$ (applied after 1 month)
Finish & services $= 1.5\,\text{kN/m}^2$ (applied after 3 months)
Design live load $= 2.5\,\text{kN/m}^2$ (applied after 6 months)
Sustained live load $= 0.3 \times 2.5 = 0.75\,\text{kN/m}^2$

The slab is prestressed by tendons following the profile shown in Figure 9.4(a), giving a prestressing force after immediate losses of 600 kN. Further losses after 3 months are 20%, and after several years are 30%. Cylinder strength $f_c' = 40\ \text{N/mm}^2$.

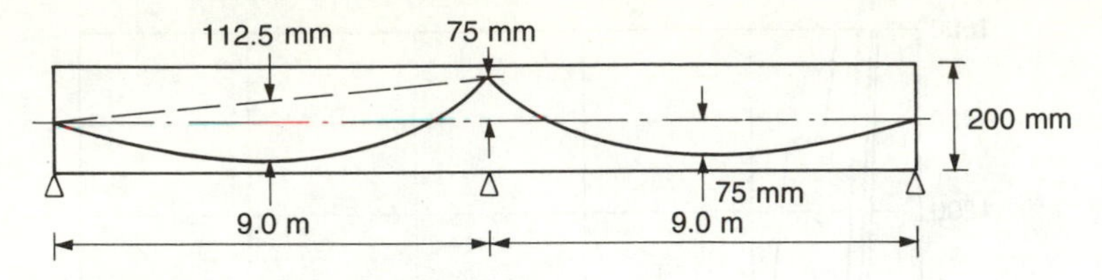

(a) Elevation showing tendon profile

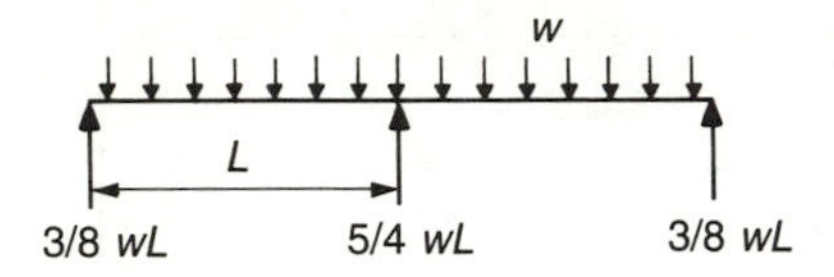

(b) Support reactions for a uniformly loaded two-span beam

Figure 9.5 *Example 9.1*

Solution

Section depth $= 200$ mm
Self-weight $= 0.2 \times 24 = 4.8$ kN/m^2 (applied after 1 month)
Moment of inertia $I_c = 1000 \times 200^3/12 = 6.667 \times 10^8$ mm^4.

Using the BSI recommendations given in Chapter 2,

$E_c = 28$ kN/mm^2, rising to 32.2 kN/mm^2 after 1 year.
$C_c = 1.8$ for loads applied after 1 month
$C_c = 1.2$ for loads applied after 6 months

Equivalent uniform load

$$w_e = 8Ps/L^2 = 8 \times 600 \times 0.1125/9.0^2 = 6.67 \text{ kN/m}$$

For a two-span beam carrying a uniformly distributed load w per unit length, the reactions are as shown in Figure 9.5(b) and the bending moment expression for the left-hand span is:

$$M = 0.375wLx - 0.5wx^2 = E_cI_c[d^2y/dx^2]$$

Integrating twice gives an expression for the deflection y at any point in the span:

$$E_cI_cy = wLx^3/16 - wx^4/24 + Ax + B$$

Substituting $y = 0$ at $x = 0$ and at $x = L$ gives

$$A = -wL^4/48, \quad \text{and} \quad B = 0$$
$$\delta = 0.00521wL^4/E_cI_c \quad \text{at midspan}$$

A slightly larger deflection occurs just away from midspan, but the difference is sufficiently small to be neglected.

Similarly, in a two-span beam with a uniform load w on one span only:

$$\delta = 0.00915wL^4/E_cI_c \quad \text{at midspan}$$

Using these results, the deflections at various stages during the life of the structure can be calculated. Use units of N and mm, and take downwards deflections as positive.

(*a*) *Initial elastic deflections*:

After 1 month, the props are removed and the slab carries its self-weight, a construction load and the prestress.

Self-weight and construction load:

$$\delta = \frac{0.00521 \times (4.8 + 1.5) \times 9000^4}{28 \times 10^3 \times 6.667 \times 10^8} = 11.5\,\text{mm}$$

Prestress equivalent load:

$\delta = 11.5 \times (-6.67/6.3) = -12.2\,\text{mm}$

Net deflection $= -0.7\,\text{mm}$

(*b*) *After 3 months (just prior to installation of services and finishes)*:

The deflections calculated above must be modified to allow for early creep and prestress loss. Assume that 50% of the long-term creep has taken place.

Self-weight and construction load	$\delta = 11.5 \times (1 + 1.8 \times 0.5)$	$= 21.9\,\text{mm}$
Prestress equivalent load	$\delta = -12.2 \times (1 + 1.8 \times 0.5) \times 0.8$	$= -18.5\,\text{mm}$
	Net deflection	$= 3.4\,\text{mm}$

(*c*) *After several years (no transient loads present)*:

Full creep and prestress loss have now occurred. Note that only the sustained part of the live load is used in the creep calculations.

Self-weight, services, finishes	$\delta = 11.5 \times (1 + 1.8)$	$= 32.3\,\text{mm}$
Prestress equivalent load	$\delta = -12.2 \times (1 + 1.8) \times 0.7$	$= -23.9\,\text{mm}$
Sustained live load	$\delta = (1 + 1.2) \times 11.5 \times (0.75/6.3)$	$= 13.1\,\text{mm}$
	Net deflection	$= 11.5\,\text{mm}$

(*d*) *Short-term deflections*:

In addition to the sustained loads considered above, the full design load may be active for short periods. The largest deflection will occur when only one of the two spans is loaded:

Additional long-term deflection

$$\delta = \frac{0.00915 \times (2.5 - 0.75) \times 9000^4}{32.2 \times 10^3 \times 6.667 \times 10^8} = 4.9\,\text{mm}$$

Comparing with the BS 8110 requirements (Table 9.1):

Maximum total deflection $= 11.5 + 4.9 = 16.4\,\text{mm} < L/250$ acceptable.

Maximum total deflection occurring after the installation of finishes
$= 16.4 - 3.4 = 13.0\,\text{mm} < L/500$ acceptable.

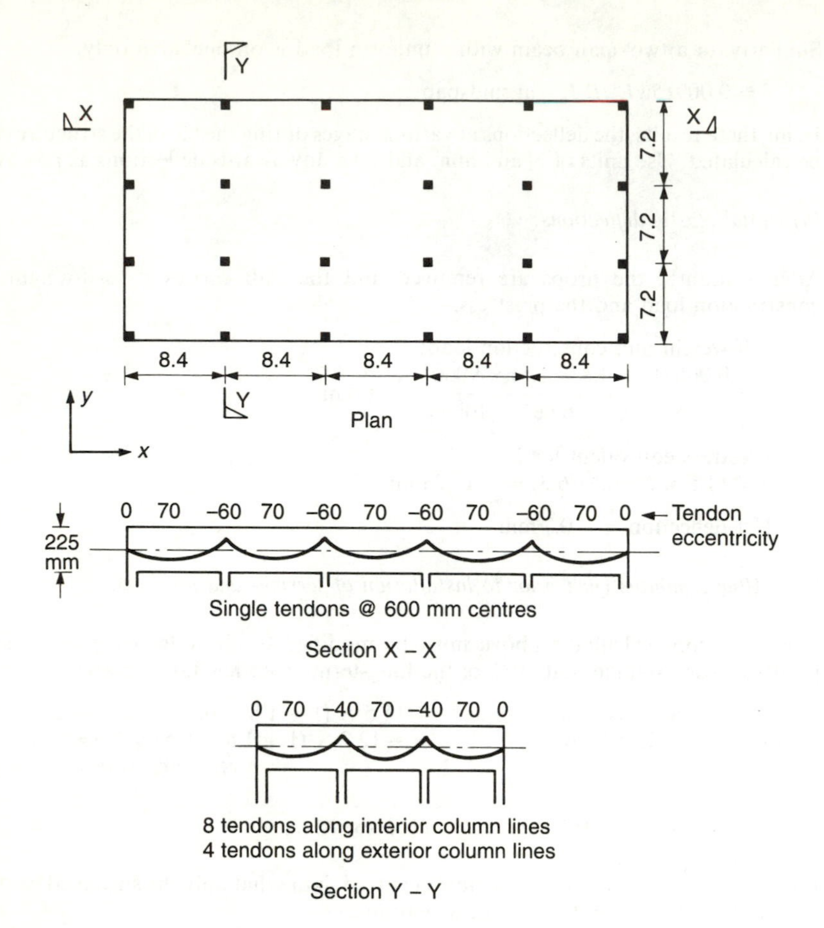

Figure 9.6 *Example 9.2: Two-way post-tensioned flat slab*

Example 9.2 Deflection calculation for a two-way slab

The various approaches are illustrated for panel B3 of the flat slab shown in Figure 9.6. Deflection is calculated for self-weight only.

Assume E_c:	28 kN/mm^2	
P_x:	400 kN/m	prestressing force in x-direction
P_y:	1600 kN	prestressing force along interior column lines in y-direction
:	800 kN	prestressing force along end column lines in y-direction.

Self-weight = 24 × 0.225 = 5.4 kN/m^2

Solution

Equivalent loads produced by the prestress:

Along X-X direction, tendons are evenly spaced, giving a uniformly distributed load:

Spans 1,5 : $w_e = 8Ps/L^2 = 8 \times 400 \times 0.10/8.4^2 = 4.54\ \text{kN/m}^2$
Spans 2,3,4: $w_e = 8 \times 400 \times 0.13/8.4^2 = 5.90\ \text{kN/m}^2$

In the Y-Y direction, the bunched cables give a line load along the column lines:

Spans A,C: $w_e = 8 \times 1600 \times 0.09/7.2^2 = 22.22\ \text{kN/m}$
Span B : $w_e = 8 \times 1600 \times 0.11/7.2^2 = 27.16\ \text{kN/m}$

(*a*) *Classical plate theory*:

A uniform load over the panel is required in order to compute the deflection using Equation (9.6). Therefore, distribute the effect of the tendons in the *y*-direction across the full width of the panel—this is crude, but better than neglecting their effect entirely.

Net distributed load $= 5.4 - 5.9 - (27.16/8.4) = -3.73\ \text{kN/m}^2$

Interpolating from Table 9.2 for $b/a = 8.4/7.2 = 1.17$, $\alpha_2 = 0.00835$

Plate bending parameter $H = 28 \times 10^3 \times 225^3/[12(1 - 0.2^2)] = 2.7685 \times 10^{10}$ Nmm

Using Equation (9.6),

$$\delta = \alpha_2 wa^4/H = \frac{0.00835 \times (-3.73) \times 10^{-3} \times 7200^4}{2.7685 \times 10^{10}} = -3.0\ \text{mm}$$

(*b*) *Crossing beam method*:

Consider a 4.2 m wide slab strip along the columns in the *y*-direction.

$I_c = 4200 \times 225^3/12 = 3.987 \times 10^9\ \text{mm}^4$.

Using a load distribution factor of 1.2 for a column strip, the net loads per unit length are:

Spans A, C: $5.4 \times 4.2 \times 1.2 - 22.22 = 5.0\ \text{kN/m}$
Span B : $5.4 \times 4.2 \times 1.2 - 27.16 \approx 0$

The vertical reactions due to these loads are shown in Figure 9.7(a). Then, setting up and integrating the moment-curvature expression gives the deflection δ_1 at the centre of span B

$\delta_1 = -0.75$ mm.

Now consider a 3.6 m wide middle strip in the *x*-direction.

$I_c = 3600 \times 225^3/12 = 3.417 \times 10^9\ \text{mm}^4$.

Using a load distribution factor of 0.8 for a middle strip, the net loads are:

Spans 2,3,4: $(5.4 - 5.9) \times 3.6 \times 0.8 = -1.44\ \text{kN/m}$.

The deflections of an interior span of a multispan beam closely resemble those of a fixed-ended beam, so at the centre of span 3:

$$\delta_2 = wL^4/384E_cI_c = -1.44 \times 8400^4/(384 \times 28 \times 10^3 \times 3.147 \times 10^9)$$
$$= -0.20\ \text{mm}$$

Therefore, the total displacement is $\delta = \delta_1 + \delta_2 = -0.95$ mm.

(*c*) *Frame and slab method*

Width of column strip in *y*-direction $= 0.4 \times 8.4 = 3.36$ m

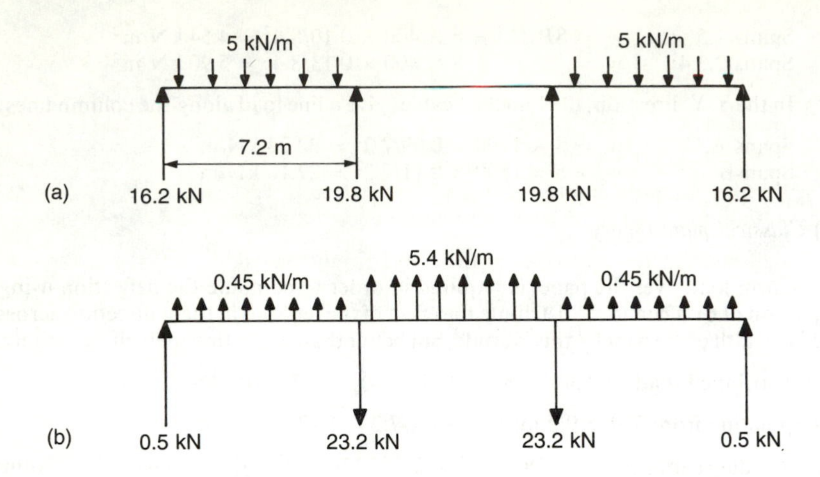

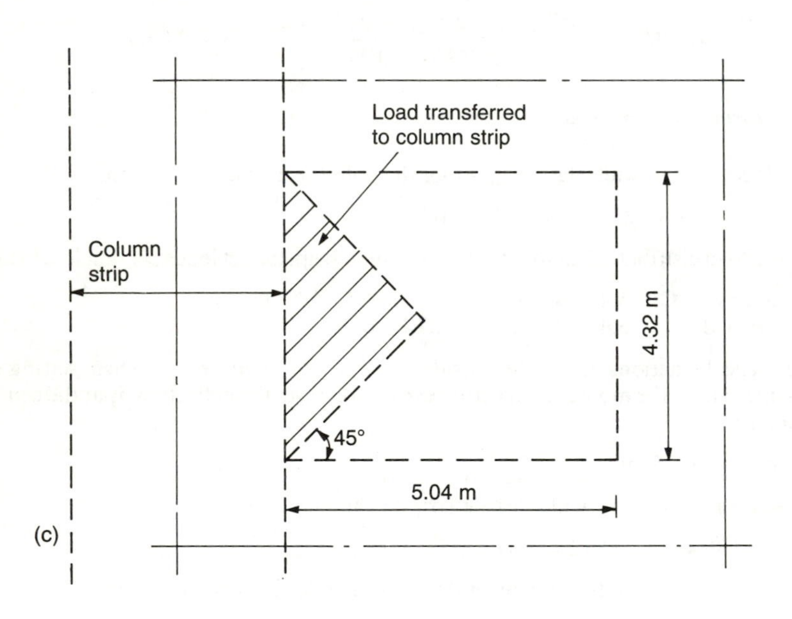

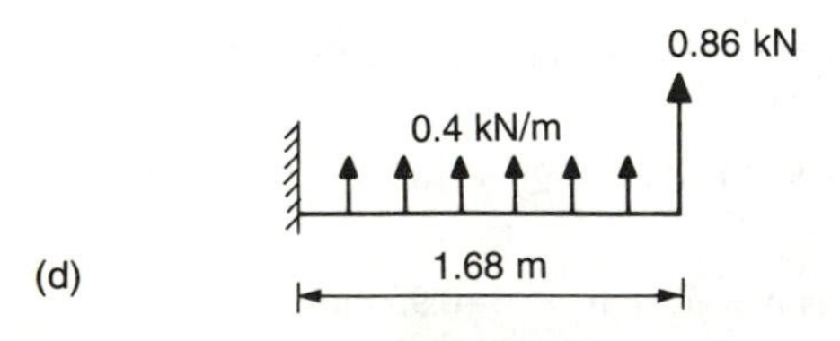

Figure 9.7 *Example 9.2 (a) Support reactions for crossing beam example (b) Support reactions for frame-and-slab example (c) Subdivision of slab load for frame-and-slab example (d) Cantilever loads for frame-and-slab example*

$I_c = 3.189 \times 10^9$ mm^4

Loads per unit length:

Spans A, C: $5.4 \times 3.36 \times 1.2 - 22.22 = -0.45$ kN/m
Span B : $5.4 \times 3.36 \times 1.2 - 27.16 = -5.4$ kN/m

The vertical reactions under these loads are shown in Figure 9.7(b). Again, integrating the resulting moment-curvature expression gives:

deflection at the centreline of the slab strip in span B, $\delta_1 = -1.02$ mm.

Now consider the interior panel between slab strips. This has the dimensions shown in Figure 9.7(c), and carries a load of:

$(5.4 - 5.9) \times 0.8 = -0.4$ kN/m^2.

Interpolating from Table 9.2, for $b/a = 5.04/4.32 = 1.17$, $\alpha_1 = 0.00540$

$$\delta_3 = \alpha_1 wa^4/D = 0.0054 \times (-0.4) \times 10^{-3} \times 4320^4/(2.7685 \times 10^{10})$$
$$= -0.03 \text{ mm}$$

The proportion of the panel load which is transferred into the adjoining slab strip is given by the shaded area in Figure 9.7(c). Assuming the load is uniformly distributed along the panel edge gives a line load of -0.86 kN/m. Now analyse a 1 m wide segment of the slab strip as a cantilever with its root at the column line, Figure 9.7(d).

Tip deflection

$$\delta_2 = wL^4/(8E_cI_c) + WL^3/(3E_cI_c)$$
$$= [(-0.4 \times 1680^4)/8 + (-860 \times 1680^3)/3]/(28 \times 10^3 \times 9.492 \times 10^8)$$
$$= -0.07 \text{ mm}$$

Total deflection

$$\delta = \delta_1 + \delta_2 + \delta_3 = -1.02 - 0.03 - 0.07 \approx -1.1 \text{ mm.}$$

Comment

Clearly the three methods used here yield quite different results. Since they all contain significant approximations, it is hard to say with confidence which is the most accurate. As stated in Section 9.1.4, the simplicity and wide applicability of the crossing beam approach make it probably the most suitable method for everyday design use. It should be noted that the central panel examined here may not be the most critical location for this structure, requiring a number of deflection checks to be performed. Lastly, it will be necessary to calculate long-term deflections using a similar approach to that illustrated in Example 9.1 for one-way spanning slabs.

Example 9.3 Vibration assessment of slabs

The two floors whose deflections have been calculated in Examples 9.1 and 9.2 will now be assessed for vibrations.

(a) One-way spanning slab

Consider the floor slab of Example 9.1. Assume that only the sustained part of the live load is present. Since the slab panel is continuous only at one end, no reduction in span length is made to account for the continuity. The natural frequency is found from Equation (9.7), and the acceleration response from Equation (9.16).

For a 1 m wide strip, $I_c = 6.667 \times 10^{-4}$ m^4

$$w = (4.8 + 1.5 + 0.75) \qquad = 7.05\ \text{kN/m}^2$$

$$f_0 = (\pi/2) \times \left[\frac{28 \times 10^9 \times 6.667 \times 10^{-4} \times 9.81}{7.05 \times 10000 \times 9^4}\right]^{0.5} = 3.1\ \text{Hz}$$

$$a_0 = \frac{\pi \times 3.1 \times 70 \times 9.81}{8 \times 7.05 \times 1000 \times 9 \times 0.2} = 0.67\%\, g$$

Assuming 3% damping, the limiting acceleration given by the CSA scale, Figure 9.3, is 1.05%g, so the calculated acceleration is acceptable.

(b) Two-way spanning slab

The flat slab of Example 9.2 is assessed using the Concrete Society approach.

For the x-direction,

$$\begin{aligned} D &= 225\ \text{mm},\ E_c I_x = E_c I_y = 2.6578 \times 10^7\ \text{Nm}^2/\text{m} \\ \lambda_x &= 5 \times 8.4 / 7.2 = 5.833 \\ k_x &= 1 + 1/5.833^2 = 1.029 \\ w &= 5.4 + 25\%\ \text{finishes} = 6.75\ \text{kN/m}^2 \end{aligned}$$

A first estimate of the natural frequency is given by Equation (9.11):

$$f'_x = \frac{1.029\pi}{2}\left[\frac{2.6578 \times 10^7 \times 9.81}{6.75 \times 1000 \times 7.2^4}\right]^{0.5} = 6.13\ \text{Hz}$$

As this is a flat slab, this value must be modified by the calculation of a second frequency, Equation (9.12), and the slab frequency is then found from Equation (9.14).

$$f_b = \frac{\frac{\pi}{2}\left[\frac{2.6578 \times 10^7 \times 9.81}{6.75 \times 1000 \times 7.2^4}\right]^{0.5}}{(1 + 7.2^4/8.4^4)^{0.5}} = 3.53\ \text{Hz}$$

$$f_x = 6.13 - (6.13 - 3.53)(\tfrac{1}{5} + \tfrac{1}{3})/2 = 5.44\ \text{Hz}$$

With 4% damping, the coefficients required for the calculation of response factor are $N_x = 2.04$ and $C_x = 321.2$. The x-direction response factor is given by Equation (9.20).

$$R_x = \frac{1000 \times 2.04 \times 321.2 \times 9.81}{6.75 \times 1000 \times 5 \times 3 \times 8.4 \times 7.2} = 1.05$$

Repeating the calculations for the y-direction gives $f_y = 4.63$ Hz and $R_y = 1.08$

The overall response factor is $R_x + R_y = 2.13$.

This is well below the most strict of the specified limits, making the floor acceptable for all kinds of use. This example shows that even floors of very low natural frequency can give acceptable vibration performance, as their high mass means that only low accelerations result from typical everyday loadings.

10 SHEAR

In this chapter, the mechanisms of shear failure are discussed, methods for calculating shear strength are presented and recommendations are given for reinforcing members to resist shear loading. The chapter first deals with shear in beams and one-way slabs, then with punching shear in two-way spanning members. Methods of analysis are given which comply with both British and American codes of practice.

Failure of floors in shear is an ultimate strength criterion, which is usually checked after the flexural design is complete. In beams, and in one-way slabs spanning between beams or walls, excessive shear stresses result in the formation of diagonal tension cracks. In flat slabs, punching shear failure around a column or under a very large concentrated load is the principal concern. In practice, punching shear is the more important criterion; one-way slab shear is rarely critical in design. For instance, beam strips, although in other respects designed as one-way spanning, should be checked for punching shear. Nevertheless, since an understanding of one-way shear is an essential prerequisite to the assessment of punching shear, the former topic is dealt with in some detail here.

A major difference between prestressed and reinforced concrete is that the vertical component of the prestressing force will, in nearly all cases, oppose the shear due to the applied loads, thus reducing the shear force which the concrete section is required to withstand. For example, Figure 10.1 shows the forces acting on a section of a simply supported beam under uniform loading. The resultant shear on the right-hand face is

$$V = -(wL/2) + wx + P \sin \alpha$$

The magnitude of this force is smaller than the value in the absence of the prestress so long as the tendon is sloping downwards (i.e. α is between 0° and 90°). In a floor, however, both the average prestress and the inclination of the tendons are quite low, so that the contribution of the prestress to the shear strength is small. Additionally, tests have shown that this beneficial effect is normally only effective at locations where the concrete section is uncracked. At cracked sections, therefore, the effect of inclined tendons is assumed to occur only if it increases the effective shear force on a section.

The normal procedure for shear design is to compare the capacity of the concrete section, including rod reinforcement and prestressing tendons, with the maximum applied shear force, inclusive of ultimate load factors. Values for load factors recommended by BS 8110 and ACI 318 are given in Chapter 8. If the

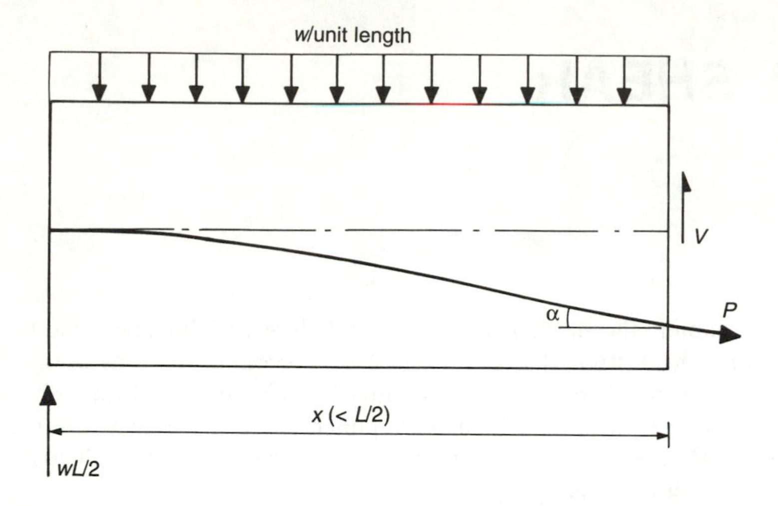

Figure 10.1 *Effect of inclined tendon on effective shear force*

capacity is adequate then no further action is required. If the applied shear force is excessive then some additional capacity must be provided, either by inclusion of shear reinforcement or by increasing the cross-section. Shear reinforcement may take the form of conventional stirrups, proprietary reinforcing cages (*shearhoops*) or structural steel shearheads. These alternatives are discussed further in Section 10.4.

For ribbed or waffle slabs, increases in cross-section are achieved by converting to a solid section close to the supports. For solid slabs, drop panels or column heads can be provided, but these require special formwork and so can cause significant increases in construction time and cost. This solution is, therefore, normally used only when it is not possible to provide sufficient shear strength using reinforcement.

10.1 Shear strength of concrete

The behaviour of concrete in shear is complex and remains poorly understood. Plain concrete develops shear strength primarily by the mechanism of aggregate interlock and friction between the constituent particles. Longitudinal reinforcement increases the shear strength by dowel action, and by acting as a tie across shear cracks, preventing them from opening. Additionally, the shear behaviour of concrete elements is closely related to the tensile strength of concrete, since a vertical shear force usually causes failure by the development of diagonal tension cracks.

Codes of practice give guidance on the appropriate material strengths to be used in shear calculations. For uncracked sections, BS 8110 relates the shear

Table 10.1 *Values of concrete shear strength, N/mm² (BS 8110)*

$\frac{100A_s}{b_v d}$	Effective depth (mm)							
	125	150	175	200	225	250	300	≥ 400
≤ 0.15	0.53	0.50	0.48	0.47	0.46	0.44	0.42	0.40
0.25	0.62	0.60	0.57	0.55	0.54	0.53	0.50	0.47
0.50	0.78	0.75	0.73	0.70	0.68	0.65	0.63	0.58
0.75	0.90	0.85	0.83	0.80	0.77	0.76	0.73	0.67
1.00	0.98	0.95	0.91	0.88	0.85	0.83	0.80	0.74
1.50	1.13	1.08	1.04	1.01	0.97	0.95	0.91	0.84
2.00	1.24	1.19	1.15	1.11	1.08	1.04	1.01	0.94
≥ 3.00	1.43	1.36	1.31	1.26	1.23	1.19	1.15	1.06

Notes:
*In calculating the reinforcement ratio, BS 8110 recommends that the areas of rod reinforcement and prestressing steel should simply be summed. However, this fails to take adequate account of the greater strength of the tendons. Some engineers, therefore, multiply the area of prestressing tendons by the ratio f_{pu}/f_y prior to combining it with the rod reinforcement.

*The values shown are for a concrete strength $f_{cu} = 40$ N/mm² or greater. For lower strengths the values in the table should be multiplied by $(f_{cu}/40)^{1/3}$.

*For members prestressed by unbonded tendons, the value obtained should be multiplied by 0.9.

strength to the tensile strength of the concrete as outlined in Section 10.2.2. The tensile strength is, in turn, related to the cube strength by

$$f_t = 0.24 f_{cu}^{0.5} \text{ N/mm}^2 \text{ units or}$$
$$f_t = 2.89 f_{cu}^{0.5} \text{ psi units.}$$

For sections cracked in flexure, as for reinforced concrete, the shear strength is found using an empirical formulation in terms of the section depth, the reinforcement ratio and the cube root of the compressive strength. Tabulated values of shear strength for concrete having compressive strength 40 N/mm² (5800 psi) are given in Table 10.1.

ACI 318 defines the shear strength as a linear function of the square root of the compressive strength f_c', the value of the coefficient, β_p, varying with the nature of the shear loading. Guidance on the choice is given in the relevant sections later in this chapter.

When lightweight concrete is used, the shear strength assumed in the calculations should be modified appropriately. BS 8110 recommends that the design shear stress v_c should be taken as 0.8 times the value obtained from Table 10.1, with the exception of grade 20 lightweight concrete, for which the values should be taken from Table 10.2. No recommendation is given for the tensile strength, but it is reasonable to assume the same reduction factor as for shear strength.

ACI 318 recommends that all values of $\sqrt{f_c'}$ affecting the calculation of shear capacity or cracking moment should be multiplied by 0.75 for concrete in which

Table 10.2 *Values of v_c for Grade 20 lightweight concrete*

$100A_s/b_vd$	*Values of v_c*	
	N/mm²	*psi*
0.15	0.25	36
0.25	0.30	44
0.50	0.37	54
0.75	0.43	62
1.00	0.47	68
1.50	0.53	77
2.00	0.59	86
≥ 3.00	0.68	99

all the aggregates are lightweight, and by 0.85 for sand-lightweight concrete.

Lastly, the shear strength of a prestressed section is to an extent influenced by whether the tendons used are bonded or unbonded. Bonded tendons act in much the same way as normal reinforcement, providing dowel action and load transfer across cracks, but unbonded tendons are less effective in this respect. The Concrete Society (1994) recommends that the shear strengths found from Table 10.1 should be reduced by a factor of 0.9 for members prestressed by unbonded tendons.

10.2 Beams and one-way slabs

The nature of the shear failure in a beam or one-way slab is dependent on the interaction between the bending moment and the shear force in the member. If the bending stress at a given location is below the modulus of rupture, then the concrete section remains uncracked. The stresses on a small element near the neutral axis are as shown in Figure 10.2(a). From a simple Mohr's circle analysis, Figure 10.2(b), the principal stresses are found to be as shown in Figure 10.2(c), so that failure occurs by the development of diagonal tension cracks. If, on the other hand, the bending stress is sufficient to cause some flexural cracking, the shear stresses will cause these cracks to grow and to become increasingly inclined towards the neutral axis of the section. The shear behaviour in this case is similar to that in a non-prestressed member.

For simply supported spans under predominantly uniform loads, the peak bending moment occurs at midspan and the peak shear force at the supports, making the development of combined flexure/shear cracks unlikely. For continuous spans, however, the situation is rather more complex, since both the bending moment and the shear force are maximum at the supports. Figure 10.3 illustrates the most likely failure modes in various parts of an interior span of a continuous beam. Usually, both modes of failure must be considered, with the shear strength taken as the lower of the two values thus calculated.

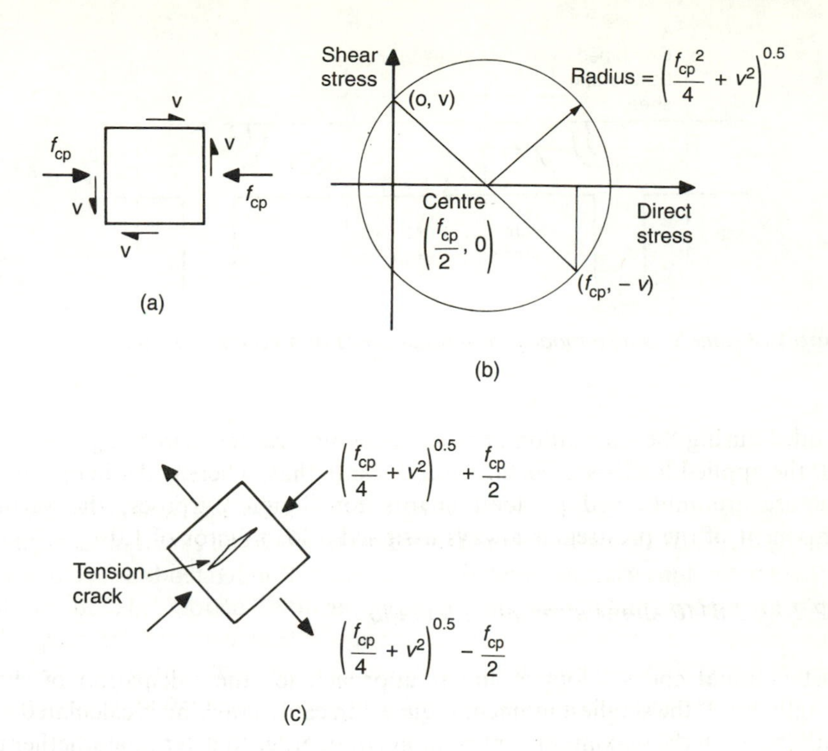

Figure 10.2 *Mohr's circle analysis for a concrete element near the neutral axis*

Very often, the critical location for shear in continuous beams is approximately one-fifth of the way along the span, where the moment is very low, or even negative, so that combined flexure/shear cracking dominates.

10.2.1 Calculation of applied shear force

The shear forces to which a one-way spanning member is subjected can be calculated by any simple elastic method. For simply supported spans the calculation is trivial. Continuous spans are normally analysed in unit widths, with beams and walls treated as simple supports. The maximum shear is likely to occur when the full live load is present on adjacent spans. Since excessive shear can cause structural collapse, the loads must be multiplied by appropriate safety factors, as discussed in Chapter 8.

For uncracked sections, the usual practice in the UK is to modify the shear force thus calculated by inclusion of the vertical component of the tendon force. In nearly all cases this will lead to a reduction in the shear which the section must carry, though situations may occasionally arise in which the shear force is increased. For cracked sections, the vertical component of the prestress is included only if it increases the applied shear force. The approach adopted in the USA is slightly different, in that the vertical component of the tendon force is

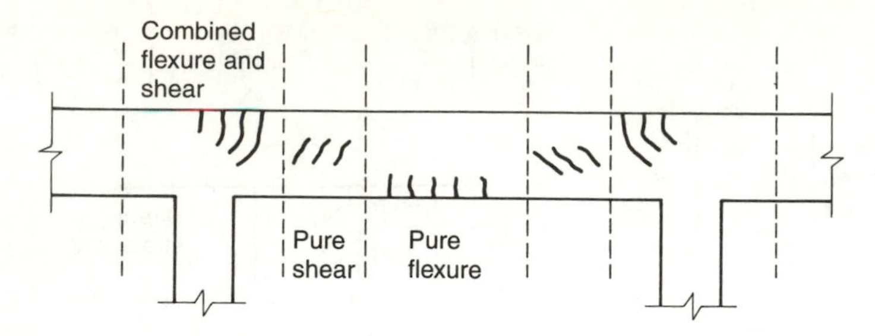

Figure 10.3 *Likely failure modes in various parts of a continuous span*

included during the calculation of shear resistance, rather than being combined with the applied load—see Section 10.2.3. Note that, whereas the loads on the structure are multiplied by load factors for design purposes, the vertical component of the prestress is always assigned a load factor of 1.0.

10.2.2 BS 8110 shear strength calculation

Most national codes adopt a similar approach for the calculation of shear strength. First, the smallest moment required to cause cracking is calculated and compared with the maximum applied moment in order to determine whether the section under consideration is likely to be cracked in flexure. If the section is uncracked, then the shear strength can be calculated using elastic theory. For combined flexure/shear cracking no theoretical formulation is available, so an empirical formula is used. The shear strength of a cracked section is then taken as the smaller of the combined flexure/shear and the uncracked values.

BS 8110 stipulates upper limits on the average shear stress over the cross-section, regardless of any reinforcement present. For normal weight concrete the shear stress may not exceed the smaller of 5 N/mm^2 (725 psi) and $0.8\sqrt{f_{cu}}$ (units of N and mm) or $9.6\sqrt{f_{cu}}$ (units of psi). For lightweight concrete the corresponding values are 4 N/mm^2 (580 psi) and $0.63\sqrt{f_{cu}}$ (units of N and mm) or $7.6\sqrt{f_{cu}}$ (units of psi). It is worth checking this requirement before proceeding with a more detailed strength check, as, if the average shear stress is excessive, there is no option but to increase the cross-section.

In BS 8110 the moment M_0 required to produce zero stress in the extreme fibre of the section is calculated as:

$$M_o = 0.8(PZ_{ten}/A_c + M_p) \tag{10.1}$$

where P = prestressing force, including an allowance for losses
Z_{ten} = elastic modulus of fibre which would normally be in tension under applied loads
M_p = total moment due to prestress (i.e. sum of primary and secondary moments)

Under normal circumstances, that is, with the tendon eccentricity towards the tension face, both of the terms within the brackets will be positive. The multiplier of 0.8 is a safety factor.

If the applied moment M is less than M_0 then it can safely be assumed that no flexural cracking has occurred. The applied shear force V may then be adjusted to take account of the vertical component of the prestress. As discussed above, this will usually lead to a reduction in the effective shear force that the section must carry.

The shear strength of an uncracked section, V_{c0}, can be found from the elastic analysis already introduced in Figure 10.2. The maximum tensile stress in the section is shown in Figure 10.2(c), acting at right angles to the tension crack. Equating this stress to the tensile strength of the concrete, f_t, and again introducing a safety factor of 0.8 on the prestress, the maximum allowable shear stress on the section can be expressed as

$$v_{all} = (f_t^2 + 0.82p_{av}f_t)^{0.5}$$

Now for a rectangular section it can be shown using classical elastic theory that the average shear stress is 2/3 of this peak value. Hence the shear force that the section can carry can be found by taking the average shear stress and multiplying by the shear area to give

$$V_{co} = 0.67b_vD(f_t^2 + 0.8p_{av}f_t)^{0.5} \qquad (10.2)$$

Calculation of the uncracked shear strength can be simplified by using the values of the average stress V_{co}/b_vD given in Table 10.1. These have been calculated using Equation (10.2). The concrete strengths and prestress levels given cover the ranges likely to be encountered in post-tensioned floors. Values outside these ranges can be calculated directly from Equation (10.2).

For the case of combined flexural/shear cracking, an empirical formulation is used:

$$V_{cr} = [1 - 0.55f_{pe}/f_{pu}]v_cb_vd + M_oV/M \qquad (10.3)$$

where d = depth to centroid of all steel, tendons and rod reinforcement
v_c = concrete strength

In this equation, the applied shear force V and bending moment M should be taken as positive, as should the cracking moment M_0. In post-tensioned floors the ratio f_{pe}/f_{pu} is normally between 60 and 70%, so that Equation (10.3) can be simplified to

$$V_{cr} = 0.65v_cb_vd + M_oV/M \qquad (10.4)$$
$$\leq 0.1b_vdf_{cu}^{0.5} \quad \text{N-mm units}$$
$$\text{or } 1.2b_vdf_{cu}^{0.5} \quad \text{lb-in units}$$

The shear capacity of the section, V_c, is taken as V_{c0} for an uncracked section, and as the lower of V_{c0} and V_{cr} for a cracked section.

The concrete strength v_c, given in Table 10.3, is related to the rod reinforcement ratio and the section depth. In calculating the reinforcement ratio, both rod

Table 10.3 *Uncracked shear strength ($V_{co}/b_v d$), N/mm², calculated using Equation (10.2)*

Average prestress p_{av} *(N/mm²)*	*Concrete strength* f_{cu} *(N/mm²)*					
	25	*30*	*35*	*40*	*50*	*60*
1.0	1.04	1.12	1.19	1.26	1.38	1.49
2.0	1.23	1.31	1.39	1.46	1.58	1.70
3.0	1.39	1.48	1.56	1.63	1.77	1.89
4.0	1.54	1.63	1.72	1.79	1.93	2.05
5.0	1.67	1.77	1.86	1.94	2.08	2.21
6.0	1.80	1.90	1.99	2.07	2.22	2.36
8.0	2.02	2.13	2.23	2.32	2.48	2.63
10.0	2.23	2.34	2.45	2.55	2.72	2.87

reinforcement and prestressing steel should be included, with the area of prestressing steel multiplied by the factor f_{pu}/f_y, to account for its greater strength.

10.2.3 ACI 318 shear strength calculation

The approach given in ACI 318 is based on similar principles to the BS 8110 method, but the exact formulae differ slightly. One significant difference is that material safety factors are not implicitly included in the ACI formulae. Instead, a factor of 0.85 is introduced after the calculation of shear strength—see Section 10.2.4. The shear strength in the absence of flexural cracking is defined as

$$V_{co} = [\beta_p (f_c')^{0.5} + 0.3 p_{av}] b_v d + V_p \qquad (10.5)$$

where, for a one-way spanning member,

$\beta_p = 0.3$ N-mm units
or 3.5 lb-in units.

The term in square brackets in Equation (10.5) is an approximation to the square root expression in Equation (10.2), with the term $\sqrt{f_c'}$ representing the tensile strength. Note that the vertical component of the prestress, V_p, is included explicitly in the equation for the shear strength, rather than being treated as part of the applied shear force.

The expression for combined flexural/shear cracking is

$$V_{cr} = 0.05 (f_c')^{0.5} b_v d + M_{ct} V/M \text{ in N-mm units} \qquad (10.6a)$$

$$\text{or } = 0.60 (f_c')^{0.5} b_v d + M_{ct} V/M \text{ in lb-in units} \qquad (10.6b)$$

where M_{ct} = live load moment required to cause cracking

$$= Z_t [0.5 (f_c')^{0.5} + f_{ct}] \text{ N-mm units} \qquad (10.7a)$$

$$\text{or } Z_t [6.0 (f_c')^{0.5} + f_{ct}] \text{ lb-in units} \qquad (10.7b)$$

f_{ct} = stress at extreme tension fibre due to dead loads.

Note that the inclusion of the square root terms in Equations (10.5) to (10.7) means that they are not dimensionally consistent, and so require the use of appropriate units. Care must, therefore, be taken to use the appropriate form of each equation, and to use consistent units. The shear strength of the section should be taken as the smaller of the two values calculated using Equations (10.5) and (10.6).

10.2.4 Shear reinforcement

If the applied shear force exceeds the calculated shear strength, then some remedial action must be taken. In beams, the most usual method is to provide reinforcement in the form of closed vertical shear links. It is possible to provide shear links in one-way slabs, but in practice this is not normally done. If the shear is excessive then usually the slab depth is increased and the calculation repeated.

The shear reinforcement in a beam must be sufficient to carry the diagonal tension induced by the shear loading, and so prevent cracking. This leads to a simple formula relating the number and size of links to their strength and to the shear forces acting. The BS 8110 version of the formula is:

$$A_{sv}/s_v \geq (V - V_c)/(0.87 f_{yv} d) \qquad (10.8)$$

where A_{sv} = total cross-sectional area of the two vertical legs of a link
s_v = longitudinal spacing of links
f_{yv} = yield strength of shear steel
= 460 N/mm (66.7 ksi) for high tensile steel
= 250 N/mm (36.3 ksi) for mild steel

The multiplier of 0.87 is a material partial safety factor for the shear reinforcement.

In calculating shear reinforcement using this approach, attention must be paid to the minimum shear reinforcement requirements given in BS 8110. For prestressed beams, no shear reinforcement is required if the applied shear force V is less than $0.5V_c$. If V exceeds $0.5V_c$ but is less than $V_c + 0.4b_v d$, then a nominal amount of shear reinforcement is chosen by setting the term $(V - V_c)$ in Equation (10.8) equal to $0.4b_v d$; this is done as using the actual value of $(V - V_c)$ would result in an unacceptably large link spacing. For larger values of V, Equation (10.8) is applied as it stands. The spacing of links along the length of a member should not exceed $0.75d$ under normal circumstances, and should not exceed $0.5d$ when V is greater than $1.8V_c$. There are no minimum shear steel requirements for slabs.

Because of its slightly different approach to the inclusion of safety factors the ACI 318 formula is:

$$A_{sv}/s_v \geq (V - 0.85V_c)/(0.85 f_{yv} d) \qquad (10.9)$$

Again, there is no minimum shear steel requirement for slabs or for beams in which V is less than $0.5V_c$. Otherwise, the minimum shear link area to spacing ratio may be taken as the smaller of

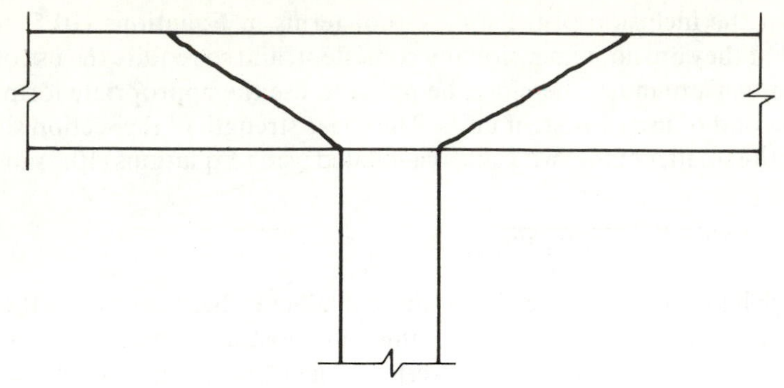

Figure 10.4 *Punching cone formation at a column connection*

$$A_{sv}/s_v \geq b_v/(3f_{yv}) \tag{10.10a}$$

or

$$A_{sv}/s_v \geq [(A_p f_{pu})/(80 f_{yv} d)](d/b_v)^{0.5} \tag{10.10b}$$

10.3 Two-way slabs

In two-way spanning floors the critical design case is punching shear around the columns or under very large concentrated loads. The exact form of the failure mechanism varies with the reinforcement details, but in general it consists of cracking through the slab, usually at an angle considerably flatter than 45°, so that an inverted, truncated cone of material is dislodged, as shown in Figure 10.4.

In design, rather than trying to assess the cone angle, it is usual to calculate the shear carried by a vertical failure zone whose edges are a fixed distance from the face of the column. The zone thus defined is known as the *critical perimeter*. Since the prestressing arrangements in the two perpendicular span directions are rarely the same, the shear capacities of each pair of parallel edges are likely to be different and so must be calculated separately.

10.3.1 Applied punching shear force

Obviously, the maximum punching shear force in a slab occurs on a perimeter at the face of the column, and equals the total vertical load being transferred from the floor into the column. For a critical perimeter at some distance from the face of the column, this maximum shear force is reduced by subtracting the vertical load acting on the area of slab inside the perimeter.

For uncracked sections, the punching shear can be further reduced by the vertical component of the prestressing force, V_p, within the critical perimeter. For example, Figure 10.5 shows a typical prestressing tendon passing over an interior column and the resulting equivalent vertical loads. Note that the entire reverse curvature of the tendon lies within the critical perimeter. The contribution of this

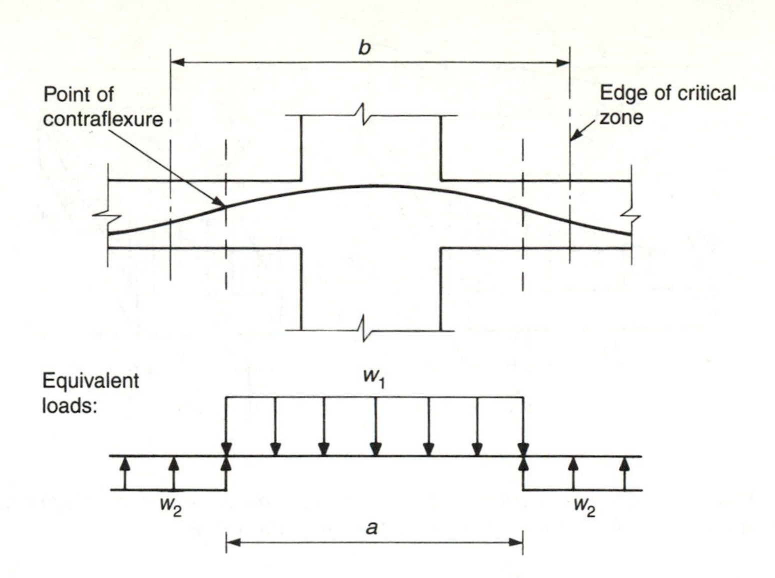

Figure 10.5 *Tendon over a support*

tendon is $w_1a - w_2(b - a)$ and the total value of V_p is found by summing the contributions of all tendons (in both span directions) which pass within the critical perimeter. It should be remembered that placing tendons so as to achieve a high reverse curvature over a column is difficult, and that small variations in the cable profile can cause quite large changes in V_p. It is, therefore, prudent to evaluate this term conservatively, and it is normal to take V_p as zero for tendons whose points inflexion lie outside the critical zone.

As with one-way shear, appropriate load factors as stipulated in the national codes should be applied to the external loads causing the punching shear, while the reduction term due to the vertical component of the prestress should always be assigned a load factor of 1.0.

10.3.2 Moment transfer

If the total shear force acting on a perimeter around a column is V and the length of the perimeter is u, then the average shear stress v acting on the assumed failure plane is simply given by V/ud. However, in many floors the slab-column connection is required to transmit a moment M_t from the floor slab into the column. The value of this moment can be determined from the structural analysis carried out for the flexure design. The transfer of moment across the critical perimeter occurs by a combination of flexure and eccentricity of shear, resulting in a non-uniform shear distribution around the perimeter, with a maximum shear stress which may be considerably larger than the average value. Figure 10.6 shows the commonly assumed forms of shear stress distributions around both internal and edge column connections where moment transfer occurs. The various national codes of practice give approximate formulae, based on experimental

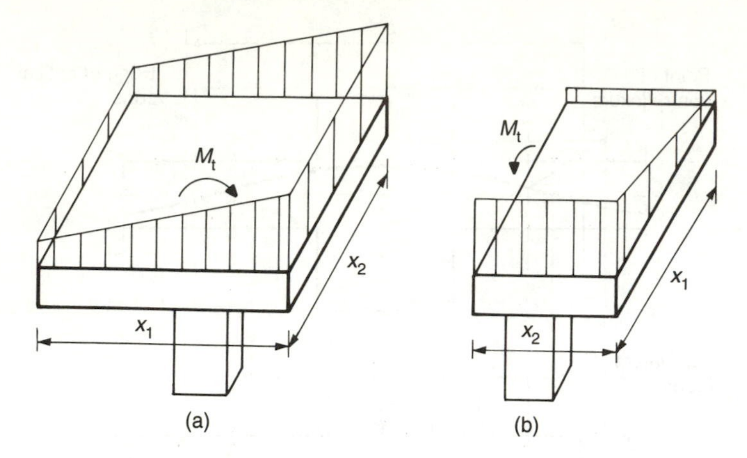

Figure 10.6 *Effect of moment transfer on shear stress distributions (a) Internal column (b) Edge column, with moment transfer parallel to the edge*

testing, from which the effect of moment transfer on the shear stress distribution can be calculated. The details of these approaches are discussed further in Sections 10.3.3 and 10.3.4.

10.3.3 BS 8110 strength calculation

Punching shear calculations are dealt with in slightly different ways in Britain and the USA. The usual British practice is to sum the shear capacities of the four edges of the critical perimeter and compare the value thus obtained with the total shear load on the perimeter. The shear strength in a given direction is calculated by using the formulae for one-way spanning members given in section 10.2.

The area around a column is divided into a series of critical perimeters, the first $1.5d$ from the face of the column, with subsequent perimeters spaced at intervals of $0.75d$, as shown in Figure 10.7. The shear failure associated with a given perimeter is assumed to occur within an area extending $1.5d$ inwards from the perimeter towards the column. For example, the notional failure zone associated with the first perimeter is shown shaded in Figure 10.7. If free edges or openings in the slab exist in the vicinity of the column, then these will significantly alter the shear distribution around the perimeter, since the shear stress at a free edge must be zero. This is taken into account by reducing the length of the critical perimeter as shown in Figure 10.8.

The shear strength is first assessed on the innermost perimeter. If the strength is adequate, then no further action is required. If the shear stresses are excessive, then shear reinforcement or an increase in slab thickness must be provided. The process is repeated for successive perimeters, until a perimeter is reached where the strength is adequate.

Rather than assessing in detail the variations in shear stress around the critical perimeter caused by moment transfer, BS 8110 simply recommends the use of an

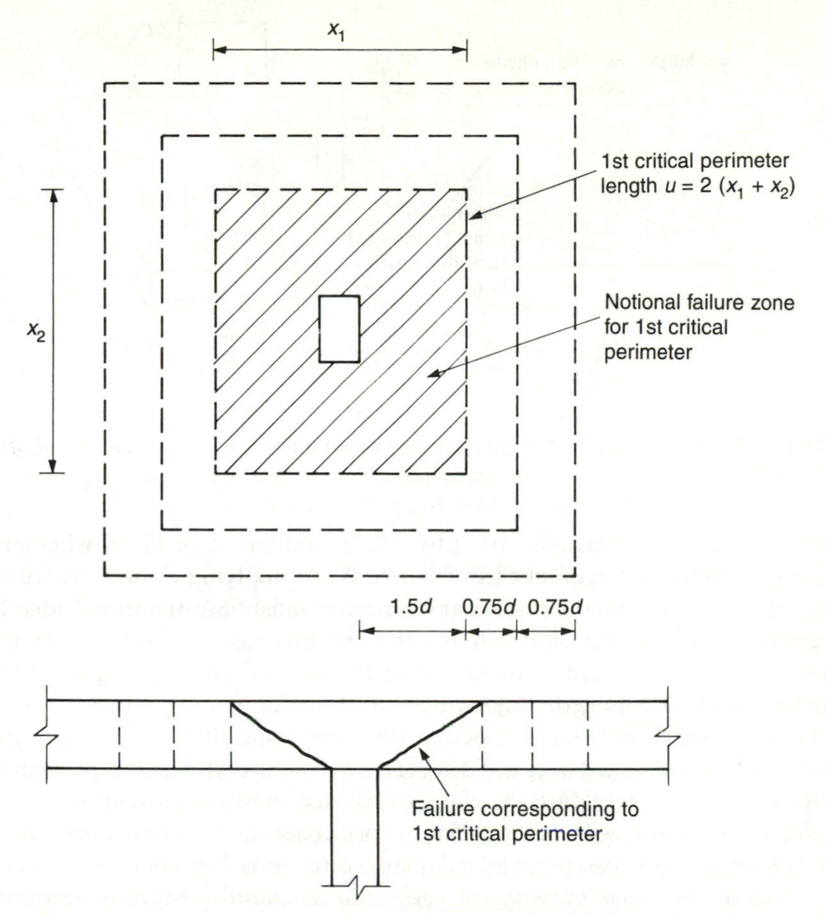

Figure 10.7 *BS 8110 definition of critical perimeters*

effective shear force, V_{eff}. If prior analysis has yielded the value of the moment M_t transferred to the column, then V_{eff} may be calculated as, for internal columns:

$$V_{\text{eff}} = V[1 + 1.5M_t/(Vx_2)] \tag{10.11}$$

or, for edge columns where bending is about an axis perpendicular to the free edge

$$V_{\text{eff}} = V[1.25 + 1.5M_t/(Vx_2)] \tag{10.12}$$

where x_2 = length of the edge of the critical section parallel to the moment axis – see Figure 10.6.

In the absence of calculated values of M_t, V_{eff} may be taken as $1.15V$ for interior columns, $1.4V$ for edge columns where moment transfer is about an axis perpendicular to the free edge, and $1.25V$ for edge columns where moment transfer is about an axis parallel to the free edge.

The value of V_{eff} thus calculated must be compared with the design shear

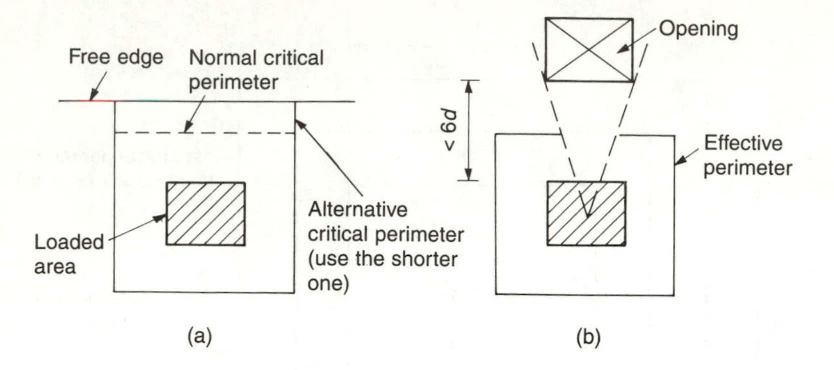

Figure 10.8 *Effect of (a) free edges and (b) openings on assumed length of critical perimeter*

strength given by Equations (10.1) to (10.4), and will depend on whether the section is cracked or uncracked in flexure. When applying Equations (10.3) or (10.4), the value of the concrete shear strength v_c must be found from Table 10.1, as a function of the section depth and the reinforcement ratio A_s/b_vd. In many slabs, tendons are placed in bands along the column lines, giving a very high reinforcement ratio immediately adjacent to the column, but a much lower value in the nearby slab. The Concrete Society (1994) recommends that the reinforcement ratio should be averaged over a width of twice the side of the critical perimeter, in order to avoid exaggerating the effect of banded prestressing tendons.

Nearly all slabs will have different prestressing and rod reinforcement configurations in the two perpendicular span directions. It is, therefore, necessary to calculate the shear capacity of each pair of parallel edges of the critical perimeter separately. It is then sufficiently accurate to sum the shear resistances around the perimeter and compare the value thus obtained with the total applied shear force, V_{eff}.

If the total shear resistance around a perimeter exceeds V_{eff} then the slab is adequate. If, on the other hand, it is less than V_{eff} some strengthening is required. For slabs greater than 200 mm (8 in) deep, strength is normally provided by the inclusion of links. Links in thinner slabs are extremely difficult to fix and are likely to be too short to allow the development of adequate anchorage. Links must be provided around at least two perimeters within the notional failure zone corresponding to the critical perimeter being checked. The innermost perimeter of reinforcement should be approximately $0.5d$ from the face of the column, with subsequent perimeters not more than $0.75d$ apart. The link spacing along a perimeter should not exceed $1.5d$. The total amount of shear reinforcement provided must compensate for the difference between the applied shear force and the concrete shear strength, i.e.:

$$\Sigma A_{sv} \geqslant (V_{eff} - V_c)/(0.87 f_{yv}) \tag{10.13}$$

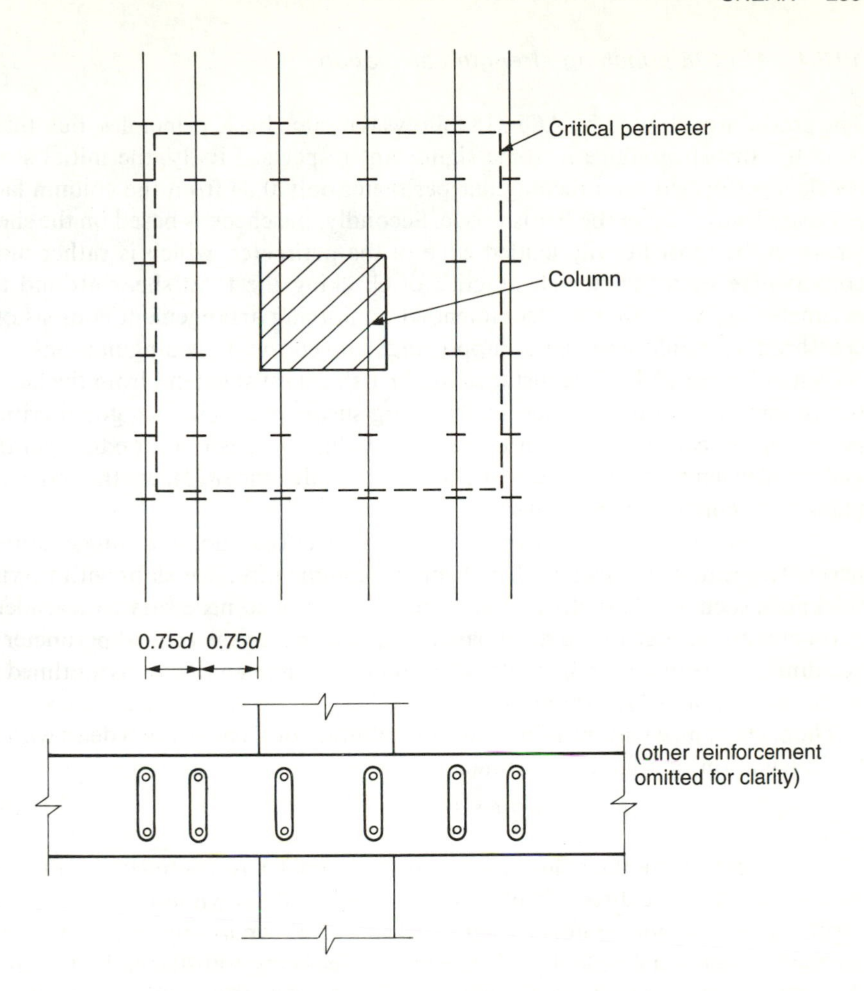

Figure 10.9 *Typical arrangement of links and lacing bars to resist punching shear in accordance with BS 8110*

Obviously the failure zones corresponding to successive critical perimeters overlap. In assessing the reinforcement requirements at a particular perimeter, BS 8110 allows account to be taken of any shear reinforcement lying within the relevant zone, even if it was provided to reinforce other zones. For example, Figure 10.9 shows a suitable arrangement of shear links corresponding to the innermost critical perimeter. As well as reinforcing this zone, the outer set of links also lies within the zone corresponding to the next critical perimeter. Therefore, if it is subsequently found that shear is excessive on this second perimeter, then links would be required around only one additional perimeter within the second notional failure zone.

10.3.4 ACI 318 punching strength calculation

The procedure adopted by ACI 318 follows the same basic principles, but differs from the British practice in some significant respects. Firstly, the initial shear check is performed on a rectangular perimeter only $0.5d$ from the column face, compared with $1.5d$ in the British code. Secondly, the check is based on the shear stress on the most heavily loaded edge of the perimeter, which is rather more conservative than the British practice of checking the total shear around the perimeter. In ACI 318, the recommended reinforcing arrangement consists of a crosshead of reinforcing bars supporting conventional rectangular links, as shown in Figure 10.10. The distance this crosshead must extend from the face of the column is then determined by checking shear on a new, octagonal critical perimeter. As can be seen from Figure 10.10, this consists of four edges parallel and equal in length to the sides of the column, at a distance $d/2$ from the end of the main bars, connected by straight lines.

For slabs of uniform thickness, shear is checked along a single critical perimeter located at a distance $0.5d$ from the column face. For slabs with varying thickness, such as where drop panels are used, it is also necessary to consider a perimeter at the edge of the drop panel. Adjustments to the critical perimeter to account for openings or free edges are made in the same way as outlined in Section 10.3.3 and illustrated in Figure 10.8.

The effect of moment transfer on the distribution of shear stress is dealt with by the calculation of a factor γ_v defined as:

$$\gamma_v = 1 - 1/[1 + 0.67(x_1/x_2)^{0.5}] \tag{10.14}$$

where x_1 and x_2 are the sides of the critical perimeter respectively parallel and perpendicular to the direction of moment transfer, as shown in Figure 10.6. It is assumed that the non-uniform shear distribution shown in Figure 10.6 accounts for the transfer of a moment $\gamma_v M_t$, the remainder being transferred by flexure of the slab-column connection. Referring to Figure 10.6, the resulting expressions for the shear stresses on edges AB and CD of the critical perimeter are:

$$v_{AB} = (V/ud) + (\gamma_v M_t x_1/2J_c) \tag{10.15a}$$

$$v_{CD} = (V/ud) - (\gamma_v M_t x_1/2J_c) \tag{10.15b}$$

where J_c is a geometric property of the critical section, defined for the internal column shown in Figure 10.6(a) as

$$J_c = dx_1{}^3/6 + x_1 d^3/6 + dx_2 x_1{}^2/2 \tag{10.16a}$$

and for the edge column, Figure 10.6(b) as

$$J_c = dx_1{}^3/12 + x_1 d^3/12 + dx_2 x_1{}^2/2 \tag{10.16b}$$

Note that it is necessary to calculate both v_{AB} and v_{CD}, as it is the shear stress of greatest magnitude that is critical. If M_t is very large, then it is possible for v_{CD} to be more critical than v_{AB}.

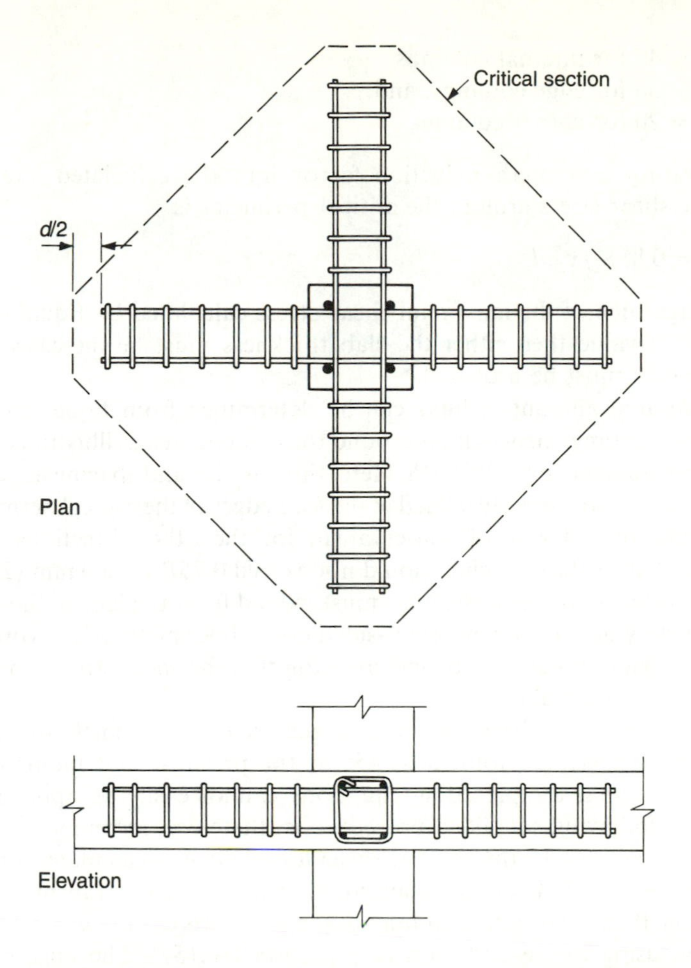

Figure 10.10 *Typical arrangement of links and lacing bars to resist punching shear in accordance with AC 318*

The maximum shear stress thus calculated must be compared to the shear strength of the slab. ACI 318 allows the punching shear capacity to be calculated on the assumption that the section is uncracked so long as certain requirements for minimum amounts of bonded reinforcement are satisfied. The shear strength V_c can, therefore, be found from the expression for an uncracked member, Equation (10.5), with the V_p term determined as outlined in Section 10.3.1. Since the edges AB and CD give the worst shear stresses, the shear width b_v in Equation (10.5) should be taken as the side x_2 of the critical perimeter. The parameter β_p is taken to be the smaller of

$$\beta_p = 0.3 \text{ or } \beta_p = (\alpha d/12u + 0.125) \quad \text{N-mm units}$$
$$\beta_p = 3.5 \text{ or } \beta_p = (\alpha d/u + 1.5) \quad \text{lb-in units}$$

where $\alpha = 40$ for internal columns
$= 30$ for edge columns, and
$= 20$ for corner columns

Incorporating a strength reduction factor into the calculated strength, the allowable shear stress around the critical perimeter is

$$v_{\mathrm{all}} = 0.85 V_c / x_2 . d.$$

If the magnitude of the maximum shear stress calculated by Equation (10.15) exceeds this value then either the slab thickness must be increased or shear reinforcement must be provided.

The required amount of links can be determined from Equation (10.9). A symmetrical arrangement of links around the column area as illustrated in Figure 10.10 is recommended by ACI 318. Here, the link size and spacing are chosen on the basis of the most heavily loaded of the four edges of the critical perimeter, and so incorporate a degree of conservatism for the other directions. ACI 318 stipulates that the link spacing should not exceed $0.75D$ or 600 mm (24 in). The required distance to which the links must extend from the face of the column is determined by defining a new, eight-sided critical perimeter $d/2$ beyond the last link, as shown in Figure 10.10, and checking that the shear stresses on this new perimeter are acceptable.

An accurate shear check on this new perimeter is extremely difficult, since several of its edges are rotated at 45° to the prestress and moment transfer directions, so that the preceding equations cannot easily be applied to these edges. The following simplified procedure is, therefore, recommended. Firstly, the shear stress due to the vertical shear force alone is calculated on the new perimeter, as V/ud. If an unbalanced moment is being transmitted by the connection, then the resulting additional shear stress is calculated for the original perimeter, using the second term in Equation (10.15a). The maximum shear stress on the perimeter is then taken as the sum of these two terms. This maximum stress must not then exceed the permissible stress of the concrete alone, neglecting the prestress, that is, $v_{\mathrm{all}} = \beta_p (f_c')^{0.5}$. This rather conservative approach is recommended on the grounds of its simplicity.

10.3.5 Decompression load method

An alternative to the approaches outlined above is the decompression load method, developed by Regan (1985). This gives a simple and convenient way of relating the punching strength of a prestressed slab to that of a reinforced concrete floor. The approach is similar to the calculation of shear strength for prestressed beams cracked in flexure, Equations (10.3) and (10.6).

The *decompression load* is defined as the force required to cancel out the initial compressive stress caused by the prestress at the face of the slab which is normally in tension under applied loads. At loads above the decompression force, the concrete section starts to crack and so behaves similarly to a reinforced section.

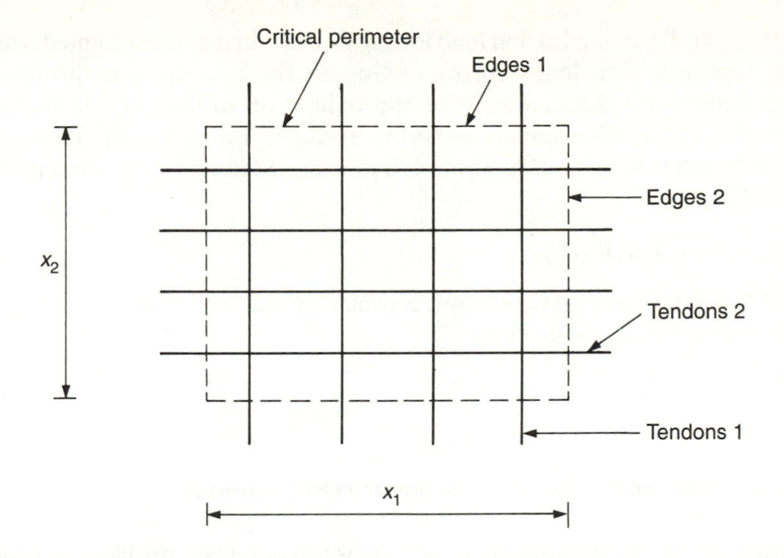

Figure 10.11 *Typical tendon arrangement through a critical perimeter*

The strength of the prestressed slab is taken as the sum of the decompression force and the shear strength of an equivalent reinforced member. (The equivalent reinforced concrete section is assumed to be geometrically identical to the slab under consideration, and to have a reinforcement quantity equal to the sum of the rod reinforcement area and the prestressing steel area multiplied by f_{pu}/f_y.)

Now for a given set of prestressing tendons, the prestress can influence only those edges of the critical perimeter perpendicular to the tendons. It is, therefore, necessary to consider the two span directions separately. Figure 10.11 shows a typical arrangement of prestressing tendons passing through a critical perimeter, with the two tendon directions and the corresponding perpendicular edges of the shear perimeter denoted by the subscripts 1 and 2. The shear strength V_r of the equivalent reinforced concrete section is simply calculated as

$$\begin{aligned} V_r &= V_{r1} + V_{r2} \\ &= 2v_{c1}x_1d_1 + 2v_{c2}x_2d_2 \end{aligned} \tag{10.17}$$

where the values of v_c are found from Table 10.1.

Next, it is necessary to calculate the decompression loads for the two span directions. For a given structural geometry and prestressing arrangement, the decompression load V_o can be related to the moment M_o given by Equation (10.1). For example, for a concentrated load at the centre of a span of length L

$$V_o = 4M_o/L$$

The punching strength of the prestressed section is then given by

$$V_c = V_r + V_{o1}(V_{r1}/V_r) + V_{o2}(V_{r2}/V_r) \tag{10.18}$$

Thus the overall decompression load for the slab is taken as the weighted average of the decompression loads corresponding to the two different prestressing arrangements. This is a conservative approximation to the exact formulation, which is difficult to solve. Equation (10.18) represents the most general case of the decompression load approach. For a slab prestressed in one direction only it can be simplified to

$$V_c = V_r + V_{01}(V_{r1}/V_r) \tag{10.19}$$

For a slab with the same prestressing arrangement in each direction $V_{01} = V_{02} = V_0$ and Equation (10.18) reduces to

$$V_c = V_r + V_o \tag{10.20}$$

10.4 Alternatives to conventional shear reinforcement

The provision of conventional stirrups in slabs can often be a problem; in order to be fully effective, the links must be anchored in the compression zone of the slab and tied into the main tension steel. For this reason, it is not normally possible to provide links in slabs thinner than about 200 mm (8 in). Even in thicker slabs, the detailing and fixing of the links can be difficult and time-consuming operations. Considerable efforts have, therefore, been expended in the development of alternative systems aimed at reducing construction time and cost while improving structural reliability.

10.4.1 Proprietary shearhoops

One attractive alternative is the use of prefabricated cages of shear reinforcement which can be simply placed on site and tied into the main bending reinforcement. The design of such systems requires considerable care; slabs reinforced by prefabricated cages which simply sit between the top and bottom reinforcing mats have been found to fail at loads lower than predicted by the BS 8110 approach, because the links are not effectively tied into the main tension reinforcement.

A system of shearhoops complying with the requirements of BS 8110, which provides full anchorage while maintaining ease of assembly, has been developed by Chana (1993). The system is commercially available in a wide range of sizes and reinforcement densities. A typical shearhoop consists of a series of specially shaped links held together by two horizontal hoop bars, Figure 10.12. The links are designed to cover the maximum possible depth without violating the cover requirements, and to be easy to tie into the main bending reinforcement, offering a barrier to the formation of shear cracks between the links and the main steel. The hoop bars hold the cage together, making it easy to handle, provide some additional anchorage to the links and offer some confinement to the slab. Structural testing of the shearhoop system suggests that it results in punching

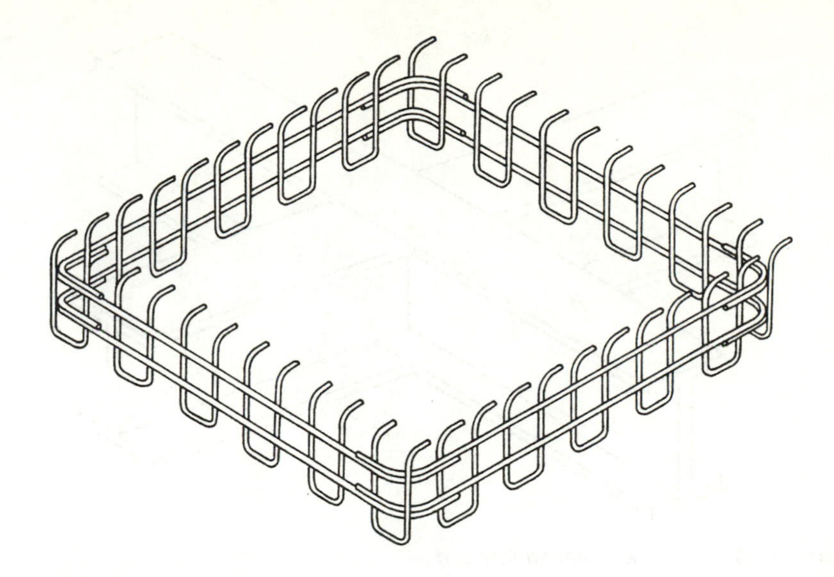

Figure 10.12 *Prefabricated shearhoop*

shear capacities around 10% greater than those of slabs reinforced by conventional links.

Shearhoops are positioned after the fixing of the bottom slab reinforcement. They sit on spacers, with the links given the same cover as the lowest layer of bending steel. The top steel is then fixed around them, the inner layer being located under and at right angles to the protruding horizontal legs of the links, thus ensuring that the links are fully tied into the tension reinforcement. Fixing trials suggest that the time taken to fix shearhoops is less than half that required for conventional shear reinforcement.

10.4.2 Structural steel shearheads

ACI 318 permits the use of shearheads made of structural steel I- or channel-sections in place of conventional shear reinforcement. Shearheads usually consist of identical sections, positioned at right-angles and connected by full-penetration welds, see Figure 10.13. These can be prefabricated away from site and quickly positioned in the slab prior to casting, leading to significant reductions in fixing times on site. Shearheads also make a significant contribution to the moment capacity of the slab at the column connection, resulting in reduced amounts of conventional bending reinforcement.

Shearheads must satisfy two design criteria. Firstly, they must extend from the column face to a perimeter at which the shear in the slab can be carried by the concrete alone. Secondly, they must have sufficient moment capacity to ensure that the required shear strength of the slab is reached before the flexural strength of the shearhead arms is exceeded. The first of these criteria can be used to assess

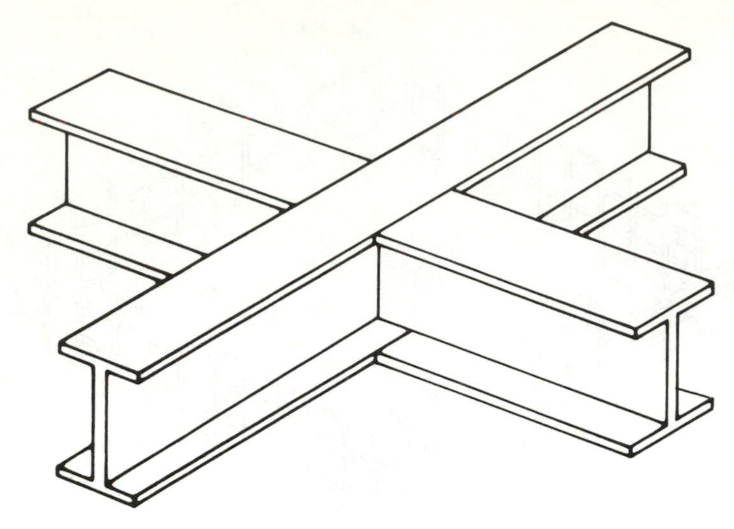

Figure 10.13 *Structural steel shearhead*

the required length of the shearhead arms by a simple iterative procedure. For a guessed arm length, the critical shear perimeter is deduced as outlined below. If the shear stress on this perimeter proves excessive, then the arm length is increased and the calculation repeated until an acceptable shear stress is achieved. Once the required length of the shearhead arms has been found, then the moment capacity criterion can be used to establish the required section size.

Consider firstly the required length of the shearhead arms. For a given length of shearhead, the critical perimeter is assumed to pass through each arm of the shearhead at three-quarters of the distance from the face of the column to the end of the arm. The shearhead must extend beyond this critical location in order to ensure adequate anchorage. The remainder of the perimeter, between the shearhead arms, is made up of straight lines, with the proviso that the perimeter need never come closer than $d/2$ to the face of the column. The resulting critical perimeters for slabs reinforced by shearheads of various sizes are shown in Figure 10.14.

The shear capacity of the concrete section can then be checked using an approximate approach similar to that described in section 10.3.4 for slabs reinforced by links. The shear stress due to the vertical shear force is calculated on the critical perimeter, while the shear stress due to any moment transfer is calculated by taking a rectangular perimeter $d/2$ from the face of the column. The total shear stress is then found by summing the two terms. This value is then compared to the allowable stress, ignoring the effect of the prestress, that is, $\beta_p\sqrt{f_c'}$. If the shear stress is excessive, then the length of the shearhead arms must be increased accordingly. The necessary adjustment is easy to calculate, since only the first of the two shear stress terms is affected by the change in critical perimeter.

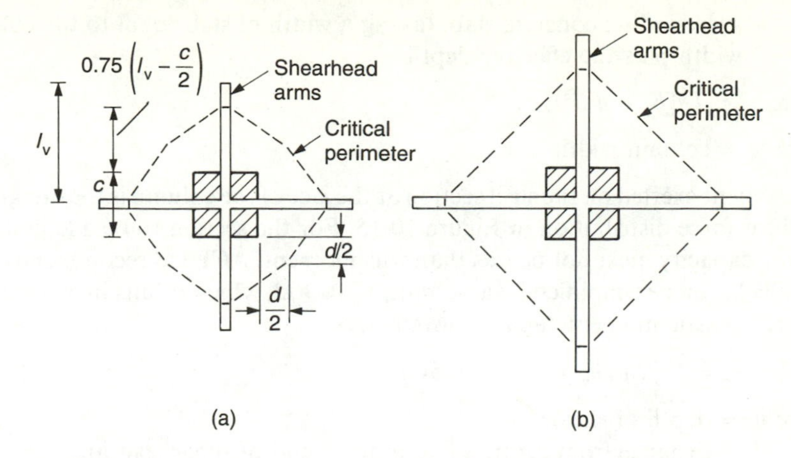

Figure 10.14 *Critical perimeters around shearheads at internal columns (a) Small shearhead (b) Large shearhead*

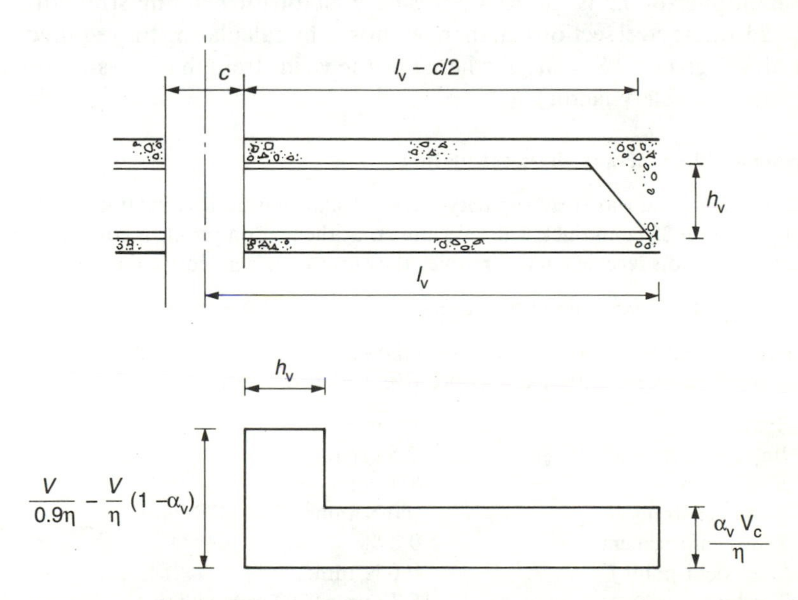

Figure 10.15 *Elevation and assumed shear force distribution along a shearhead arm*

Having determined the length of the shearhead arms, it is then necessary to choose a section having an adequate flexural strength. Figure 10.15 shows the elevation of a typical arm of a shearhead, together with the assumed distribution of shear force along the arm. Here:

η = the number of arms of the shearhead (usually three for an edge column or 4 for an internal column), and

α_v = the ratio of the flexural stiffnesses of the steel section and of the surrounding concrete slab, taking a width of slab equal to the column width plus the effective depth

$$\alpha_v = E_s I_s/[b_c + d)d^3/12]$$

where b_c = column width

The moment exerted on the steel section at the face of the column is the integral of the shear force distribution in Figure 10.15. For the section to be adequate, its flexural capacity must not be less than this moment. ACI 318 recommends that the calculation be simplified by assuming $V_c \approx V/2$, which results in a minimum required plastic moment capacity given by:

$$0.9M_p = (V/2\eta)[h_v + \alpha_v(l_v - 0.5C)] \quad (10.21)$$

where h_v = depth of shearhead
l_v = distance from centre of column to end of shearhead arm
c = column cross-sectional dimension in the direction of the shearhead arm.

The multiplier of 0.9 is the ACI 318 safety factor for bending strength.

An adequate steel section can then be chosen by calculating the required plastic modulus S, given by $S = M_p/f_y$ where f_y is the yield strength of the steel (inclusive of a material safety factor).

Example 10.1 One-way shear calculation

Calculate the one-way shear capacity of a solid slab continuous over two equal spans, supported on 250 mm thick walls. Assume that the tendon profile is concordant and that the tendon reverse curvature over the support is defined by the parabola

$$e = 0.000375x^2 - 65 \text{ in mm units}$$

relative to an origin at the intersection of the slab and the interior wall centreline, with a downward eccentricity taken as positive. Other data are:

Span	9.0 m
Imposed live load	2.5 kN/m^2
Slab depth	200 mm
Concrete strength f_{cu}	40 N/mm^2 ($f_c' =$ 32 N/mm^2)
Rod reinforcement	0.2 %
Steel yield point f_y	460 N/mm^2
Tendons	15.7 mm @ 300 mm centres
Tendon ultimate stress	1770 N/mm^2
Tendon force, final	170 kN each

Solutions

Two solutions are provided, complying with BS 8110 and ACI 318. The two approaches result in quite different values for the shear strength. The ACI method takes a rather conservative approach to the assessment of concrete shear strength, with no account taken of the reinforcement level, but it includes a more realistic assessment of the cracking moment than B 8110, which ignores the tensile strength of the concrete.

The critical section is at the face of the centre wall, at a distance of 8.875 m from the centreline of the outer support.

Section area	$A_c = 0.2\ m^2m$
Section modulus	$Z = 0.00667\ m^3m$
Eccentricity	$e = 0.000375 \times 125^2 - 65 = -59$ mm

Solution 1: Using BS 8110.

Factored load	$= 24 \times 1.4 \times 0.2 + 1.6 \times 2.5$	$= 10.72$ kN/m^2
Reaction at outer support	$= 0.375 \times 10.72 \times 9.0$	$= 36.18$ lN/m
Moment at critical section	$= 36.18 \times (8.875) - 10.72 \times (8.875)^2/2$	$= -101.1$ kNm/m
Shear at critical section	$= 0.625 \times 10.72 \times 9.0 - 10.72 \times 0.25/2$	$= 58.96$ kN/m

Cracking moment (Equation 10.1)

$$M_o = 0.8[PZ_{ten}/A_c + M_p] = 0.8(P)[Z_{ten}/A_c + e]$$
$$0.8(170/0.3) \times [0.000667/0.2 + 0.059] = 41.9 \text{ kNm/m}$$

Ultimate moment (101.1) exceeds M_0. The section is cracked, so use Equation (10.4).

Effective depth d	$= (0.2 \times 165 + 159)/0.45$	$= 162$ mm
Total steel content	$= 0.2 + 100 \times (150/0.3) \times (1770/460)/(1000 \times 162)$	$= 1.39\%$

From Table 10.1, $v_c = 1.03$ N/mm^2
From Eq. (10.4), $V_{cr} = 0.65 \times 1.03 \times 162 + 41.9 \times 58.96/101.1 = 132.9$ kN/m > 58.96

No shear reinforcement is required.

Solution 2: Using ACI 318.

Factored load $= 24 \times 1.4 \times 0.2 + 1.7 \times 2.5 = 10.97$ kN/m^2
Reaction at outer support $= 0.375 \times 10.97 \times 9.0 = 37.02$ kN/m
Moment at critical section $= 37.02 \times (8.875) - 10.97 \times (8.875)^2 2 = -103.5$ kNm/m
Shear at critical section $= 0.625 \times 10.97 \times 9.0 - 10.97 \times 0.25/2 = 60.30$ kN/m

The stress at the extreme fibre stress due to prestress $= (P/A_c) \times (1 + eA_c/Z)$

$$= \left[\frac{170 \times 1000/0.3}{200 \times 1000}\right] \times \left[1 + \frac{59 \times 170 \times 1000/0.3}{0.00667 \times 10^9}\right] = 7.90 \text{ N/mm}^2$$

From Eq (10.7), $M_{ct} = 6.67 \times 10^6 \times (0.5\sqrt{32} + 7.90) = 71.5$ kNm/m

The actual moment exceeds this value. The section is cracked, so use Equation (10.6)

$$V_{cr} = (0.05\sqrt{32}) \times 165 + 71.5 \times 60.3/103.5 = 89.7 \text{ kN/m}$$

Incorporating a shear strength factor of 0.85 gives the shear strength of 76.3 kN/m, which exceeds the actual shear force of 60.3 kN/m. Therefore, no shear reinforcement is needed.

Example 10.2 Punching shear calculation

Check the punching shear capacity of the interior slab-column connection shown in Figure 10.16, given the following data:

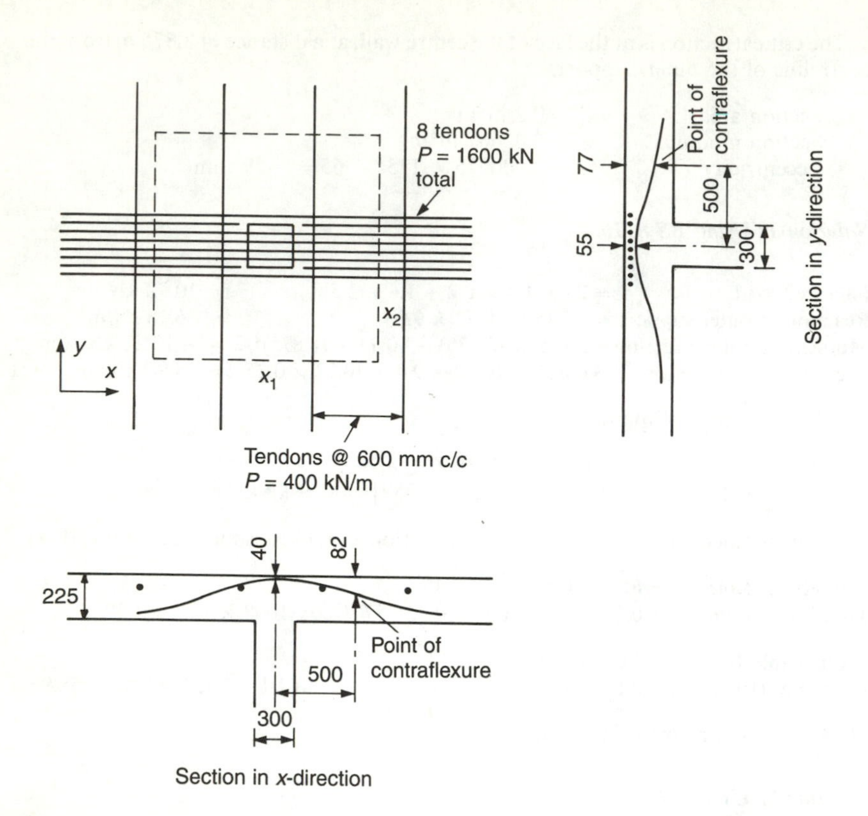

Figure 10.16 *Example 10.2*

Slab thickness	$D225$ mm
Concrete strength	$f_{cu} = 40\ \text{N/mm}^2, f_c' = 32\ \text{N/mm}^2$
Factored punching shear	$V = 700$ kN
Moments in slab at column centreline	$M(x) = 60$ kNm/m and $M(y) = 25$ kNm/m
Moment transferred to column	$M(x) = 25$ kNm/m and $M(y) = 0$ kNm/m

Tendon profiles can be assumed concordant. Eccentricities of the reverse parabolas in the two directions, relative to an origin at the intersection of the slab and column centrelines, are given by the following equations.

For the x-direction tendons, $e_x = 0.000168\ x^2 - 72$
For the y-direction tendons, $e_y = 0.000088\ x^2 - 57$

SOLUTION 1: Using BS 8110.

For determining the critical perimeter, take d as the tendon depth at the column centreline.

x-direction tendons: $d = 225 - 40 = 185$ mm, $x_2 = 300 + 3 \times 185 = 885$ mm
y-direction tendons: $d = 225 - 55 = 170$ mm, $x_1 = 300 + 3 \times 170 = 810$ mm

Using Equation (10.11),

$$V_{eff} = V[1 + 1.5M_t/(Vx_2)]$$
$$= 700[1 + 1.5 \times 25/(700 \times 0.855)] = 744 \text{ kN}$$

The x_2 edges are at a distance 405 mm from the column centreline, and the x_1 edges are 428 mm from the centreline. So the tendon eccentricities are:

x-direction $e = 0.000168 \times 405^2 - 72 = -44$ mm
y-direction $e = 0.000088 \times 428^2 - 57 = -41$ mm

For the x-direction, where the tendons are concentrated in a narrow band, assume the prestress is distributed over a width equal to twice the edge of the critical perimeter, so that the force per metre width is $1600/(2 \times 0.855) = 936$ kN. From Equation (10.1), the cracking moments per metre width are:

$$M_o(x) = 0.8(936 \times 0.00844/0.225 + 936 \times 0.044) = 61.0 \text{ kN/m}$$
$$M_o(y) = 0.8(400 \times 0.00844/0.225 + 400 \times 0.041) = 25.1 \text{ kN/m}$$

The section is uncracked in both directions, and shear capacity can be found from Table 10.3. Since the tendon points of inflexion lie outside the critical perimeter, the vertical component of the prestress is not included in the calculations.

Edges x_2 (x-direction tendons):
$p_{av} = 936 \times 1000/(1000 \times 225) = 4.16 \text{ N/mm}^2$
$V_{co} = 1.81 \times (2 \times 855 \times 225)/1000 = 696$ kN

Edges x_1 (y-direction tendons):
$p_{av} = 400 \times 1000/(1000 \times 225) = 1.78 \text{ N/mm}^2$
$V_{co} = 1.42 \times (2 \times 810 \times 225)/1000 = 518$ kN

Total shear resistance = 696 + 518 = 1214 kN

The shear resistance exceeds the actual shear, so no shear reinforcement is needed.

Solution 2: Using ACI 318.

Using the same values of d as in Solution 1, the edges of the critical perimeter are:

$x_1 = 300 + 170 = 470$ mm
$x_2 = 300 + 185 = 485$ mm

Moment transfer is accounted for using the parameter

$$\gamma_v = 1 - 1/[1 + 0.67(470/485)^{0.5}] = 0.40$$

J_c is found conservatively by using the smaller of the two d values.

$$J_c = 170 \times 470^3/6 + 470 \times 170^3/6 + 170 \times 485 \times 470^2/2 = 12.43 \times 10^9 \text{ mm}^4$$

From Equation (10.15),

$$v_{AB} = \frac{700 \times 10^3}{2(470 + 485) \times 170} + \frac{0.4 \times 25 \times 10^6 \times 470}{2 \times 12.43 \times 10^9} = 2.16 + 0.19$$
$$= 2.35 \text{ N/mm}^2$$

$$v_{CD} = 2.16 - 0.19 = 1.97 \text{ N/mm}^2$$

β_p is the smaller of:

0.3 and $(40 \times 170)/[12 \times 2 \times (470 + 485)] + 0.125 = 0.42$. Take $\beta_p = 0.3$

$p_{av} = 4.16\ \text{N/mm}^2$ as for Solution 1

Using Equation (10.5),

$$V_{co} = [\beta_p(f_c')^{0.5} + 0.3p_{av}]b_v d + V_p$$
$$V_{co}/(b_{vd}) = 0.3\sqrt{32} + 0.3 \times 4.16 = 2.95\ \text{N/mm}^2$$

Multiplying by a strength reduction factor of 0.85 gives a design shear strength of $2.5\ \text{N/mm}^2$, which is greater than 2.35. So no shear reinforcement is needed.

11 SLABS ON GRADE

Ground-bearing floors, commonly used in commercial and industrial buildings, are normally required to carry a variety of load-patterns, such as wheel loads from fork trucks, handling machinery, concentrated loads from racking and stacking systems, and uniformly distributed loads separated by unloaded aisles. Any fixed machinery is usually founded on its own separate base.

The design process is based on empirical rules backed by elastic analysis, and the construction techniques have largely evolved through experience over a period. Design guides and background data are published by a number of organizations in different countries; these are based on experience, and on the loadings relevant to the particular country and should be referred to when designing a ground bearing floor. The publications also contain useful guidance on the properties of the various soils, the loads and loading patterns, construction techniques and details of joints in use in the country. The organizations include:

The Portland Cement Institute—USA
The Post-tensioning Institute—USA
The British Cement Association
The Concrete Society—UK

This chapter deals with the design aspects pertinent to post-tensioning, though there is considerable overlap with conventional floors constructed in reinforced or plain concrete. It also gives a comparison between conventional reinforced and post-tensioned floors, and data on the elastic stresses produced under a concentrated or line load, and uniform loading with unloaded aisles.

A ground slab fundamentally differs from a suspended slab by virtue of its being continuously supported by the ground. Its function is to distribute the load concentrations to the ground in a safe manner. A slab carrying a uniform load over the whole of its area serves only as a separation medium between the ground and the load; structurally it is redundant. The performance of a ground slab is, therefore, strongly linked with the characteristics of the sub-base and ground, and, of course, the loading.

Reinforced ground floors tend to develop a variety of defects. The most common problem is that of uncontrolled cracking of the slab panels. In time, the concrete begins to spall at the cracks. In order to minimize such cracking, the floor is provided with a grid of joints—consisting of saw-cuts, contraction joints or expansion joints—filled with a sealant. Under continual wheel traffic, the joints gradually deteriorate. Another problem is that of *curling* of the edges at joints, caused by variation of strain in the depth of the slab, which may be due to shrinkage or temperature gradient.

From the user's point of view, cracks and joints in a floor are highly undesirable. They generate dust, cause damage to the vehicles, and slow down or impede the working of the stacking systems.

Post-tensioning allows large areas of the floor, exceeding 100 × 100 m (330 × 330 ft), to be constructed without any joints. This appreciably reduces the problems associated with cracks and joints, and allows the floor to be finished to the flatness accuracy required by the various mechanized stacking systems. A post-tensioned floor is also more tolerant of occasional overloading, because minor cracks tend to close and heal in time. By comparison, a non-post-tensioned floor normally relies on the tensile strength of concrete for distributing concentrated loads; once the concrete has cracked, its strength is lost for ever.

To a much smaller extent, post-tensioned floors can suffer from the same problems as the normal non-prestressed floors. The problems and the remedies are equally important in both types of construction, and it is, therefore, useful to discuss the performance of non-prestressed floors first.

11.1 The design process

In the past, the strength requirement of a ground floor was often specified in terms of a uniformly distributed loading. This is structurally meaningless, because under a uniform loading, applied over the whole of the floor area, the slab has no structural function; it acts only as a separation medium between the load and the subgrade. The current trend is to specify the magnitude of concentrated loads, such as the wheel load or the leg load from a racking system, or to specify aisle widths between areas of block stacking.

The long-term settlement behaviour of a floor depends on the characteristics of the subgrade and the average intensity of the sustained loading. Except for local effects, it is independent of the strength of the slab. Advice on soil properties should be sought from specialist consultants.

The design process is most commonly concerned with the stresses developed by the concentrated loads and the short term strength of the slab. The calculation of stresses is based on an elastic analysis of the floor-subgrade system, assuming the floor to be a thin elastic plate continuously supported by an elastic medium. The theory developed by Westergaard in the 1920s forms the basis of the current design methods. In the future, finite element analysis techniques are likely to have an increasing role.

The procedure for the design of a ground bearing floor using elastic theory consists of the following steps:

1. Determine the characteristics of the subgrade
2. Determine the loading pattern and magnitude of loads
3. Determine the appropriate safety factor for the number of load movements

4. Assume thicknesses of the sub-base and the concrete slab
5. Calculate the slab strength
6. Calculate the stresses produced by the loads
7. Revise step 3 and repeat if necessary

The stress at a point may be influenced by several discrete concentrated loads in the vicinity. Methods of assessing this are given in C&CA (now BCA) Technical Report 550 (Chandler, 1982) and by Fatemi-Ardakani, Burley and Wood (1989).

Experimental and theoretical work in the United States and in the UK has led to the derivation of semi-empirical rules and nomographs for determining the thicknesses of the various layers of the floor construction without the need for iteration.

A ground slab can be analysed on a computer as a grid or using finite elements, replacing the soil with equivalent springs. The analysis can be extended to predict the long-term settlement of the floor. Specialized design software is also available from various sources.

The Westergaard equations, the moments produced by partition loads and the uniform loading of a floor with unloaded aisles are given later in this chapter.

11.2 Factors affecting the design

The factors which influence the stresses produced in a floor and determine its strength are:

- loading, pattern and intensity
- loaded area
- slab thickness
- modulus of elasticity of concrete
- flexural tensile strength of the concrete
- amount of prestress
- friction between the slab and the sub-base/grade
- and elastic characteristics of the sub-base/grade

Some of the above factors, and the safety factor used in the design, are discussed below.

11.2.1 Loading

The loading patterns on a floor can be classified in four categories.

1. Wheel loads from fork lift trucks and stacking machinery. The trend towards high stacking has necessitated a minimum of elastic springing of the machine and solid tyres. The load is applied over a small area of the floor; the average contact pressure has been measured as 13.9 N/mm^2 (2000 psi), and the maximum single wheel static load registered was 5.6 tonnes, (Fatemi-Ardakani, Burley and Wood, 1989). The static load is magnified by the irregularities in

the floor surface, particularly by open joints. The dynamic load can be up to twice the static value, depending on the quality of the floor. High stacking mobile equipment is particularly detrimental to open joints in the floor.

Typical examples of laden tyre contact pressures are:

3 tonne counterbalance lift truck	2.4 N/mm^2	350 psi
2 tonne reach trucks	5.6 N/mm^2	800 psi
2 tonne pallet trucks	11.0 N/mm^2	1600 psi

Wheel loads in the region of 8.5 tonnes are not unusual.

2. Leg loads from racking systems. Leg loads of the order of 20 tonnes are common. Each leg is provided with a baseplate, which may vary from a minimum of about 100 mm square (4 in) up to 500 mm square (20 in) or more.

   The baseplate often consists of a steel plate without adequate ribs or gussets to enable it to distribute the load over the whole of its area. In order to be effective, the baseplate should be stiffer than the concrete slab. A baseplate of stiffness equal to that of the slab is only about 50% efficient. Even for this efficiency, assuming an E_s/E_c ratio of 8, the required thickness of a steel plate (without gussets) is half that of the concrete slab as the following simple calculation shows.

   $$E_s t^3 = E_c D^3$$

   therefore,

   $$t = (E_c/E_s)^{1/3} D = (1/8)^{1/3} D = D/2.$$

3. Areas of floor carrying uniformly distributed loading, separated by aisles. The maximum tensile stress develops in the middle of the aisle at the top of the slab.

4. Line loads. Occasionally, a floor slab is required to support a partition wall. The loads are usually permanent and affect the long-term settlement of the floor. The implication of supporting heavy partitions on the floor should be studied in conjunction with the settlement characteristics of the subgrade.

The wheel load and the leg load from a racking system are both concentrated loads acting on a relatively small area of the floor. Rack loading may vary in intensity but the legs stay in position for long durations; they directly affect the long-term settlement. The wheel load is transient and its influence on the long-term settlement is negligible; the repetitious wheel loading may, however, affect the strength of the floor through fatigue. The suggested load factors are discussed in Section 11.2.4.

11.2.2 Modulus of rupture

The modulus of rupture is expressed in terms of a power of the concrete strength. The expressions are of the form

$$f_t = C(f_c)^n \qquad (11.1)$$

Table 11.1 *Moduli of rupture* (N/mm^2), *UK practice.*

f_{cu}	25	30	35	40	45	50
$f_{t\ 28}$	3.36	3.79	4.21	4.60	4.97	5.33
$f_{t\ 90}$	3.70	4.17	4.63	5.06	5.47	5.87

where C = a coefficient
n = a constant
f_c = 28-day concrete strength f_{cu} or f_c'

In the UK practice, C is usually taken as 0.393 and n as $\frac{2}{3}$. A period of 90 days is expected to lapse before the floor is loaded. At 90 days, the modulus of rupture is assumed to have increased by 10% over the 28-day value. Therefore

$$\begin{aligned} f_t &= 0.393(f_{cu})^{2/3} \text{ at 28 days} \\ &= 0.432(f_{cu})^{2/3} \text{ at 90 days} \end{aligned} \tag{11.2}$$

Table 11.1 gives the moduli of rupture for a range of 28-day concrete strengths. The values shown on the second line are at 28 days and those on the third line at 90 days.

In American practice, the modulus of rupture is expressed as $C(f_c')^{0.5}$ psi. Ringo and Anderson (1992) state that the coefficient C commonly varies from 9 to 11 for bank-run gravel and crushed stone aggregate respectively. The Portland Cement Association recommends a coefficient of 9. The designer either assumes one of these coefficients (7.5, 9 or 11), or specifies tests to be run from the trial concrete mix to determine an appropriate value for the modulus of rupture. A default value of 7.5 is normally used for the coefficient C.

The Imperial and metric equivalents of the coefficient C are:

Imperial	7.5	9.0	11.0	psi units
Metric	0.62	0.75	0.91	N-mm units

The Imperial values of the moduli of rupture for various cylinder strengths f_c' corresponding to the three values of the coefficient C are shown in Table 11.2.

Table 11.2 *Moduli of rupture* (*psi*), *US practice*

C				f'_c			
	3000	3500	4000	4500	5000	5500	6000
7.5	410	444	474	503	530	556	581
9.0	493	532	569	604	636	667	697
11.0	602	651	696	738	778	816	852

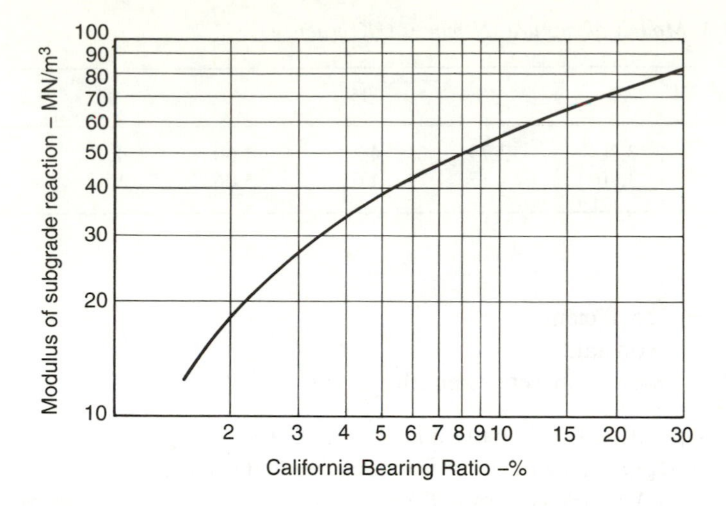

Figure 11.1 *Relationship between subgrade modulus and CBR*

11.2.3 Subgrade modulus k

The modulus of subgrade reaction is a measure of its elastic compressibility. It is defined as the stress on the surface of the subgrade which causes it to deform by one unit of length, and is expressed in MN/m^3 or pci. Its value can be determined by a plate test on the prepared grade, or an approximate value may be obtained from the California Bearing Ratio (CBR); the relationship, as given by Chandler (1982), is shown in Figure 11.1.

General guidance on the value of k can be obtained from Table 11.3. For more detail refer to the Concrete Society Technical Report 34 (1988).

The value of k on the surface of the sub-base differs from that on the subgrade. Chandler and Neal (1988) give the enhanced values shown in Table 11.4 for granular sub-base.

The stresses generated in a slab by a load are not very sensitive to the value of k, as can be judged from Section 11.6. Little advantage in the slab thickness is gained by taking values higher than 80 MN/m^3.

Table 11.3 *Subgrade moduli*

Description	*k* MN/m^3	*k pci*
Coarse grained gravelly soils	54–82	200–300
Coarse grained sand and sandy soils	54–82	200–300
Fine grained soils–silts and clays	27–54	100–200
High plasticity clays	14–27	50–100

Table 11.4 *Enhancement of* k *(MN/m^3) with sub-base*

k value for subgrade alone	*Sub-base thickness, mm* 150	200	250	300
13	18	22	26	30
20	26	30	34	38
27	34	38	44	49
40	49	55	61	66
54	61	66	73	82
60	66	72	81	90

11.2.4 Safety factor F_s

In the UK, a safety factor of 1.5 is consistently applied to the loads where the loading patterns are known. This is approximately in line with a partial safety factor of 1.3 for concrete and 1.2 for loading. For repetitive wheel loading, fatigue may affect the strength of the concrete slab. Chandler (1982) has suggested an increase in the factor of safety, shown in Table 11.5. One cycle of loading consists of a vehicle travelling laden and returning unladen along the aisle.

In the USA, a range of values for the safety factor F_s is used, depending on the particular conditions as judged by the designer. According to Ringo and Anderson (1992):

- A conservative value of 2.0 is commonly used. This is appropriate where loadings are not accurately known at the time of design, or where support

Table 11.5 *Safety factors for fatigue loading*

Cycles	*Safety factor*
>400,000	2.00
400,000	1.96
300,000	1.92
240,000	1.87
180,000	1.85
130,000	1.82
100,000	1.79
75,000	1.75
57,000	1.72
42,000	1.70
32,000	1.67
24,000	1.64
18,000	1.61
14,000	1.59
11,000	1.56
8,000	1.54
<8,000	1.50

conditions, or any other key items, are either not accurately known or are suspect.

- A value of 1.7 is acceptable consistent with the load factors used in other concrete design applications. This can be used where loading is frequent and input values (design parameters) are reasonably well known.
- A value of 1.4 is acceptable for use under certain conditions. For example, where impact loadings do not exist, or where the loading intensities and/or the frequency of load application are less, then values between 1.4 and 1.7 may be appropriate.

11.2.5 Restraints to lateral movement

The design of a ground slab assumes that, apart from the friction between the slab and the subgrade, it is free to undergo changes in length arising from the variation in temperature, moisture content of the concrete and humidity. If the restraint induces tensile stresses higher than the strength of the slab, then it cracks. Restraint to lateral movement may be caused by a number of factors, including the following.

- Friction between the subgrade and the floor slab. The conventional reinforced concrete slab is normally provided with sufficient reinforcement to avoid cracks when shrinkage takes place at the early age of the concrete. This is discussed in more detail in section 11.3.

  Changes in temperature or humidity may cause higher tensile stresses in a fully loaded slab than in an unloaded one, because of the additional friction due to the load itself. The tension in a loaded slab is much more severe than in an unloaded slab. The additional friction due to the load is, however, normally ignored in the design process.
- Where a floor is required to carry permanent line loads, such as at a partition, it is often made locally deeper to increase its stiffness. Since the soffit of the floor is no longer level, the subgrade has to have a local trough, whose vertical or sloping sides hinder the movement of the floor. Either the subgrade must move with the slab or the slab has to ride the slope if it is not to crack. In the case of a single line load, the trough may form an anchor point for all movement. Some compressible material is usually provided along the vertical, or sloping, faces of the troughs to reduce the tensile forces which would otherwise develop. The compressibility of the material should be studied with reference to the expected movement to assess the magnitude of the tensile force which would develop.

  An alternative is to provide shallow slopes, say of the order of 5%.
- The penetrations for the internal columns form a grid of rigid points in a large floor.

  A similar restraint may occur along the outside edges of a floor if the external columns project into the building so that the edge of the slab has to be notched around the columns. The notches prevent the shortening of the slab along the particular edge.

  In these locations too, the magnitude of the tensile stress that may develop is controlled by providing a compressive layer between the floor and the columns. The remarks above, about the compressibility of the material, also apply to these cases.

The mass of a column foundation represents a hard spot under the slab. If the subgrade has not been properly compacted, or the columns are founded on stiffer foundations such as piles, the floor gradually settles but the small area of the slab immediately above a base does not. This causes flexural tension at the bottom surface of the slab around the column base.

Alternatively, if the ground pressure under the foundations is higher than the general floor loading, as is often the case, then the foundations settles, leaving a gap between the base and the floor. In this case the slab develops tension at its top surface in flexure.

11.3 Traditional RC floors

Traditionally, ground floors have consisted of an unreinforced, or lightly reinforced, concrete slab laid on a slip membrane on prepared grade. The slip membrane usually consists of one or two layers of heavy polythene. A sub-base, consisting of compacted ballast or hardcore, and sand blinded, may also be provided below the concrete. Figure 11.2 shows the traditional construction.

For its capacity to distribute load concentrations, a floor relies upon the flexural tensile strength of the concrete. The floor is assumed to behave as a fully elastic thin plate, supported by a semi-infinite elastic grade. If the floor cracks, its physical properties change drastically, grossly reducing its load carrying capacity.

Direct shear has not been found to be critical in determining the load capacity of a floor; punching shear may need to be checked under heavy concentrated loads. Shear is transferred across fine cracks through aggregate interlock.

It is recognized that for a floor to perform in a satisfactory manner it must not be allowed to crack. Apart from overloading and deficiencies in the grade, a floor can crack from a number of causes, including heat generation during concreting, drying shrinkage and subsequent temperature changes. Of these, the most important is shrinkage. As a floor is allowed to dry up, possibly months after its construction, the loss of moisture is accompanied by a reduction in the volume of

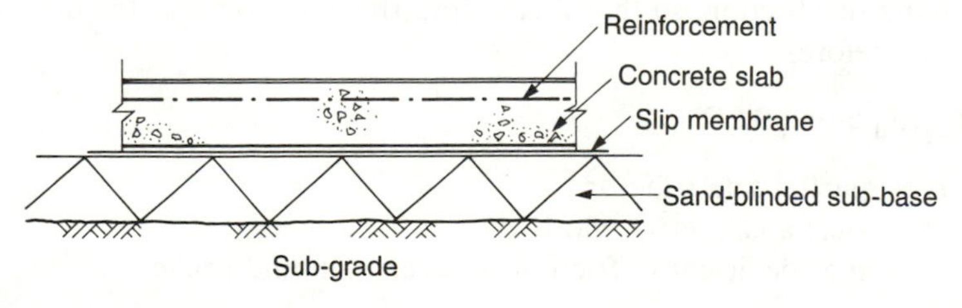

Figure 11.2 *A traditional concrete floor*

concrete. The horizontal movement associated with the reduction in the plan size of the floor is resisted by friction between the concrete and the grade below, which sets up tensile stresses over the concrete section. If the tensile stress exceeds the strength of the concrete, then it cracks.

The slip membrane, being usually waterproof, acts as a vapour barrier. The moisture from the body of the concrete migrates upwards and evaporates from the top surface. The surface is, therefore, dryer than the body of the concrete; and consequently, the surface zone has a higher strain and a higher tensile stress. The cracks, therefore, start from the top surface and penetrate into the concrete.

Seasonal and daily temperature variations superimpose further tensile stresses, causing the cracks to travel deeper into the concrete. A freeze-thaw cycle is particularly detrimental.

The measures taken to control the cracks may include one or more of the following.

- Reduce total shrinkage strain. Use plasticizers to reduce the water content. High-strength concretes usually contain a greater quantity of cement, which needs more water. Therefore, prefer a low strength concrete if possible.
- Reduce the rate of shrinkage. Most of the sprayed curing membranes lower, and delay, the rate of loss of moisture. This reduces the intensity of the tensile stresses, and they occur later in the life of the concrete, when it is stronger.
- Provide crack control joints at frequent intervals. Joints, however, are weak spots deliberately introduced at strategic locations. Movement of a floor is often concentrated at the joints, which leads to a gradual deterioration of the surface. Traversing wheels accelerate the process. A joint may eventually fail as the arrises get worn out, thereby widening it, or even causing a step to develop across it.
- Use steel fibre to enhance the tensile strength of concrete. Floor areas of around 1000 m^2 have been cast without any joints using this approach.
- Provide a reinforcement mesh fabric near the top surface, where the tensile stresses are highest. This is a far from ideal solution as the mesh often gets displaced to a lower than intended level and it restricts the movement of the concreting crew.

The amount of rod reinforcement is normally based on the *drag theory*. In a floor of length L, frictional resistance to shrinkage will build up linearly to a maximum at the mid-point of the floor length. The reinforcement must be capable of overcoming this friction, so that it can drag the floor towards the middle of its length. Therefore,

$$A_s f_y / F_d > \mu w_c D L / 2 \tag{11.3}$$

where F_d = partial safety factor
A_s = steel area mm^2m width
μ = the coefficient of friction between slab and grade

In order to ensure that the reinforcement remains within the elastic range, the

Table 11.6 *Coefficient of friction*

Material	μ
Sand and gravel	2.0
Granular sub-base	1.7
Layer of sand	1.0
Polythene	0.9

value of F_d is usually taken between 1.5 and 1.15. The coefficient of friction depends on the firmness and roughness of the grade, and on the nature of the slip membrane. Typical values of the coefficient of friction against the initial movement are given in Table 11.6.

Taking $F_d = 1.15$, $f_y = 460$ N/mm^2 and $\mu = 1.5$, and measuring D in millimetres and L in metres, for concrete of normal density,

$$A_s = 0.045DL \qquad \text{mm}^2\text{/m width} \qquad (11.4a)$$

In imperial units, with $F_d = 1.15$, $f_y = 60\,000$ psi, and measuring D in inches and L in feet,

$$A_s = 0.00016DL \qquad \text{in}^2\text{/ft width} \qquad (11.4b)$$

The problem with the drag steel concept is that, working at a stress in the range $f_y/1.5$ to $f_y/1.15$, the strain is high enough to cause the concrete to crack—the very defect it is supposed to guard against. Admittedly, the cracks are very fine and well distributed, nevertheless the concrete cannot remain intact. While the presence of reinforcement does not stop the concrete from cracking, it does hold the concrete tightly together, maintaining good aggregate interlock for transfer of shear.

However, the concrete is assumed to be uncracked and capable of withstanding a moment of

$$M_{cr} = f_t Z/F_s \qquad (11.5)$$

where F_s = Factor of safety

11.4 Post-tensioned ground floors

In a post-tensioned floor the practice is to provide a concentric prestress, which enables the floor to cope with stress reversals. The amount of prestress in a ground-bearing floor is dependent on the length of the floor, the loading pattern and intensity, and the quality of the subgrade. Prestress is applied in both directions and the average stress is usually much smaller than that in a suspended floor. An average final stress of 1.0 N/mm^2 (150 psi) after losses is quite common, though it may be as low as 0.4 N/mm^2 (60 psi) in floors of short length.

The amount of prestress is chosen to satisfy two requirements:

1. Concrete must not crack with the drag
2. Concrete must not crack under the loading

The first requirement is satisfied for the unloaded slab if:

$$P > \mu w_c DL/2. \qquad (11.6)$$

Using $\mu = 1.5$, $w_c = 24\ \text{kN/m}^3$, and measuring D in millimetres and L in metres,

$$P = 0.018DL \qquad \text{kN/m width, and}$$
$$p_{av} = P/A_c = 0.018L \qquad \text{N/mm}^2 \qquad (11.7a)$$

In Imperial units, with a concrete density of 144 pcf, D measured in inches and L in feet, the equivalent equations are:

$$P = 9DL \qquad \text{lb/ft, and}$$
$$p_{av} = P/A_c = 0.75L \qquad \text{psi} \qquad (11.7b)$$

If the loaded slab is considered, then the transient wheel loads may be ignored and the rack loading translated into an equivalent uniformly distributed loading. Then

$$P = (0.018D + 0.75w)L \qquad \text{in metric units}$$
$$P = (9D + 0.75w)L \qquad \text{in Imperial units} \qquad (11.8)$$

where w = the equivalent uniformly distributed load

As regards the second requirement, that of the strength, the concrete section does not suffer from any cracks when post-tensioned to satisfy the first requirement. The concrete being intact, the total tensile strength available is that due to the modulus of rupture and the prestress, less the friction loss. Therefore,

$$M_u = (f_t + p_{av} - \mu w_c x)Z/F_s \qquad (11.9)$$

where x = distance from the slab edge $\leqslant L/2$
F_s = the partial safety factor applied to stresses

Note that the prestress is maximum some distance away from the edge; it is lower near the anchorage because of the draw-in and it reduces to a minimum at the far end in a slab stressed from one end. In a slab stressed from both ends, or where the

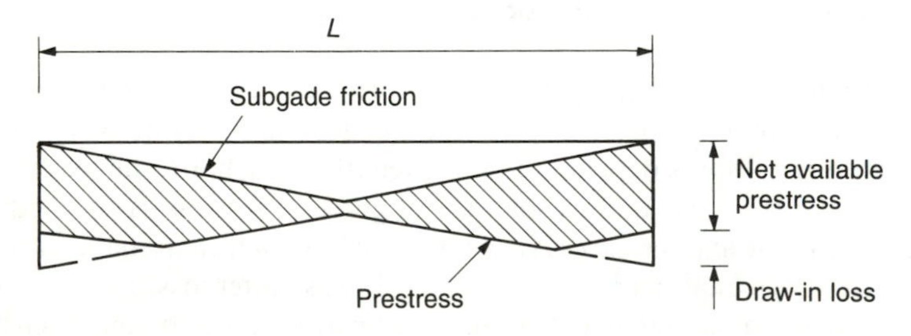

Figure 11.3 *Stresses in an unloaded post-tensioned floor*

tendons are stressed alternately from each end, the prestress is minimum at mid-length of the slab. The strength of the floor varies correspondingly, see Figure 11.3.

In a bonded floor this state of stresses is locked in after grouting. Factors affecting the length of a floor, such as temperature and humidity changes, may tend to even out the prestress in a floor using unbonded tendons, but it is considered safer to use the distribution shown in Figure 11.3 in the design.

11.5 Elastic analysis

This section deals with the elastic analysis of concentrated, line and uniformly distributed loads. In designing a floor on the basis given here, a floor construction is initially assumed, and the load, including dynamic effect if any, is multiplied by the appropriate factor of safety. The resulting tensile stress, or moment, is checked against the permissible stress, or the cracking moment. If necessary, the floor construction is revised.

The formulas given in this section are based on Westergaard's equations, which are now believed to over-estimate the stresses produced by concentrated loads. As stated in Section 11.1, others have devised methods based on elastic theory and experimental measurements, which yield more economical designs. Some of these are referred to in this section but not discussed in detail.

11.5.1 Concentrated loads

The elastic analysis of ground-bearing floors carrying isolated concentrated loads is usually based on the equations of Westergaard (1948) and their modifications, which assume a thin elastic plate bearing on an infinite or semi-infinite elastic medium.

The basic equations do not take account of the separation between the slab and the grade which may occur because of curling and temperature gradient across the slab thickness. The stresses given by these equations are immediate and do not consider the long-term settlement of the subgrade. No reliable correlation exists between the elastic properties of a soil and its long-term deformation characteristics. For poor soils, the long-term settlement may be up to 40 times the immediate deflection. Specialist soils engineers should be consulted for predicting the long-term behaviour of a floor.

The relative stiffnesses of the slab and the ground are represented by a *radius of relative stiffness* r, measured in units of length. Its value is given by:

$$r = \{E_c D^3/[12(1 - \mu^2)k\}^{0.25}$$

where ν = Poisson's ratio for concrete
k = modulus of subgrade reaction

For $\nu = 0.2$, the expression reduces to $r = 0.543\{E_c D^3/k\}^{0.25}$

Table 11.7a *Values of radius of relative stiffness (mm)*

k (MN/m^3)	Slab thickness (mm)						
	150	175	200	225	250	275	300
13	887	996	1101	1203	1301	1398	1492
27	739	830	917	1002	1084	1164	1243
54	621	498	771	842	912	979	1045
82	560	628	695	759	821	882	942

Table 11.7b *Values of radius of relative stiffness (in)*

k (pci)	Slab thickness (inches)						
	6	7	8	9	10	11	12
50	34.8	39.1	43.2	47.2	51.1	54.9	58.6
100	29.3	32.9	36.4	39.7	43.0	46.2	49.3
200	26.8	27.7	30.6	33.4	36.1	38.8	41.4
300	22.3	25.0	27.6	30.2	32.7	35.1	37.4

With E_c in kN/mm^2, D in mm and k in MN/m^3,

$$r = 17.17\{E_c.D^3/k\}^{0.25} \text{ mm} \qquad (11.10a)$$

and, with E_c in ksi, D in inches and k in pci

$$r = 3.054\{E_c D^3/k\}^{0.25} \text{ inches} \qquad (11.10a)$$

Tables 11.6a and 11.6b give the values of the radius of relative stiffness for different combinations of subgrade modulus and slab thickness in metric and Imperial units respectively. These are based on a concrete modulus of elasticity of 28 kN/mm^2 (4000 ksi).

The maximum values of stresses in the slab are given by the following expressions at an interior point, at the edge and near a corner. The interior and edge stresses are maximum directly below the load and have tension at the bottom of the slab whereas the stress due to a corner load is maximum some distance away from the corner and has tension on the top surface. The edge and corner loads are assumed to be applied so that the slab edges are tangential to the circular loaded area. The logarithms are to base 10.

$$\sigma_i = 1000(P/D^2)[1.264 \log(r/a) + 0.3379] \qquad (11.11a)$$

$$\sigma_e = 1000(P/D^2)[3.208 \log(r/a) + 0.5518a/r - 0.026] \qquad (11.11b)$$

$$\sigma_c = 1000(P/D^2)3[1 - (a\sqrt{2}/r)^{0.6}] \qquad (11.11c)$$

where P = load in kN

a = radius in loaded area mm
σ_i = stress in N/mm^2 under an interior load, away from the edges
σ_e = stress in N/mm^2 under a load at an edge N/mm^2
σ_c = stress in N/mm^2 due to a load at a corner N/mm^2

The empirical observations of Teller and Sutherland (1943) show that the stress due to a corner load is given more accurately by

$$\sigma_c = 1000(P/D^2)3[1 - (a\sqrt{2}/r)^{1.2}] \quad (11.11d)$$

Equations (11.12a) to (11.12d) are the Imperial versions of the above four equations, (11.12d) being the Teller and Sutherland equation.

$$\sigma_i = (P/D^2)[1.264 \log(r/a) + 0.3379] \quad (11.12a)$$

$$\sigma_e = (P/D^2)[3.208 \log(r/a) + 0.5518a/r - 0.026] \quad (11.12b)$$

$$\sigma_c = (P/D^2)3[1 - (a\sqrt{2}/r)^{0.6}] \quad (11.12c)$$

$$\sigma_c = (P/D^2)3[1 - (a\sqrt{2}/r)^{1.2}] \quad (11.12d)$$

Equations (11.11a), (11.11b) and (11.11d) can be expressed in the convenient forms of Equations (11.13) and Figure 11.4 gives the values of the coefficients for a range of a/r values.

$$\sigma_i = 1000(P/D^2)K_i$$
$$\sigma_e = 1000(P/D^2)K_e \quad (11.13)$$
$$\sigma_c = 1000(P/D^2)K_c$$

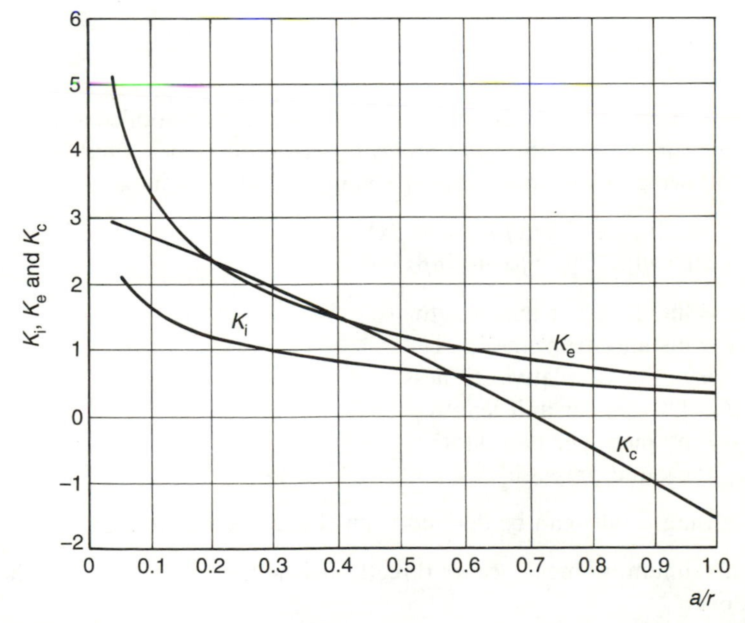

Figure 11.4 *Values of* K_i , K_e *and* K_c

Table 11.8 *Increase in concentrated load near joints and edges*

Joint description	*Increase*
Tied joints	Nil
Dowelled contraction joints, opening 1 to 6 mm	33%
Dowelled contraction joints opening more than 6 mm, induced contraction joints opening more than 1 mm, free edges and isolation joints	85%

The above equations give the stresses for isolated loads only, any adjacent loads are ignored. From a practical point of view, the stresses produced by any load located beyond a distance of $3r$ from the point under consideration are negligible and it can be ignored. Where a floor is considered continuous, a joint within a distance of $1.5r$ of the point under consideration interrupts the continuity.

Chandler (1982) has given methods for calculating, at a given point, the effect of a nearby load.

Much of the detailed design of the slab is concerned with stresses governed by loads close to the edges and corners. TR34 (1988) suggests that, where the load centre is within 300 mm (12 in) of an edge or joint, the concentrated load should be increased by the factors given in Table 11.8. See section 11.6 for a definition of joints.

11.5.2 Line loads

Line loads, such as those from partitions, are sometimes applied at a slab edge. For an interior line load, the *beam on elastic foundation theory* equations are given by the following expressions; angle βx is measured in radians.

$$m = (0.25w/\beta)e^{-\beta x}(\sin \beta x - \cos \beta x)$$
$$p_g = (0.5w\beta)e^{-\beta x}(\cos \beta x + \sin \beta x) \qquad (11.14)$$

where w = line load per unit length
x = distance from the load
r = radius of relative stiffness
$\beta = 1/(\sqrt{2}r)$ per unit length
m = moment per unit width
p_g = ground pressure

The following results can be derived from the above equations:

- The maximum moment occurs directly below the load and takes the value $0.3536wr$.

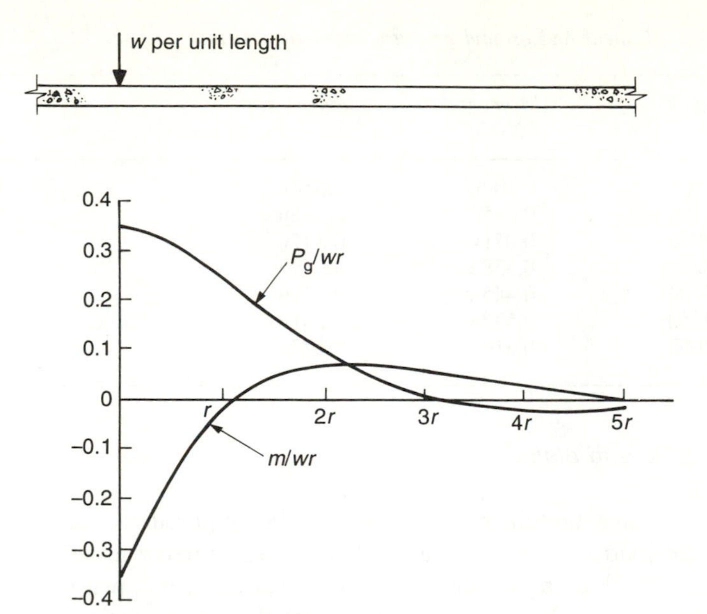

Figure 11.5 *Moment and ground pressure under a line load*

If r is measured in millimetres and w in kN/m units then the moment in kNm/m is given by $0.3536wr/1000$.

The moment drops to zero at a distance of $1.1r$ from the load, and then reverses its sign, peaking at a distance of $2.2r$ from the load where its magnitude is 21% of that under the load.

- The ground pressure is maximum under the load, its magnitude is $0.3536w/r$.

  If w is measured in kN/m and r in millimetres then the ground pressure in kN/m^2 is given by $353.6w/r$.

  The ground pressure is zero at a distance of $3.33r$ from the load, becoming negative beyond that point, peaking at a distance of $4.4r$ from the load where its value 4.3% of that below the load.

Figure 11.5 shows the distribution of the moment and the ground pressure for a uniform line load. The moment diagram is drawn on the tension face of the concrete. Note that the parameter m/wr is non-dimensional, and p_g/wr has the unit of per unit area.

The above values are based on the load being applied over an infinitely small width, which is not the case with partitions. The peak values of the moment and ground pressure, directly below the load, are reduced to the values shown in Table 11.9 if the load is assumed to be applied over a notional width of 600 mm. Such a width would result from a 200 mm partition supported on a 200 mm slab, assuming a spread of load at an angle of 45°.

Table 11.9 *Moment and ground pressure under a partition*

r (mm)	*Moment* (*kNm/m*)	*Ground pressure* (*kN/m*^{2})
500	0.102w	0.658w
750	0.185w	0.456w
1000	0.271w	0.347w
1250	0.358w	0.279w
1500	0.445w	0.234w
1750	0.532w	0.201w
2000	0.619w	0.177w

11.5.3 UDL and aisles

In some storage buildings, the floor may be required to carry a uniformly distributed load over large areas with relatively narrow aisles, which remain unloaded. The loading is equivalent to a floor carrying an upward uniformly distributed load on the aisle widths only, with the rest of the floor unloaded. The maximum moment in a floor thus loaded develops in the middle of the aisle, and has tension on the top of the slab. For an infinitely long thin elastic plate on an elastic medium, loaded as shown in Figure 11.6, the moments and the ground pressure at point O are given by the following expressions:

For $x < a$,

$$m = -(w/4\beta^2).[e^{-\beta(a+b)}\sin\{\beta(a+b)\} - e^{-\beta a}\sin(\beta a)]$$
$$p_g = -(w/2)[e^{-\beta(a+b)}\cos\{\beta(a+b)\} - e^{-\beta a}\cos(\beta a)] \qquad (11.15)$$

Moments and ground pressures within the width b can be calculated if the distance a is made zero and the load on the two sides is considered in two separate calculations. Values in the middle of length b are given by:

$$m = -(w/2\beta^2)e^{-\beta b/2}\sin(\beta b/2)$$
$$p_g = w[1 - e^{-\beta b/2}\cos(\beta b/2)] \qquad (11.16)$$

Figure 11.7 shows the relationship between the radius of relative stiffness r and the moment for aisle widths of 1.2 and 1.5 metres; also shown for comparison is the relationship between r and moment/w for partition loads applied over a 600 mm width. The figure is based on Equation (11.16). The units are kNm/m for

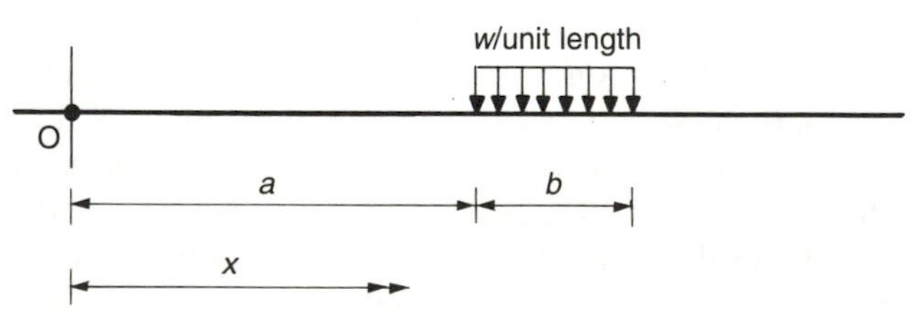

Figure 11.6 *Partial uniformly distributed load*

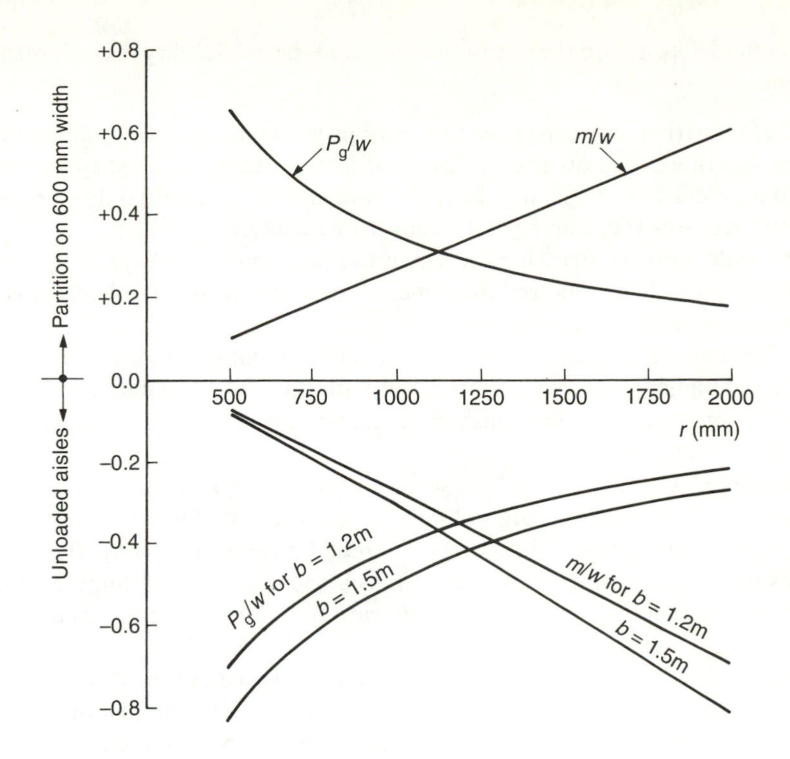

Figure 11.7 *Moment and ground pressure due to a partition, and an unloaded aisle*

moment (m), kN/m^2 for uniform loading (w), kN/m for partition load (w), and millimetres for r.

With reference to Figure 11.7, note that the partitions produce tensile stress at the bottom of the slab whereas the unloaded aisles have tension at the top. The ground pressure shown for the unloaded aisles represents a reduction from that under a uniformly loaded slab.

11.6 Construction

Unbonded tendons are often preferred in post-tensioned ground floors because of the ease of construction, though bonded tendons have been used. The amount of prestress required being much smaller than that in a suspended floor, 13 mm strand ($\frac{1}{2}$ in) is often used.

The floors are usually prestressed in both directions. In the long direction the prestress is applied within 24 to 48 hours of casting the concrete. In the transverse direction, the time of application of the prestress depends on the method of construction, with early prestressing preferred.

The currently favoured methods of construction are briefly described below. They are similar to the methods used for reinforced concrete floors. The choice of the method depends on the size of the floor and the availability of equipment and labour:

- The long strip method has become established as the conventional method of constructing floors on grade. The floor is cast in a series of strips up to 6 m (20 ft) wide. The strips may be laid consecutively or alternately, though the latter requires transverse prestressing to be delayed.
- The wide strip construction, in which the floor is cast in bays of up to 20 m (65 ft) in width. Specialized machinery and early stressing in both directions are required for such wide bays.

  The main advantages are those of the reduced number of construction joints and the higher productivity compared with the long strip method. However, the larger floor area to be finished in one operation requires more skilled and attentive finishers.
- The large bay method, in which several thousand square metres of a floor are cast in one operation. High slump concrete is discharged directly on the prepared grade (or membrane), and spread and screeded manually, followed by surface floating when the concrete has sufficiently hardened. Superplasticizers are used to achieve the high slump; vibrators or other mechanical compacting devices are not used.

  The advantages of the system are the ease and speed of construction, because no compaction is required. Very large areas of floor can be cast in a day. The disadvantage is that the large area requires much more skilled and attentive finishers. Also, in the interest of speed of construction, and for convenience, the fabric reinforcement is sometimes placed near the bottom of the slab, where it may be less effective.

The current trend is towards the use of high slump concrete containing plasticizers and other admixtures to give the setting and hardening times to suit the contractors. The high slump makes it much easier to achieve good compaction, often the concrete is described as *self-compacting*. Grade 40 concrete ($f_c' = 3000$ psi) is most commonly used. A higher strength may be used if it is more economical.

The slab is usually cast after the external walls have been erected. Access to anchorages would, therefore, be very restricted if they were positioned along the edge of the slab. The usual procedure is to construct the post-tensioned floor about two metres (say six feet) shorter in length than the clear distance between the external walls, leaving a metre at each end. The gap is filled at a late stage to allow time for as much shrinkage and creep to take place as the programme allows. The gap strip is constructed in reinforced concrete; reinforcement projecting from the post-tensioned floor is bonded into the gap strip.

A post-tensioned floor has a much smaller number of joints than a reinforced concrete floor, and a properly designed and constructed post-tensioned floor should be more stable and relatively crack free. Joints being undesirable from the user's point of view, the opinion is often expressed that, in floors where hygiene and a dust-free atmosphere are not particularly important, an unexpected crack

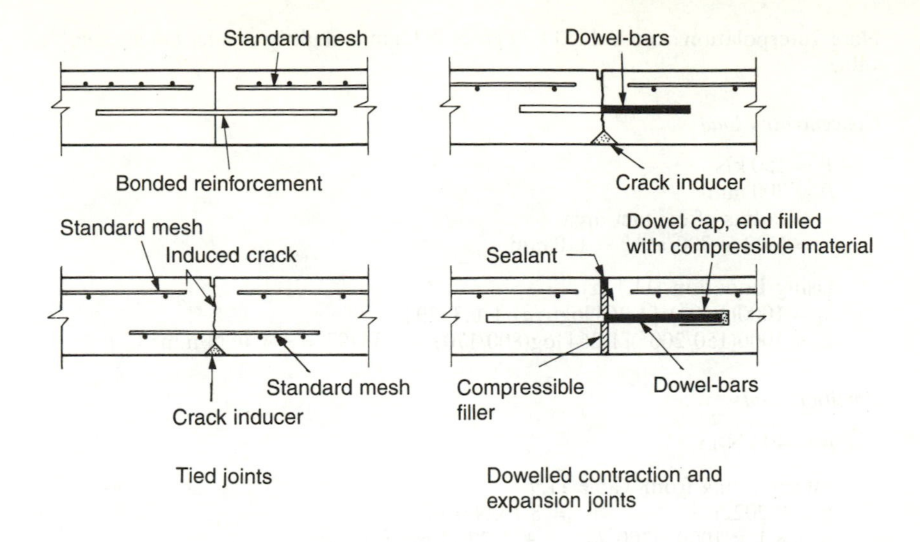

Figure 11.8 *Movement joints*

is more stable and causes fewer problems than a purpose-made joint. Cracking is, of course, unsightly.

The joints, if required in a post-tensioned floor, are similar to those in reinforced concrete. The only exception is the joint between the walls and the edge of the floor, which should be designed to cater for the relatively larger contraction of the floor due to shrinkage and creep.

Typical tied and dowelled joints, as used in reinforced concrete floors, are shown in Figure 11.8.

Example 11.1

Determine the prestress in a 100 m long 200 mm thick ground slab for the following loads, which include the safety factors:

Concentrated load of 150 kN applied over a 300 × 300 mm area
Uniform loading of 30 kN/m^2 with 1.25 m wide aisles
Partition loading of 40 kN/m applied over an effective floor width of 600 mm

Assume f_{cu} = 40 N/mm^2 and k = 50 MN/m^3

Solution

The Westergaard equation is used for calculating flexural stress under the concentrated load.

From Table 11.1, $f_{t90} = 5.06$ N/mm^2
From Table 2.3, $E_c = 30.3$ kN/m^2
From Equation (11.10a), $r = 17.087\{E_c D^3/k\}^{0.25}$
$= 17.087\{30.3 \times 200^3/50\}^{0.25} = 800$ mm

Note: Interpolation from Table 11.7a gives 793 mm, a slightly smaller but acceptable value.

Concentrated load

$P = 150$ kN
$D = 200$ mm
a = radius of contact area
$= (300 \times 300/\pi)^{0.5} = 170$ mm

Using Equation (11.11a),
$\sigma_i = 1000(P/D^2).[1.264 \log(r/a) + 0.3379]$
$= 1000(150/200^2)[1.264 \log(800/170) + 0.3379] = \pm 4.46$ N/mm^2

Partition load

$w = 40$ kN/m

Interpolating from Table 11.9,
$m = 0.2022w$ $= 8.1$ kNm/m
$\sigma = 8.1 \times 1000 /(200^2/6) = \pm\ 1.22$ N/mm^2
$P_g = 0.434w$ $= 17.36$ kN/m^2

UDL with aisle

$w = 30$ kN/m^2

From Figure 11.7,
$m = 0.2 \times w$ $= 6.0$ kNm/m
$\sigma = 6.0 \times 1000 /(200^2/6) = \pm\ 0.90$ N/mm^2
$P_g = 0.6w$ $= 18$ kN/m^2

Prestress

Of the three loading cases, the concentrated load is the critical one,
stress $= \pm 4.46$ N/mm^2

Using $\mu = 1.5$, from Equation (11.7a), p_{av} the average stress required to overcome drag friction,
$p_{av} = 0.018L$ $= 1.8$ N/mm^2

Therefore, total required prestress after allowing for f_t
$= 4.46 + 1.8 - 5.06 = 1.2$ N/mm^2

P = required final prestressing force
$= 1.2 \times 200$ $= 240$ kN/m width

Assuming a final force of 170 kN per 15.2 mm superstrand,
strand spacing $= 1000 \times 170/240 = 708$ mm centres

Provide 15.2 mm superstrand at 700 mm centres.

12 DETAILING

This chapter is concerned primarily with the production of post-tensioning drawings. It also deals with some related aspects which may need to be considered but are not necessarily part of the calculation process.

12.1 Drawings and symbols

The specialist prestressing contractor requires shop drawings, showing the information required to assemble all rod reinforcement, post-tensioning hardware and tendons, and for stressing the tendons. Often, separate drawings are produced for the prestressing items and rod reinforcement. The drawings showing rod reinforcement conform with the practice for reinforced concrete.

The post-tensioning shop drawings are usually prepared by the specialist, to suit his products and method of working. They should show:

anchorage locations on plan and section
anchorage pocket dimensions

tendon layout in plan
bundling of tendons
horizontal and vertical profiles
points of inflexion

tendon heights at support points
order of assembly

tendon forces
calculated extensions

Anchorages should be clearly identified with regard to the number and types of strand they are designed for. The actual number of strands being anchored in an anchorage assembly may, of course, be less than the capacity of the anchorage. The live end and the type of dead anchorage should be identified.

Tendons are usually colour coded for identification. The drawing should show the number and type of strands in a tendon, and its colour coding, length, jacking force and extension. This information is given at the live end in case of tendons to be stressed from one end.

The symbols, proposed by the Concrete Society, for identifying the number of

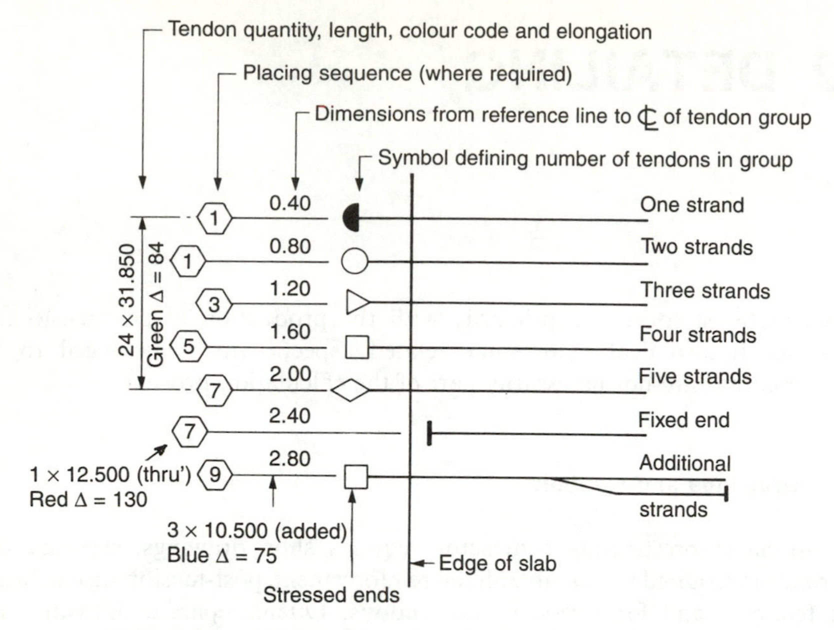

Figure 12.1 *Tendon notation*

strands in a tendon, and the data to be given at the live end, are shown in Figure 12.1.

When the design is carried out by other than the specialist contractor, the information given to him by the designer varies according to the local practice. The minimum practical information required by the specialist to enable him to produce his shop drawings should include:

- Drawings identifying the structural elements to be post-tensioned and their sizes
- Concrete quality, minimum covers and
  either: guidance on the strength required at stressing
  or: stressing stages
- Tendon profile type for each element—harped or parabolic
  High and low points of the profile
- Either: final prestress in concrete
  or: number and type of tendons and jacking forces.

Figure 12.2 shows part of a typical drawing giving the consultant's requirements for the prestress and rod reinforcement for a continuous two-way spanning solid floor. Tendon heights are shown on a section in the longitudinal direction; a similar section is required in the transverse direction. Note that the single strand tendons are not marked with the symbols proposed in Figure 12.1

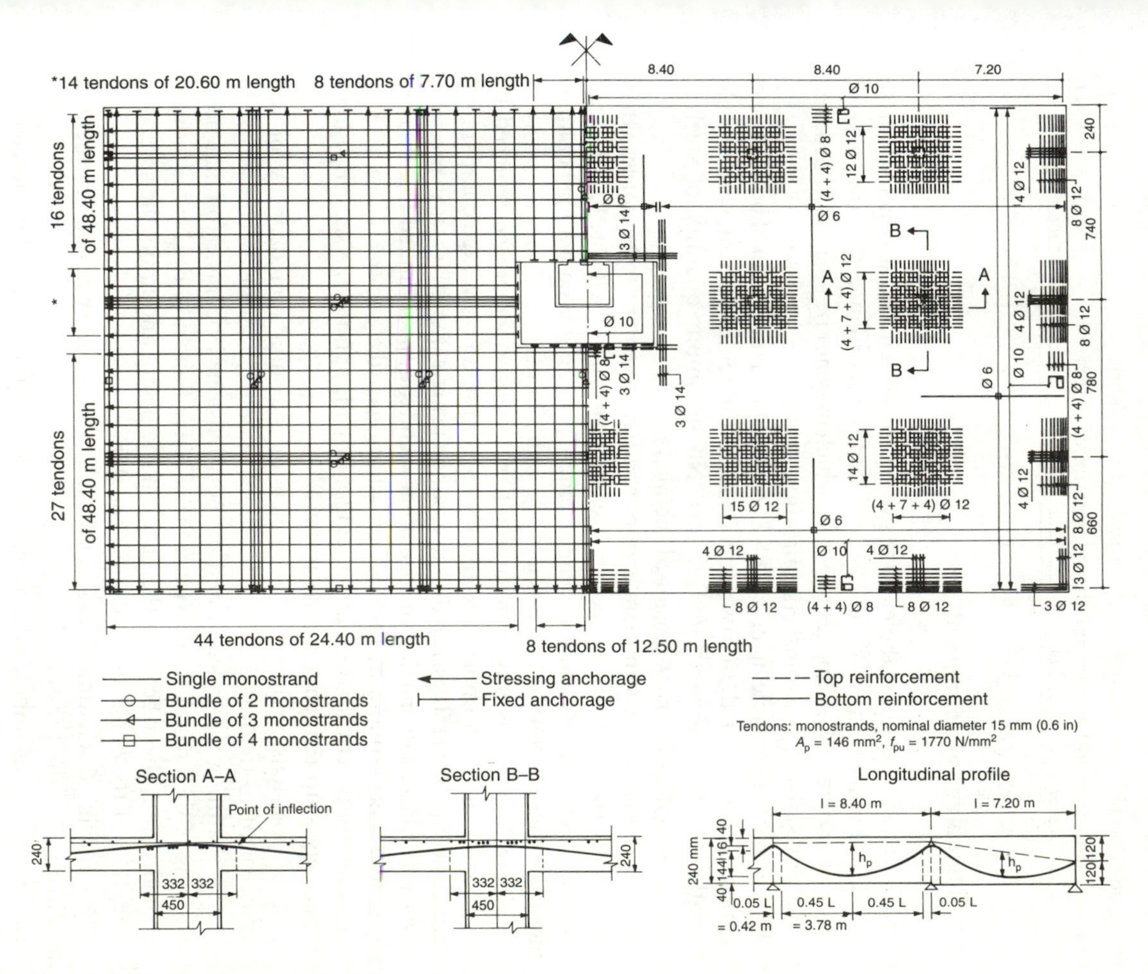

Figure 12.2 *Tendon and reinforcement layout*

Table 12.1 *Minimum steel proportions in reinforced concrete* (*ACI 318*)

Description	*Percentage*
For grade 300 (45 ksi) deformed bars	0.20
For grade 400 (60 ksi) deformed bars or welded wire fabric	0.18
For steel with f_y >400 N/mm^2 (60 ksi) at a strain of 0.35%	$0.18 \times 400/f_y$

12.2 Minimum reinforcement

In a post-tensioned floor, the elements designed as reinforced concrete, or so implied, rod reinforcement should be provided in accordance with the rules for reinforced concrete. This applies to beams which may not be post-tensioned, and to one-way slabs in the transverse direction.

For reinforced concrete members BS 8110 specifies a minimum of bonded steel quantity of 0.13% of the gross concrete area for reinforcement with a yield strength (f_y) of 460 N/mm^2 (66 000 psi). No minimum quantity of steel is, however, specified for post-tensioned members.

ACI 318 requires the minimum area of rod reinforcement to be not less than that given by the ratios of gross concrete section shown in the Table 12.1, but not less than 0.14% in any case. This reinforcement is to be spaced not further apart than five times the slab thickness, nor 500 mm (20 in).

ACI 318 allows this minimum amount of reinforcement to be replaced by prestressed tendons proportioned to provide a minimum average comparative stress of 1.0 N/mm^2 (150 psi) on the gross concrete area after all losses in prestress have taken place. The maximum spacing of such tendons is limited to 2.0 m (78 in). Where their spacing exceeds 1.4 m, additional steel, as per Table 12.1, is required between the tendons at slab edges extending from the slab edge for a distance equal to the tendon spacing.

Use of post-tensioning tendons in lieu of the minimum quantity of rod reinforcement in the non-stressed directions should be seriously considered in one-way spanning floors supported on post-tensioned beams. Nominal post-tensioning of the area of slab between the effective flanges of the beams reduces the longitudinal shear between the slab and the beam, and limits the spread of prestress from the beam into the adjacent slab.

The following discussion relates to the minimum quantity of reinforcement in the direction of the prestress.

In members using unbonded tendons, ACI 318 requires bonded reinforcement to be provided, distributed uniformly along the tension face. The area of this reinforcement should not be less than 0.4% of the area of the concrete on the tension side of the section centroid. This reinforcement is required regardless of the service load stress condition.

In two-way slabs, ACI 318 allows the bonded reinforcement in the positive moment zone to be omitted if the computed tensile stress in concrete at service loads does not exceed $[(f_c')^{0.5}]/6$. If this limit is exceeded, then the minimum amount of bonded reinforcement required is

$$A_s = N_c/(0.5f_y) \qquad (12.1)$$

In the above expression, N_c is the tensile force in concrete under service load, calculated on the basis of an uncracked section. f_y should not exceed 400 N/mm^2 (58 000 psi). This steel should be uniformly distributed over the tensile zone as close as practicable to the tension surface.

In negative moment areas at column supports, the minimum area of bonded reinforcement in each direction should be computed by

$$A_s = 0.00075DL \qquad (12/2)$$

where L is the span length in the direction of the reinforcement being considered. This reinforcement should be distributed within a slab width between lines that are 1.5D outside opposite faces of the column. At least four bars should be provided in each direction and their spacing should not exceed 300 mm (12 in).

Links in post-tensioned beams are normally governed by requirements of shear capacity. Some links are, however, required in the beams and in the ribs of a floor to support the tendons; the normal spacing for such supports is 1000 mm (40 in). Figure 12.3 shows typical arrangements

In the case of a number of slab panel tendons getting damaged, the concrete may be able to sustain some of the load provided that arch action can develop. This may be possible in the inner spans of a continuous two-way floor with a series of spans in each direction. The edge spans may also develop arch action along the direction of the edge, but not across it. The corner panels would be very unlikely to develop any arch action. In such floors it may be prudent to provide the minimum quantity of reinforcement in corner panels in both directions and in edge panels at right angles to the free edge, particularly if the prestress is provided by unbonded tendons. A small amount of tie reinforcement would, of course, be

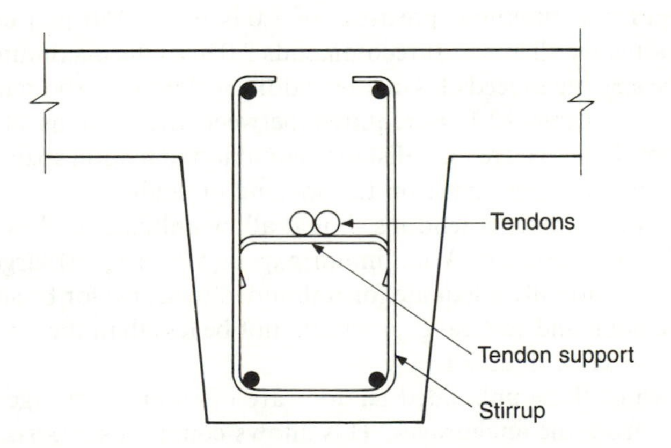

Figure 12.3 *Tendon supports in beams and ribs*

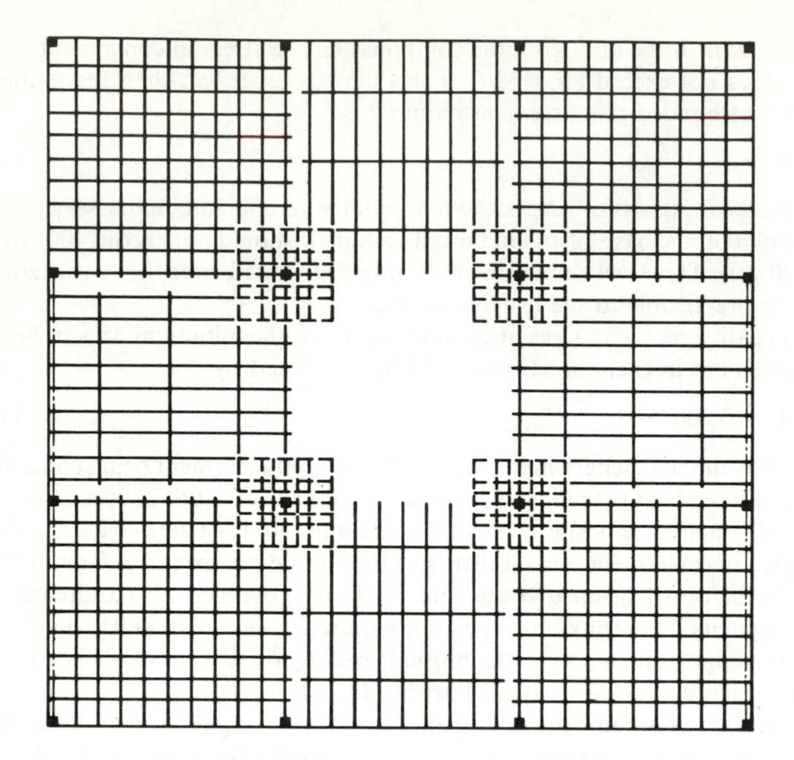

Figure 12.4 *Suggested arrangement of minimum reinforcement*

required in the edge panels parallel to the free edge. Figure 12.4 indicates the suggested position of such reinforcement.

12.3 Tendon spacing and position

ACI 318 requires a minimum prestress of 1.0 N/mm^2 (150 psi) on the gross concrete section after all losses. It recommends 2.0 m as the maximum spacing of tendons. If the spacing exceeds 1.4 m, then additional bonded rod reinforcement, of area given by Table 12.1, is required between the tendons at slab edges extending from the slab edge for a distance equal to the tendon spacing. BS 8110 does not specify any upper limit on the spacing of tendons.

The gap between adjacent tendons should allow unhindered flow of concrete through it during vibration. A minimum gap of 5 mm ($\frac{1}{4}$ in) larger than the aggregate size is normally adequate for unbonded tendons; for bonded tendons the clear horizontal and vertical gaps should not be less than the size of the duct measured in the same direction.

In slabs, two or three unbonded tendons are often bundled together a short distance away from the anchorages. This allows common supports to be used, and a greater gap is made available between the tendon groups for service holes.

The angular deviation, over the distance between the anchorages and the point where they touch each other, should be taken into account in calculating the friction loss in prestress.

It is desirable to run a few tendons through the internal columns, and to anchor some within the width of the external column if possible. Such tendons act as ties between columns and transfer some of the shear load directly to the columns. In practice, however, the internal columns are often so congested with steel that tendons cannot be accommodated within their boundaries. It is often not possible to provide anchorages in the external columns, because of congestion, and because the loss of concrete due to the anchorage pocket may not be acceptable. In such circumstances, there is no alternative but to locate all of the tendons, and the anchorages, outside the columns; the requirement of adequate rod reinforcement, to satisfactorily transfer the shear load from the tendons to the columns, should be checked.

12.4 Deflection and cladding

In normal reinforced concrete, the deflection is always downwards, unless the formwork was provided with a camber. In a post-tensioned floor, the prestress may induce an initial camber, which may change to a downward deflection when the particular element is fully loaded. The creep effect would, over a long period, increase the deflection, or the camber if under full load the deflection was still upwards. This can cause difficulties in fixing precast panels, particularly if heavy precast units are seated on thin edges of the slab, or if the cladding derives vertical support from such a slab. The alternatives and possibly their combinations, to be considered in such a case, include:

- Increase the stiffness of the slab edge. This would normally require a local increase in the thickness or the provision of a downstand; both are in conflict with the buildability aspect of a flat soffit.
- Design the cladding to be independent of the slab edge. The precast panels can be designed to bay lengths, supported at columns; the panels would, of course, be large, heavy, and difficult to handle and manoeuvre on site.

  Curtain wall type of cladding can be attached to the slab edge for horizontal support but not vertical. Either its load must be transferred to the columns by structural members incorporated within the cladding system, or it must be self-supporting.
- Design the cladding to accommodate the expected movement of the slab edge.

12.5 Movement joints

In post-tensioned floors it is often necessary to delay the connection between a slab and a stiff vertical element, such as a lift core or shear wall. The delay allows some of the shortening in the slab length to take place before the connection is established. Without this delay, the shortening due to the axial elastic and creep strains would be restrained by the vertical element, with the result that some of the axial prestress would be lost from the floor to the vertical element.

A long delay of several months would, of course, be desirable, because it would allow more of the movement to take place, but the construction programme is unlikely to allow this; a period of one month is generally considered to be the minimum. The need for delayed connection has been discussed in Chapter 5. Figure 12.5 shows two arrangements for joints between a slab and a wall. Both allow the wall to be constructed above the particular floor level while leaving the floor temporarily free to move laterally.

In arrangement (a), a gap of about one metre is left unconcreted between the slab and the wall, and is concreted after a suitable delay. The arrangement is called a *gap* or an *infill strip*. The slab anchorages, whether live or dead, are cast in the slab; they are located away from the wall and so do not transfer any of the shear to the wall. The slab must remain propped until the connection with the wall is fully effective. The reinforcement in the gap must be sufficient to enable the gap strip to act as reinforced concrete member for the combination of the bending moment and residual axial tension resulting from the restraint offered by the wall.

Arrangement (b) consists of a toothed support to the slab, provided at short lengths of the wall left unconcreted. The slab is supported directly by the wall, thus dispensing with the props. Meticulous attention to the construction of the sliding joint is essential to ensure freedom of lateral movement. The vertical tube through which the dowel passes may consist of a piece of prestressing sheath if bonded tendons are being used. The quantity of concrete required to fill the tubes and make good the gaps in the wall is very small compared with that needed in the gap strip. The anchorages are located in the *teeth* of the slab, as near the wall as they would have been in the absence of the joint. This arrangement is slightly more difficult to construct than (a), but the absence of props should allow a longer delay.

A gap strip can be used in the middle of a large slab area, to reduce the movement. If the strip runs along the direction of span of a one-way slab then no propping is required.

A post-tensioned floor requires the same considerations for a movement joint as a reinforced concrete floor, but with one major difference. A reinforced concrete floor reduces in length because of the shrinkage of concrete; the action is resisted by the reinforcement content of the floor. By comparison, a post-tensioned concrete has a smaller quantity of steel to resist shrinkage, and it undergoes further long-term shortening in length due to creep. It is, therefore, extremely unlikely that a post-tensioned floor will ever in its life reach the same length as it had at the time of its construction. Therefore, post-tensioned floors only need contraction joints, where the gap is likely to widen. The joint filler is likely to become loose in the joint and come out if it is not restrained. The sealant chosen for such a joint must be capable of withstanding the tension which may develop at the contraction joint.

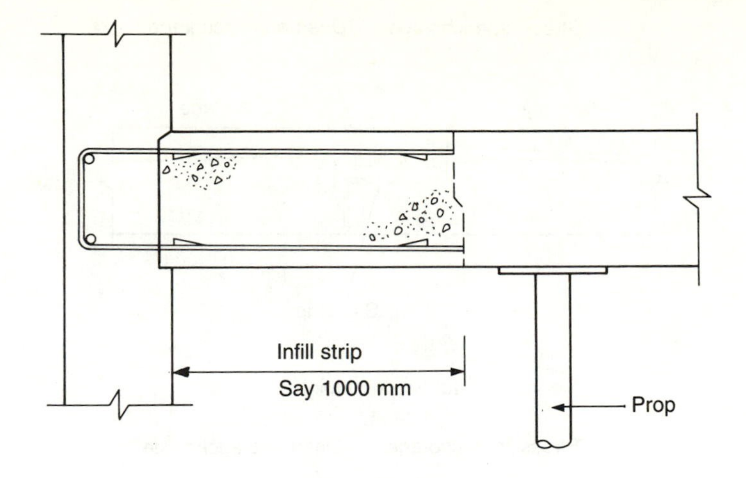

(a) Infill or gap strip

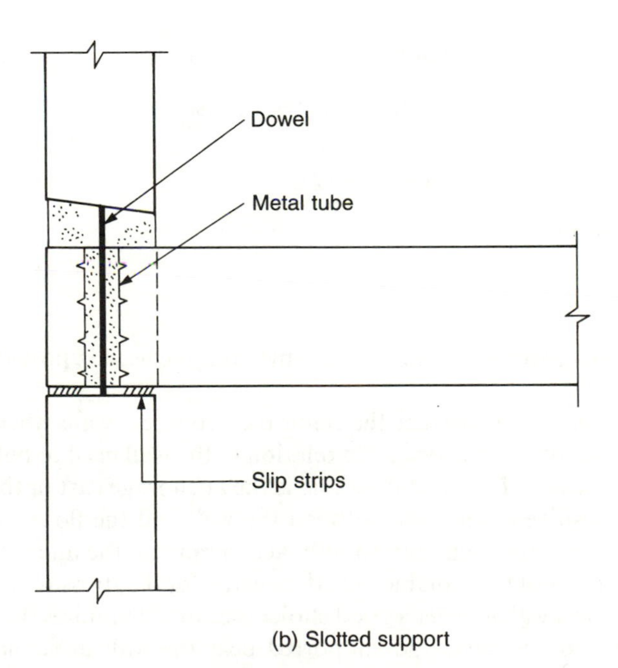

(b) Slotted support

Figure 12.5 *Temporary release joints*

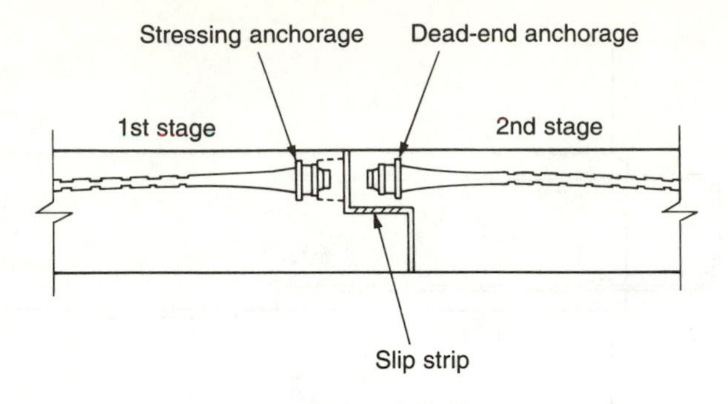

(a) Stepped joint

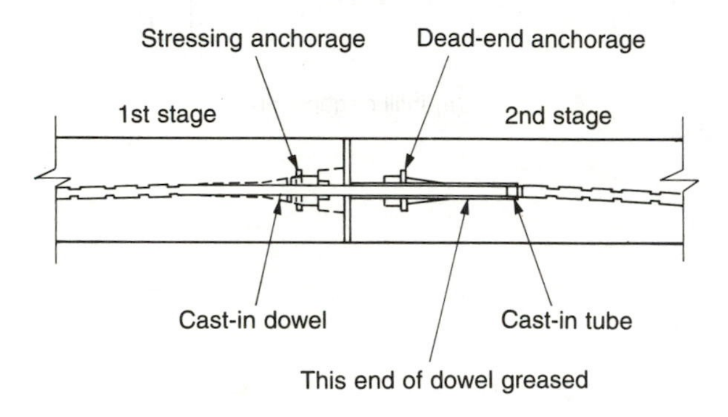

(b) Dowel joint

Figure 12.6 *Contraction joints*

Figure 12.6 shows two contraction joints which may be used in post-tensioned floors.

It is often expedient to construct the reinforced concrete walls ahead of the floors. The positions of the anchorages in relation to the wall need to be carefully considered in such cases. The most desirable is the anchorage cast in the wall; it then provides a positive connection between the wall and the floor. However, often the walls are not thick enough to fully accommodate the anchorage, and space may sometimes not be available outside the wall for the stressing operation. A dead anchorage in a wall is never a good choice, because it requires the whole of the tendon length to be coiled and supported near the wall until the floor is constructed. Figure 12.7 shows the construction joints at the floor-wall contact. Details (a) and (b) show the dead ends, and (c), (d) and (e) the live ends. In all cases the rod reinforcement at the joint must be capable of coping with the tension, and any flexure, which may develop.

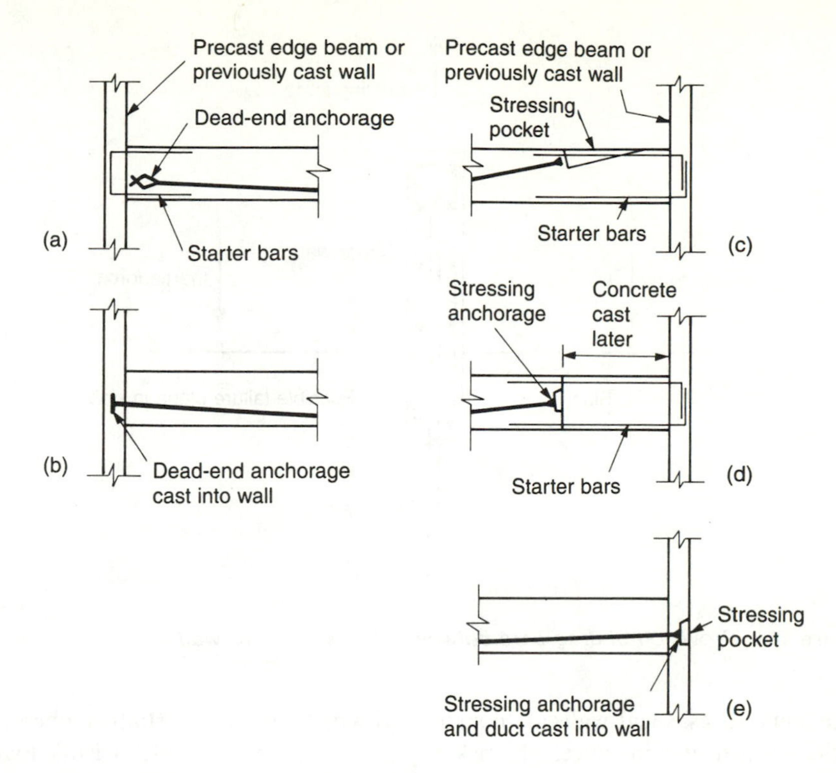

Figure 12.7 *Wall-floor construction joints*

12.6 Detailing for seismic resistance

This section outlines some additional detailing considerations for slabs constructed in areas of high seismicity. It was stated in Section 4.13 that no special seismic design procedures are generally required for floors. There are, however, a few issues relating to seismic resistance which should be borne in mind during the detailing of the floor.

As was discussed in Section 4.13, the principal role of a floor under seismic loading is to act as a horizontal diaphragm. In order to ensure adequate diaphragm action and prevent cracking, holes in the floor should be framed with additional unstressed rod reinforcement or with edge beams. However, the framing necessary to ensure diaphragm action is unlikely to exceed that required for the resistance of gravity loads and the anchorage of the tendons.

Cantilever slabs, being very flexible, are likely to undergo particularly large displacements under seismic loading. Bending reinforcement should be provided in both the top and bottom faces of such slabs, since the combination of frame bending and upwards seismic acceleration can cause a reversal of the normal curvature. Eccentricity of the prestress should be limited so that no tension is produced in the absence of gravity loads.

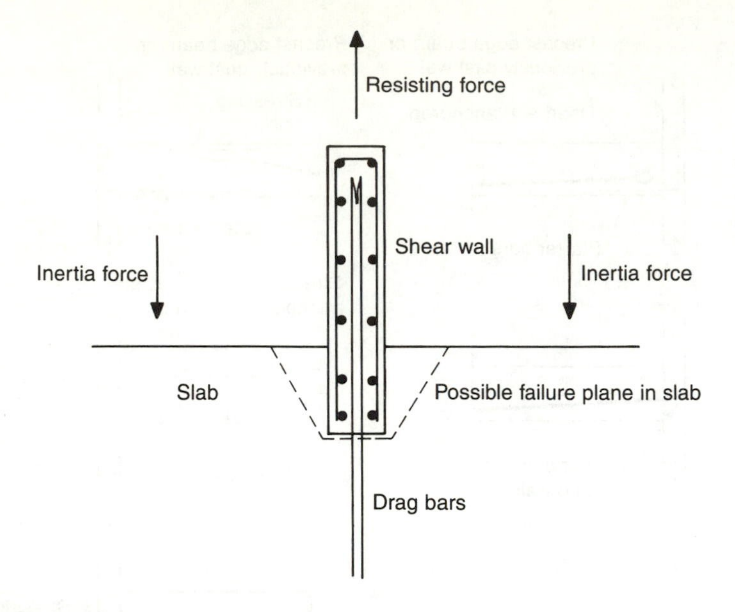

Figure 12.8 *Provision of drag bars between slab and shear wall*

In some cases, shear walls are constructed which are only partially embedded in the slab. In such instances, the risk of pull-out of the wall under seismic loads can be prevented by the provision of drag bars, as shown in Figure 12.8.

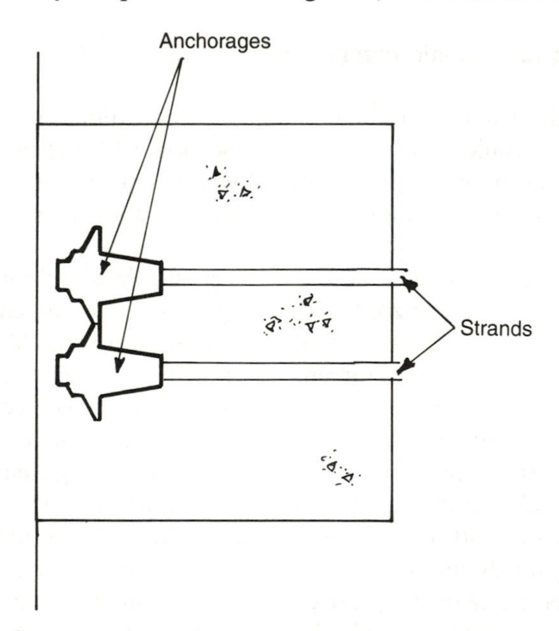

Figure 12.9 *Tendons as column ties*

Tendons can be detailed so as to tie columns into the slab as shown in Figure 12.9. The disadvantage of this approach is that the anchorages take up a considerable amount of space at the column's most congested section; Figure 12.9 shows two monostrand anchorages in a column approximately 300 mm wide (12 in). However, if the room can be spared, then this measure provides a highly beneficial level of structural continuity.

13 SITE ACTIVITIES AND DEMOLITION

This chapter deals with those aspects of construction and site activities of particular relevance to post-tensioned floors. It deals with the care of the proprietary materials and the activities of the specialist subcontractor on a project containing post-tensioned floors. It is not meant for the specialist—though he may derive some benefit from it; its purpose is to give some insight into the site procedures to other disciplines involved in the project. Also, recommendations are given for the safe demolition of post-tensioned concrete structures.

The sequence of operations performed in the construction of a post-tensioned building is given below. The important operations are discussed under separate sections.

1. Erect soffit formwork
2. Erect edge boards
3. Attach anchorages to vertical edge boards
4. Place and secure bottom reinforcement
5. Place and secure anchorage zone reinforcement
6. Place and secure tendons (or sheaths)
7. Place and secure top reinforcement
8. Place concrete and cure
9. Remove edge boards and pocket formers
10. Test concrete cube or cylinder for strength required for stressing
11. Thread bonded tendons, if only the sheaths were assembled in stage 6
12. Stress tendons
13. Strip formwork and backprop if required
14. Grout sheaths for bonded tendons
15. Cut surplus strand lengths
16. Apply rust-inhibitant, and place grease caps
17. Fill anchorage pockets.

Other than the specialist prestressing crew, the main workforce on a post-tensioned floor site consists of three major categories: formwork erectors, steel fixers and concretors. For maximum efficiency, the site operations would be so organized that the three crews have continuous work, with no or minimum idle time. The site programme would, therefore, depend on a number of factors, including the

Table 13.1 *A typical programme cycle*

Ref	*Description*	*Day*	*1*	*2*	*3*	*4*	*5*	*6*	*7*	*8*	*9*
	Columns formwork										
	Columns reinforcement										
	Concrete columns										
	Strip columns										
1 2	Slab formwork										
3	Fix anchorages										
4–7	Steel & tendons										
8	Concrete										
9 10	Strip edges										
12	Stress tendons										
13	Strip forms & backprop										
14	Grout & finish										
	Columns formwork										

size and plan shape of the building. Table 13.1 shows a construction programme consisting of a five-day cycle, which may be adapted to suit the particular needs of a project.

13.1 Storage of materials

All of the specialized prestressing components should be handled with care and stored under cover, away from chemicals that may attack steel, and protected from excessive moisture. Particular care is needed to avoid damage to plastic extrusion on unbonded strand.

The anchoring wedges and their seatings in the castings should not be allowed to get rusty, dirty, or greasy. Rust or dirt on the teeth of the wedges, or the conical surfaces of the wedges and the anchorages, weakens the grip of the anchorage on the strand. Grease may have a similar weakening effect on the grip. Grease on the conical anchoring surfaces also reduces friction, which results in an increase in the lateral bursting force on the anchorage casting.

The strand should be protected from rusting and from attack by chemicals, particularly deleterious are chlorides, nitrates, sulphides, acids and hydrogen releasing agents. The grease-packed extrusion around the strand for use in unbonded tendons provides adequate protection. Care should be taken in transporting, handling and storing the tendons to avoid damage to the protection. The extrusion can also be damaged by vibrators during concreting. The plastic extrusion and grease are removed from the strand at the anchorage, so that the wedges can grip it. The strand in this zone is, therefore, exposed to possible attack, and an unbonded tendon relies completely on the integrity of the anchorage which, of course, is not protected in the manner of strand. It is, therefore, particularly important to ensure that the anchorage assembly is well protected by a rust-inhibitant and a grease cap, and that the anchorage recess is properly made good.

Prepared tendons should be clearly marked to identify the location where they are to be used.

Tendons and strand are usually supplied in coils; extreme care is necessary to ensure that the coil does not unwind uncontrollably, as this is likely to cause accidents.

Metal sheathing for use with bonded tendons is usually not very sturdy, and is prone to mechanical damage. Any holes or tears may allow ingress of mortar paste during concreting which would hinder the stressing of the tendons. Multiple handling of the sheath increases the risk of damage. Open ends of the sheath should be protected to avoid any material, rainwater, or water from the site finding its way in. Water in the sheath is likely to render the grout ineffective and cause rusting of the tendon, and any deleterious matter would also attack the tendon.

Anchorages, tendons and sheaths should be stored in the order in which they are to be used. This procedure reduces multiple handling and the risk of damage.

All equipment should be stored in a clean and dry location, and it should be accessible only to the operatives. Maintenance and repair, which may affect the calibration of the equipment, should not be allowed on site.

13.2 Installation

To avoid split responsibility and possible delays, the same crew should be responsible for the assembly and installation of rod reinforcement, tendons and anchorages. The installers, the stressing operatives and the personnel responsible for the finishing operations—grouting, cutting strand, making good the pockets—should be controlled by one supervisor.

The installers should study the shop drawings and other relevant documents before starting on site. They should work out a sequence of operations, allowing for the incorporation of any non-structural elements in the slab, such as conduits or service runs.

Holes in vertical edge boards, for the strand to project through, should be drilled accurately so that the live anchorage can be attached in the specified position. The live anchorage should be set square to the board, unless otherwise specified. It should be firmly attached to avoid being dislodged during concreting. No mortar paste should be allowed to reach any of the anchoring surfaces. Dead anchorages should be securely supported on chairs, with the required end cover; they must not be attached to the vertical edge board of the formwork.

Pocket formers should be rigid, they should fit into the anchorage without any gaps, and should be firmly attached so as not to be damaged or displaced during concreting. They should be set true in position. Expanded polystyrene formers are often difficult to remove. Burning or cleaning with chemicals leaves the surface of the pocket in too poor a state for the concrete plug to make a good seal.

Anchorages should remain accessible for subsequent operations, unobstructed by scaffolding or other construction.

For bonded tendons, the strand may be threaded in the sheath before the sheath is laid in position. Threading of the strand after the sheaths have been placed in position may cause them to be displaced. The operation also requires access to the sheath end, which may not be available on a building site, particularly at the upper floor levels.

Tendons may also be threaded after concreting. In this case it is important to prove that the duct is unobstructed by drawing a dolly through it.

It is important that joints between lengths of sheathing are watertight. If water gets in then it may cause corrosion of the tendon and, worse, if cement grout gets in during concreting then the tendon would become bonded and stressing may not be possible. Subsequent remedy can only consist of cutting out the damaged portion and re-concreting which can be time consuming and expensive.

Make sure that the requisite length of strand is available for jacking at the live end.

After laying the unbonded tendons in position, a length of the plastic extrusion should be removed at the live end and the strand inserted through the anchorage casting. No more than 25 mm (one inch) of bare strand should be exposed behind the anchorage. A watertight sleeve should be placed on this length, so that none of the bare strand remains exposed behind the anchorage; this is required to prevent bond.

Tendons should be supported on properly designed chairs, or on rod reinforcement, and secured in position at the specified heights so that none get displaced during concreting. The tendon supports are normally spaced about one metre (40 in) apart. Care should be taken to ensure that the support arrangement is stable and that the ties are not too tight around an unbonded tendon; else the plastic extrusion may get damaged.

Horizontal deviation of the tendons from the intended position should be minimized as the wobble causes a loss in the prestressing force. This is particularly important in ribbed and waffle floors, where the horizontal deviation of a tendon may set up high lateral tensile stresses in the thin web. This may cause local bursting of the concrete if adequate reinforcement has not been provided in the anchorage zone. The horizontal wobble in a ribbed, or waffle floor, may be minimized by securely tying the tendons to a reinforcement cage in the rib.

To ensure that a tendon develops the expected prestressing force, and works as envisaged in the design, it is essential that its high and low points are placed accurately. Vertical deviation of the tendon from its intended profile between the high and the low points should be kept within the specified tolerance. If not specified, the tolerances in Table 13.2 may be used.

13.3 Concreting

The concreting operation for a post-tensioned floor is very similar to that for a reinforced concrete floor; the concretor must be aware of the following:

- Sheaths for bonded tendons and the plastic extrusion on unbonded tendons can be easily damaged during concreting, this would cause problems in stressing and grouting operations.
- Anchorages and tendons must not get displaced during concreting. Vibrators and pump pipeline should not be allowed to contact the tendons.
- It is essential that the concrete immediately behind and around the anchorages is well compacted.
- The edge boards should be removed as soon as possible after the concrete has

Table 13.2 *Vertical tolerances for tendon profile*

Depth		*Tolerance*	
Up to 200 mm	(up to 8 in)	± 6 mm	(±¼ in)
200–600 mm	(8–24 in)	±10 mm	(±⅜ in)
over 600 mm	(over 24 in)	±12 mm	(±½ in)

hardened, but while it is still green. This allows the pocket formers to be easily removed, and the pockets to be inspected and any defects to be made good.

- The concrete should be protected against damage from the environment and site operations, and cured so as to avoid cracks from any cause including shrinkage.

For post-tensioned floors, it is the normal practice to test the concrete strength at about three days, when some or all of the prestress is applied. Concrete samples should be taken and cured as specified. Typical stressing systems allow 50% prestress to be applied at a cube strength of 15 N/mm^2 and full prestressing at 25 N/mm^2 (1750 and 3000 psi respectively). This must be verified for the particular system used.

13.4 Stressing

Stressing of the tendons is the central and most important operation of the post-tensioning process. It also carries an element of risk, in that potentially dangerous situations may arise when things go wrong. For these reasons, the subject is discussed in some detail.

In cold weather, the normal grease used in unbonded strand may be too stiff for the specified extension to be achieved. it may be necessary to delay full stressing until conditions are warmer. If cold conditions are anticipated then it may be possible for the strand supplier to provide a more viscous grease.

In hot weather, the reduced friction resulting from the more viscous grease may result in higher extensions and tendon forces.

13.4.1 Safety

All equipment must be checked to ensure its proper operation and calibration. Hydraulic hoses and connections should be checked for signs of leakage.

Stressing the tendons involves straining the tendons with high forces; the operation carries some risk of accident. Therefore, protective means—boards or sandbags—should be placed in line with the anchorage to arrest the projectiles in case of an accident. Only trained personnel should be allowed in the operations area.

It is advisable not to stand near a jack or the pump during stressing, or in front of an anchorage during or after stressing, until the pocket has been made good.

Hydraulic jacks should be tethered during stressing to prevent their falling down, possibly off the floor being stressed on to the ground below, in case of a tendon breaking.

Before stressing, the concrete in the pocket and near the anchorages should be inspected for any signs of weakness, such as cracks or honeycombing. If any such defect is found, or any of the projecting tendons are not at right angles to the face of the anchorage, then stressing must not proceed. The defect should be inspected by the engineer and the remedial works carried out, if necessary.

The wedge seatings in the anchorages should be inspected, and cleaned if necessary. They should be free of rust, grease, oil, dirt and other contaminants.

In case of a problem, the concrete around an anchorage has to be cut out. In cutting the concrete, attention must be given to its effect on the adjacent anchorages if they are stressed. It may be safer to de-tension them.

Don'ts of stressing

- Do not stress any anchorages with grout inside the casting. Grout in the casting may prevent proper seating of the wedges. It is safer, and less expensive, to clean out grout than to have to de-tension, repair or replace tendons, or repair the jack.
- Do not use the jack when it does not seat properly into the casting.
- Do not overstress tendons to achieve proper elongation.
- Do not allow obstructions in the path of the jack extension.
- Do not continue stressing if something is not working properly.
- Do not de-tension with loose plates, spacing shims or with two jacks in tandem.
- Do not stand near the jack, between the jack and pump, or over the anchorage during stressing or de-tensioning.
- Do not hammer or beat on the jack or jack cylinders.
- Lastly, do not do anything if you are not sure; ask someone who knows.

13.4.2 Stressing procedure and measurement of force

The tendons should be stressed in the order agreed with the designer. Stressing should take place as soon as possible after the concrete has reached the required strength.

For *simultaneous* stressing of tendons from both ends, good communication and coordination between the two teams are extremely important in ensuring simultaneity of operations. Simultaneous stressing, however, is seldom used.

The tendon force is read from a calibrated pressure gauge, and is monitored by observing the tendon elongation. A pressure cell or a proving ring may be used for a more accurate and direct measurement of the force.

Tendon elongation is measured, and recorded, at the same time as stressing takes place. Elongation is measured to an accuracy of 2 mm (to the nearest $\frac{1}{8}$ in).

Before stressing, the tendons have an unknown amount of slack. In multistrand tendons, each strand may have a different slack, but the difference cannot be allowed for if all strands are stressed together. To allow for the slack, the following procedure is used in stressing.

- The strand is gripped and stressed to about 10% of the jacking force.
- The strand is marked a set distance away from the anchorage face using a reference measuring device (which may be a piece of wood), usually by a paint spray.
- The tendon is stressed but not locked. The force is recorded. The reference measuring device is placed against the face of the anchorage, the elongation of

STRESSING RECORD

Date ____________
Job name ________________________________ Job No. ____________
Location ________________________________ Sheet ______ of ______
Floor level ________________ Zone ____________ Stressed by ________
Equipment ________________________________ Verified by ________
Remarks ____

Tendon Reference	Specified		Measured			
	Jacking force	Elongation	Force		Elongation	
			Initial	Jacking	Jacking	Draw-in

Figure 13.1 *Stressing record sheet*

the tendon is measured from the end of the device to the paint spray mark, and recorded.

- The tendon is locked, the reference measuring device is again placed against the anchorage face, and the final elongation measured. The difference between this and the previous measurement is the wedge draw-in, which is also recorded.

Figure 13.1 shows the format of a sheet for recording the stressing data.

Most of the jacks for use with post-tensioned floors have a ram movement of the order of 200 to 300 mm (8 to 12 in). For long tendons, the required elongation may be more than the ram stroke, and it may have to be stressed in two or more strokes. After the first stroke, the tendon is anchored, the jack retracted, the tendon re-gripped, and re-stressed.

13.4.3 Short members

The relatively high loss in prestress from draw-in, coupled with the uncertainty over the amount of strand slip, makes it difficult to stress short tendons. A useful procedure to follow in stressing short tendons is:

- Make a generous allowance for draw-in in the design.
- For the first few tendons, use the maximum jacking force; most standards allow 80% of the strand strength.
- Measure the actual draw-in for each tendon.
- Calculate the force in each strand from the measured draw-in
- Adjust the jacking force for the next tendon as necessary.

The procedure is slow, and requires accurate measurement of forces and extensions, and knowledge of the design. Bar systems are often preferred for short lengths, because they use a threaded anchor which has no draw-in.

13.4.4 Stressing problems

Problems may arise during the stressing operation mainly from two basic deficiencies: a poor concreting operation and malfunctioning of the equipment or wedges.

The process of rectifying the defects may require the force in a previously stressed tendon to be confirmed. This is done, if the projecting strands have not been cut off, by gripping the tendon in a jack and re-stressing until the wedges just lift off their seats. When the wedges are initially locked, part of the tendon force is supported by friction between the wedges and the conical seating. The force required to lift the wedges must overcome the residual tendon force and the friction. Therefore, the jacking force in this operation is initially high, and it drops when the wedges become free. The lower value represents the tendon force.

The force required to free the wedges may not be achievable in a tendon where the original jacking force was 80% of the tendon strength.

Repeated re-seating of the wedges may cause their teeth to fail.

Improper elongation

Any discrepancy between the measured and the calculated elongation exceeding 7%, or as specified, should be noted and reported to the designer. He may require some remedial measures to be taken. If the measured elongation consistently varies from the calculated value, then stressing must be stopped, the reason investigated, and steps taken to rectify the problem.

A smaller elongation than expected suggests the possibility of the tendon having been inadvertently bonded. The bond may be broken by repeated stressing and de-tensioning of the tendon. Care should, however, be taken not to damage the tendon in the process.

If the bond cannot be broken, then its location should be assessed from the observed extension. Several possibilities exist in this situation.

- If the remainder of the tendon length can be accepted in its unstressed state, then the tendon can be locked with only the part length stressed. It should be borne in mind that, in the event of the bond breaking later, the tendon force will reduce to the proportion of stressed to total length.

  Acceptance of a lower prestressing force in a tendon may imply placing a higher reliance on the adjacent tendons, and stressing them to a higher level if possible.
- If this is not acceptable, and the other end of the tendon is accessible, then it may be possible to stress from both ends. In this case there is no loss in the force.
- If neither of the above is acceptable, then the tendon must be exposed by cutting out the concrete and the sheath at the blockage. The tendon can then be

freed and stressed, the cut concrete made good, and the sheath grouted if the tendon is of the bonded type.

An elongation larger than expected indicates a slippage at the anchorage, or a blowout. The two are dealt with below in separate sections. The elongation may also be large if the tendon is overstressed; see 'Tendon breakage' below.

Wedge slippage

In case of excessive draw-in at the live end, the tendon should be de-tensioned, the wedges replaced, and the tendon re-stressed. Under no circumstances should more than one set of wedges and barrels be used on a strand in trying to restrict the slippage. A defective wedge assembly should always be replaced.

The problem is more difficult to rectify if it occurs on the second, or subsequent, stroke of stressing a long tendon. In this circumstance, *never use a second jack on the back of the first one*. This is a potentially dangerous situation and special equipment or procedures are called for, such as use of de-tensioning devices or de-tensioning at the far end.

Larger than expected elongation may also occur due to slip at the dead end. It is rather unlikely at a bonded dead end; the more likely is the slippage of strand in a pre-locked anchorage. In either case,

- If there is sufficient latitude in the adjacent tendons to make up the deficiency, then the damaged tendon may be abandoned and the adjacent tendons stressed to a higher level.
- If the tendon cannot be abandoned, then there is no alternative but to cut the concrete out near the dead end, thread a new tendon through if necessary, provide a new dead end anchorage—either bonded or pre-locked—and stress the tendon.

Blowout

Unexpectedly large elongation appears to occur if the concrete, near a pre-locked or a live anchorage, is not strong enough for the stresses induced. Inadequate reinforcement in the anchorage zone may also give the same impression. This condition is usually accompanied by cracks near the dead end and signs of movement of the anchorage assembly; it can lead to a blowout.

A *blowout* is failure of concrete in the anchorage zone during stressing. It is potentially a dangerous problem as the failure may be explosive. Pockets of aggregate, sand or voids in the concrete around the anchorage, and lack of anchorage zone reinforcement are the most common causes of a blowout.

The poor concrete must be cut out, made good with cement-sand mortar having a strength not less than the concrete, and the tendon stressed.

Tendon breakage

Tendon breakage can occur from misalignment of wedges, overstressing or internal damage to the tendon. It is important to determine the cause of the breakage before replacing the broken tendon.

Misalignment of wedges occurs when wedges are offset prior to stressing, in which case they can pinch one or more of the wires of the strand. If the strand has been stressed and the wedges are holding the prestress then the tendon may be accepted at the discretion of the engineer.

Overstressing of a tendon can occur if the equipment is not properly calibrated, or if the dial gauge on the stressing pump is misread. It is safer to accept an overstressed tendon if the wedges are holding; an attempt at de-tensioning may break the tendon.

Damage to a strand may be the result of lack of care in site operations. Careless handling or storage may cause pinching of the strand locally. Damage may result from improper use of concrete vibrators, and during cutting the concrete next to a tendon. Careless drilling in installing fixings can also damage the strand.

13.5 Grouting

The grout for injecting sheathing housing the bonded tendons usually consists of a neat cement paste, often containing a plasticizer, a retarder and an expanding admixture. A small quantity of fine sand is sometimes included to reduce the amount of cement, but this is more common in large ducts, which are unlikely to be used in a building. The plasticizer allows a smaller quantity of water to be used, as a result a lesser quantity of water remains unused by the chemical reaction in the sheath. The aimed setting time should be sufficient to complete the grouting of a tendon and allow for possible mishaps, such as breakdown of equipment or blockage of a sheath. Admixtures containing chlorides, fluorides, sulphites and nitrates should not be used. Expanding admixtures containing aluminium are best avoided, because they liberate hydrogen, which may cause embrittlement of the strand.

The grout is normally designed to produce a 28-day cube strength of 35 N/mm^2 (4500 psi), and a seven day strength of not less than 20 N/mm^2 (2500 psi).

Mixing is carried out for about two minutes in a specially designed mixer; modern mixers need a minute or less to deliver a good mix. Mixing time is critical for obtaining a good uniform mixture with the desired setting time and easy flow characteristics. Too long a mixing time results in unnecessary heating of the grout, which changes its properties. Too short a time may leave lumps of unmixed cement, which can cause blockage in the pump line or the sheathing.

Water should not be allowed to remain in the sheath. If a sheath is to be flushed for some reason then use of lime water should be considered because of its alkalinity. Flushing should be followed by compressed air blast to remove the water and dry the sheath.

The tendons are normally grouted as soon after stressing as possible. However, grouting should be avoided in freezing weather; if the grout freezes, its expansion may rupture the sheath and cause cracking of the cover concrete. Grouting at low temperatures should be carried out only if the member is first heated, and its temperature can be maintained above freezing for at least two days. This allows the grout sufficient time to harden.

Grouting is carried out by pumping the mix under pressure, of about 0.5 N/mm^2 (75 psi), through the inlets provided. Grout is injected at low points in a tendon profile. It should not be allowed to flow down from the high points in the profile as this is likely to leave air pockets in the sheath.

Progress of grouting is monitored at the monitoring tubes, cast in the member along the length of the sheath, usually at high points of the profile. When the grout reaches a monitoring tube and emerges without air pockets and under pressure, the particular monitoring tube is capped. Without the monitoring tubes, it is difficult to ensure that the whole of the sheath has been properly grouted. A useful check is to compare the volume of grout pumped with the theoretically required volume.

In case of a breakdown or stoppage of the equipment during grouting, or a blockage in the sheath, the extent of sheath remaining ungrouted should be determined by carefully drilling holes through the sheath so as not to damage the tendon. These holes can be used for injecting grout and monitoring its progress. Access to a spare grouting set is a useful safeguard against possible problems arising out of the breakdown of the equipment in use.

13.6 Finishing operations

On satisfactory completion of the stressing and grouting operations, surplus lengths of the strands are cut off using a disk cutter, usually within about 25 mm (an inch) of the wedge faces. Generation of too much heat, such as may occur if a flame cutter is used, may raise the temperature of the anchorage assembly and the strand and can impair the anchorage efficiency.

All exposed surfaces of the anchorage assembly are then given a rust-inhibitant spray application. When dry, the ends of the strand and the wedges in an anchorage are covered with a grease filled cap. The pocket is then made good with a cement-sand mortar containing a suitable expanding agent.

13.7 Demolition

A review of recent research into potential demolition problems shows that the additional risk associated with demolition of post-tensioned structures in comparison with reinforced concrete structures is very small. The precautions needed and the choice of suitable demolition procedures are discussed; for the most part these require only minor modifications of the methods used for reinforced concrete. In addition, considerations for the cutting of holes in existing structures are discussed.

Demolition methods for post-tensioned structures are largely similar to those for reinforced concrete. The main differences can be summarized as follows:

- Since the prestressing tendons are made of extremely tough, high-strength steel, they are difficult to sever. However, once the tendons have been cut or otherwise de-tensioned, post-tensioned elements will often be easier to demolish than reinforced concrete ones, as the total amount of steel they contain is less.

- Loss of prestress due to the cutting of tendons can create a risk of premature collapse of parts of the structure. This may require slightly more substantial propping than would be used during demolition of a reinforced slab.
- When a post-tensioned tendon is severed, there is a sudden release of stored strain energy. However, as will be shown below, in post-tensioned slabs the effects of this energy release are usually minimal and require only very minor changes to conventional demolition procedures. Transfer beams have a much higher level of stored energy and they need care in demolition.

Until quite recently, very few post-tensioned structures had been demolished. As a result, concern persisted that the sudden release of strain energy caused by the cutting of the prestressing tendons could result in an explosive failure of the structure or rapid, dangerous movement of the tendons themselves. This concern was reflected in numerous codes and guidelines (BS 6187: 1982, FIP 1982, Health and Safety Executive 1984), which warned of the potential dangers but did not recommend appropriate demolition procedures.

In the case of post-tensioned floors, since the average compressive stress in the concrete is small, the likelihood of explosive failure due to the cutting of a tendon is minimal. Far greater attention has been paid to the possibility that a severed tendon might push out its end anchorage and be ejected from the slab at high velocity, becoming a dangerous projectile. As the discussion below demonstrates, this is extremely unlikely.

13.7.1 Review of research and field testing

In recent years, considerable laboratory research and in situ monitoring has been performed in order to establish the likely levels of structural damage or tendon movement occurring when prestressed tendons are cut.

Unbonded tendons

The greatest concern over the possibility of dangerous tendon movement has been expressed with regard to slabs post-tensioned by unbonded tendons. Since the tendons are encased in greased plastic sheaths, the end anchorages are required to carry the full tendon load throughout the life of the structure. The lack of any bond with the concrete along the length of the tendons was equated with a lack of restraint to movement, the assumption being that, if a tendon was able to overcome the resistance of its recessed end anchorage, then it might be ejected from the slab edge at high velocity.

A laboratory research programme (Williams and Waldron, 1989a, 1989b) found that, for tendons in extruded plastic sheaths with a good quality grease layer, the damping and friction generated as the tendon moves within its greased sheath provide by far the greatest restraint to tendon movement. The restraint

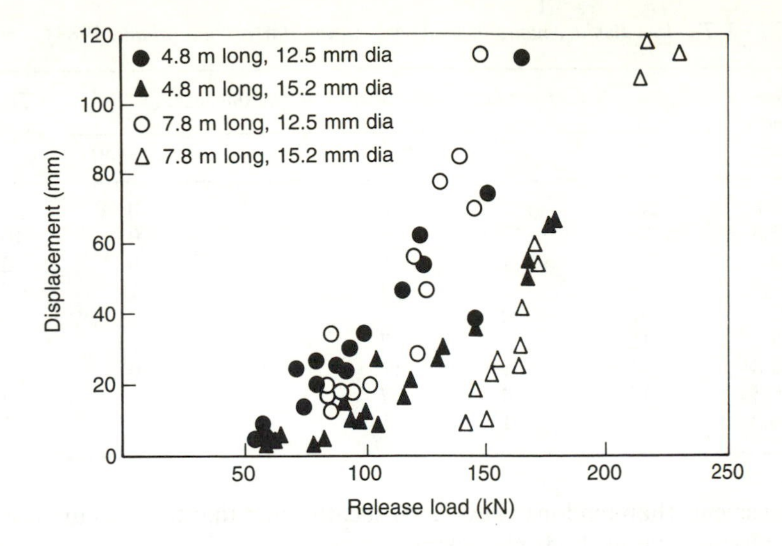

Figure 13.2 *Variation of end displacement with release load*

provided by the mortar end plugs is in most cases quite small, absorbing only a few per cent of the strain energy of a typical tendon. Figure 13.2 shows the variation of final anchorage displacement with release load from laboratory tests on tendons up to 7.8 m (26 ft) in length. For these relatively short lengths, displacements are small, of the order of 100 mm (4 in). No distinction has been made in the figure between tendons with and without plugged ends, since both gave similar results. The smaller diameter tendons show markedly higher levels of displacement at a given load, though this trend disappears if the loads are expressed as fractions of the tendon ultimate tensile strength rather than as absolute loads.

In addition to these laboratory results, numerous field studies of demolition of post-tensioned structures have been performed recently, the most comprehensive of which are those reported by Barth and Aalami (1989). They monitored movements of over one thousand 12.7 mm (0.5 in) diameter tendons of lengths varying between 1.8 m (6 ft) and 64 m (210 ft), wrapped in heat-sealed plastic sheaths. This method of encasing tendons provides a less tight fit than the more widely used extruded plastic sheathing studied by Williams and Waldron, and might, therefore, be expected to result in rather higher displacements. Barth and Aalami found that only 7% of the tendons displaced by more than 75 mm (3 in) and that 78% underwent no displacement at all. They also noted the tendency of the kern wire to shoot out ahead of the wound wires, though the kern wire displacement remained small.

In general, field studies have suggested that levels of tendon movement when severed during demolition are low, that the older types of unbonded tendon, where the sheath is wrapped rather than extruded, may give slightly higher

Table 13.3 *Tendon displacements in field tests (after Barth and Aalami, 1989)*

Released length (m)	*No. of tendons with a displacement (mm) of:*					*Total number*
	0	*≤75*	*76–150*	*151–300*	*301–450*	
1.8	248	102	0	0	0	350
15.8	89	12	5	1	0	107
16.5	17	1	5	0	0	23
20.6	36	2	4	7	1	50
22.6	32	2	5	9	0	48
28.7	15	7	7	4	1	34
40.8	21	2	11	1	0	35
50.3	17	4	7	0	1	29
64.0	324	1	1	0	0	326

displacements than tendons in extruded sheaths, and that tendon curvature has little effect on the likely displacement.

One question that arises is whether there is an optimum location for a cut so as to minimize the possibility of dangerous movement. This is hard to judge intuitively, since an increase in length will result in an increase in the strain energy released, but also in an increase in the damping and frictional restraints. There is limited evidence to suggest that very short and very long tendons give rise to the lowest displacements. For instance, the test results of Barth and Aalami are summarized in Table 13.3, which show the numbers of tendons experiencing end displacements within certain ranges. From this, it is clear that the largest displacements occurred in released lengths between 20 m (66 ft) and 50 m (164 ft), with very short tendons moving only a few millimetres and nearly all of the 326 longer tendons experiencing zero displacement. A similar trend was observed by Williams and Waldron (1990), using a finite element model validated against the laboratory tests outlined above. However, these results are too few and too disparate to allow any firm conclusions to be drawn. The exact magnitude of tendon movement will vary with tendon type, length, stress level and condition, so that it is impossible to give precise guidance on the exact displacement that would be expected in a particular situation. Nevertheless, it is clear that displacements are likely to be small in all cases.

A related, but slightly different, problem which has been observed in a number of site monitoring exercises is a tendency for tendons to burst out of the top or bottom surface of a slab at points where the cover is particularly low (Suarez and Poston, 1990, Springfield and Kaminker, 1990, Schupack, 1991). From the reports published, it appears that this is only likely to occur at locations where the cover to the tendon is very small, 20 mm (3/4 in) or less. A sensible precaution against this eventuality is to cut the tendon at a point where it is close to the mid-depth of the slab; it is likely that this will provide a more rational approach to choosing cutting points than attempts to limit the axial movement.

Bonded tendons

In general, slabs post-tensioned by bonded tendons are even less likely to give rise to demolition problems than those containing unbonded tendons. The main difference is that cutting of a bonded tendon usually only causes localized bond failure, and hence release of prestress, in a short length of tendon on either side of the cut. There is, therefore, no anchorage movement and no loss of strength in adjacent spans.

Where tendons are well grouted, the bond will be sufficient to prevent any movement at the anchorages during demolition or alteration. Some local debonding is likely around the cut point, and possibly some cracking of the concrete along the line of the tendon, caused by the increase in tendon cross-section as it contracts longitudinally. However, the resulting loss of strength will be confined to quite a small region, so that a catastrophic collapse is extremely unlikely.

The possibility of severe tendon movement only arises in cases where the grouting is extremely poor. The debonding of partially grouted tendons when cut is currently poorly understood. What little research has been carried out (Belhadj *et al.*, 1991) has concentrated on problems associated with very large tendons in beams, suggesting that the mechanisms acting are extremely complex. It is, therefore, not possible to give definitive guidance on the likely level of movement of very poorly grouted tendons.

13.7.2 Demolition procedures

The research and field observations discussed above suggest that post-tensioned slabs can be safely and easily demolished. To ensure site safety, a number of precautionary measures need to be taken, but these are unlikely to be much more onerous than those required for ordinary reinforced concrete. There are several possible methods of demolition, but a typical procedure for a post-tensioned floor is likely to consist of:

- propping of the slab being removed, and of any adjacent areas likely to be affected
- provision of shielding at locations where there is a possibility of tendons, anchorages or other debris being ejected (although, as discussed above, movement at anchorages is likely to be small for unbonded tendons and zero for bonded tendons)
- cutting or de-tensioning of tendons
- demolition of the concrete slab

The methods available for de-tensioning and demolition are discussed in sections 13.7.4 and 13.7.5 below.

Prior to commencing demolition, it is essential that a thorough preliminary investigation is carried out, to ascertain the condition of the structure and the likelihood of any problems arising during demolition. This will enable a rational demolition procedure to be devised and the necessary safety precautions to be taken.

13.7.3 Planning a demolition job

A demolition contractor should always seek guidance from a prestressing specialist before commencing the demolition of a post-tensioned structure. This is particularly important when transfer beams or unbonded tendons are involved. If possible, the original construction drawings for the building should be obtained, though it should be borne in mind that alterations may have been made during the life of the structure.

An important first step is to establish the type and location of the tendons. This can be done using the drawings and confirmed using a covermeter on site. Locations where the tendons come very close to the top or bottom face of the slab can also be identified in this way. Some shielding should be provided at these points to restrain the tendons from bursting out and prevent fragments of concrete from being ejected; a few planks held in place by props should be sufficient. Cutting of tendons should not be performed close to these locations, and all personnel should be kept well away from them during de-tensioning.

A number of other checks should be made prior to commencing demolition. The type of anchorages should be determined, since the likelihood of tendon movement will be influenced by the exact way in which the strand is held. For instance, it should be possible to establish which are dead- and which are live-end anchorages, and whether the dead-end anchorages are bonded or pre-locked. It is also important to identify any construction joints, infill strips, couplers or structural alterations, since these may have an influence on the demolition procedure. For example, cutting of an unbonded tendon on one side of a coupler will lead to a loss of tension only on the side of the cut, not over the whole tendon length.

From the number and size of the tendons, it is possible to make an estimate of the stress level in the floor. As a rough indication of the significance of the stress level, a reduction in stress of 1 N/mm^2 (145 psi) gives an energy release equivalent to a 25 mm (1 in) drop in the level of the slab. Careful consideration must be given to the method of releasing this energy, especially if the tendons are unbonded or poorly grouted; this topic is discussed further in section 13.7.4.

Transfer beams are stressed to a much higher level. They are designed to be stressed in several stages as increasing dead load is imposed by the construction of the structure above. If stressed in one operation, then the prestress will induce reverse tensile and compressive stresses of a high magnitude which can result in failure of the beam. Demolition must follow a similar multi-stage de-tensioning procedure; as part of the structure above is demolished, some tendons must be severed to reduce the prestress.

If unbonded tendons are present, it should be established whether the sheathing is of the extruded type which is now used almost universally, or of the older and looser wrapped type. The latter may give rise to slightly higher tendon displacements. Guards should be placed around the tendon ends to catch any flying debris which might be generated as the anchorages are dislodged. These need not be particularly substantial; a few planks securely held over the anchorages will suffice. No personnel should be allowed to stand near the tendon ends during demolition.

De-tensioning of an unbonded tendon will, of course, cause loss of prestress over the full length of the tendon, not just the span where the cut is made. While a floor may contain sufficient rod reinforcement to prevent complete collapse when the prestress is removed, it is still necessary to provide temporary support to a floor during de-tensioning. In general, the amount of propping required for a post-tensioned floor is slightly greater than for a reinforced concrete member. It should be noted that the props may be subjected to some horizontal forces due to the expansion of the floor when the tendons are cut.

For bonded tendons, it is vital to establish the adequacy of the grouting. The only reliable way of doing this is to break out the concrete around the tendons at a few locations and perform a visual inspection. If the tendons are well-grouted, then cutting is unlikely to cause serious problems, the bond being sufficient to prevent any movement at the anchorages. As discussed earlier, only very localized debonding of the tendon and cracking of the concrete are likely, so that any loss of strength will be confined to quite a small region. Some propping of the floor is advisable, but again this needs to be only slightly more substantial than that used when performing similar operations on a reinforced concrete floor.

13.7.4 De-tensioning of tendons

While some methods of demolition can be carried out by a single operation, several require prior cutting or de-tensioning of the tendons. This may be for safety reasons, in order to prevent the sudden, uncontrolled release of a tendon later in the demolition process, or simply because the tool being used to demolish the concrete is not able to cut through the hardened steel from which the tendons are manufactured.

For unbonded tendons, there are a number of methods by which the tension can be released. These include heating of the anchorages until slip occurs, breaking out of concrete behind the anchorages until slip occurs, and cutting of the tendons at some point along their length, away from the anchorages. This last option can be done either by breaking out the concrete around a tendon, then using a flame torch or disc cutter, or by making a single cut through the slab and tendons using a saw or thermic lance. Experience suggests that using a flame torch results in a more gradual loss of force, and, therefore, in slightly smaller tendon displacements, though the movement is unlikely to be large in any case. When choosing the cutting points, positions where the tendon is very close to the top or bottom face of the slab should be avoided. The choice between the above methods will usually depend on the particular conditions encountered on a given site; often, problems of access will govern which method is the most suitable.

For well-grouted bonded tendons, there is little risk of an uncontrolled release of energy, so cutting will only be necessary if the tools used for demolishing the concrete slab are unable to cut through the tendons. Any of the cutting methods already mentioned for unbonded tendons can be used, i.e. flame torch, disc cutter, saw or thermic lance.

While it is unlikely that the relatively small tendon forces in post-tensioned slabs would lead to complete debonding even of very poorly grouted tendons, it is recommended that a cautious approach is adopted. Probably the safest method would be a controlled de-tensioning procedure using open throat jacks, similar to the approach outlined in section 13.8 for the alteration of slabs containing unbonded tendons.

As mentioned above, all de-tensioning procedures will cause a reduction of strength in the slab, which will, therefore, need to be propped temporarily. For bonded tendons the loss of strength will be quite localized, while for unbonded tendons it may occur on the full length of the structure.

13.7.5 Demolition methods

Provided that the necessary preliminary measures outlined above have been followed, post-tensioned floors can be demolished using most of the conventional methods. The advantages and drawbacks of the various methods are briefly discussed below. The choice of method is dependent on numerous factors, including structural form, noise and dust limitations, space constraints and cost.

The traditional wrecking ball and crane approach is fast and cheap. However, it is difficult to use in a controlled way, and is constrained by height restrictions and the need for clear space around the building. It generates large amounts of noise and dust and is not able to sever the prestressing tendons, which must, therefore, be de-tensioned beforehand.

The use of large circular saws provides a slow, controllable demolition process in which the slab and tendons are cut in a single pass. The method can be expensive if there are a lot of tendons to cut, as they cause considerable wear of the saw blades. The blade must be cooled by water, creating a slurry which drips down the building; this is unimportant in demolition but can be undesirable when carrying out alterations.

Explosive demolition techniques bring the structure down very quickly and economically, but do not break up the steel tendons, which, therefore, have to be cut when clearing the demolished structure. The design of the explosive system must take account of the stored energy in the structure due to the prestressing. The method generates problems of blast, ground-borne vibrations and flying debris, but these should be no worse for post-tensioned buildings than for other structural types.

Thermic lances can be used to cut through the concrete and tendons in a single operation. This method is quiet and vibration-free, but expensive.

Percussion tools such as drills provide a cheap and controllable demolition method, but create problems of dust and noise, and require the prior de-tensioning of the tendons.

13.8 Cutting holes

In many structures, change of use or general maintenance may require the cutting of holes in a floor. Small diameter holes are in many instances easier to make in post-tensioned than in reinforced concrete floors, since the steel in a post-tensioned floor is likely to be more widely spaced. It is, therefore, quite likely that small holes can be made without cutting through any steel. Obviously, tendon locations should be ascertained from construction drawings and checked on site using a covermeter prior to commencing cutting operations.

For larger alterations, requiring the cutting of tendons, care should be taken in choosing the hole location. For instance, it is not normally possible to cut through beams, as this causes too great a disruption of the structural system. Locations where the tendons are very close to the bottom of the slab should also be avoided, as this results in an eccentric application of prestress at the newly created free edge, and because it is difficult to insert new anchorages at such locations. Sometimes a downstand beam can be added at the edge of the opening in order to alleviate the latter problem, but the eccentric application of the force remains. Cutting through bunched tendons can also present problems, since they must be splayed out in order to fit new anchorages, requiring the removal of concrete from around the tendons for some distance beyond the edge of the hole.

When making alterations involving the cutting of unbonded tendons, re-tensioning is required after the alteration, making simple cutting of the tendons unacceptable, as this may cause damage to the tendon anchorages. For these instances, special open-throat jacks are available which allow tendons to be de-tensioned in a gradual and controlled way. The normal procedure is to break out a hole slightly larger than that required, leaving the tendons intact. Open-throat jacks are then positioned over a tendon at either end of the opening and used to take up the load in the tendon, as shown in Figure 13.3(a). The length of tendon between the jacks becomes slack and is cut, and the pressure is then gradually released from the jacks, leaving the tendon undamaged and free of tension. New anchorages can now be positioned at the edges of the hole, and the two halves of the tendon re-stressed, Figure 13.3(b).

Cutting through bonded tendons presents fewer problems, as the prestress is maintained by the bond with the concrete, so that new anchorages need not be fitted. Instead, the individual wires of the tendons can simply be splayed out and concreted over when making good the edges of the hole.

When making large holes, it is likely that some additional reinforcement will be required around the perimeter, as shown in Figure 13.3(c), in order to compensate for the loss of strength caused by the hole. This will be the case in both post-tensioned and reinforced concrete floors.

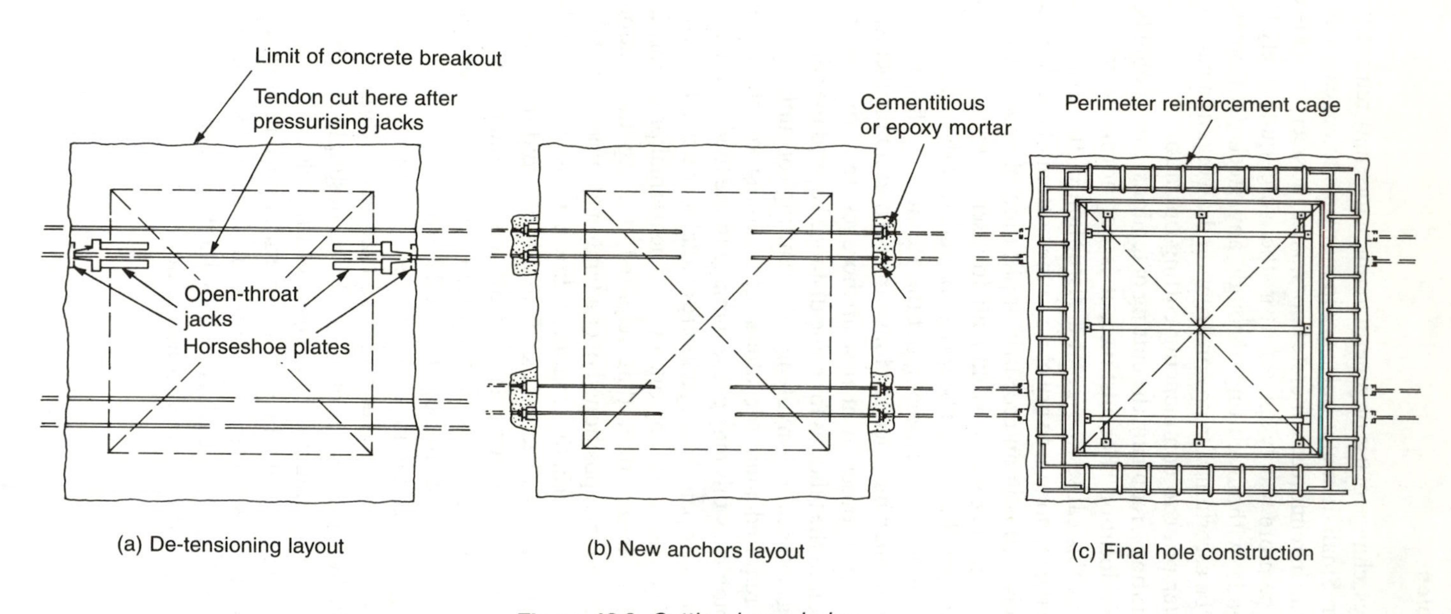

Figure 13.3 *Cutting large holes*

Figure 13.4 *VSL mono jack (twin ram), stressing slab tendon, Australia*

REFERENCES

ACI 318-89 *Building code requirements for reinforced concrete*, American Concrete Institute, Detroit.

ACI Committee 435 (1974) *Deflection of Two-Way Reinforced Concrete Floor Systems: State-of-the-Art Report*, Report No. ACI 435.6R-74 (Reapproved 1989), American Concrete Institute, Detroit.

ACI Committee 435 (1991) *State-of-the-Art Report on Control of Two-Way Slab Deflections*, Report No. ACI 435.9R-91, American Concrete Institute, Detroit.

Barth, F.G. and Aalami, B.O. (1989) *Controlled Demolition of an Unbonded Post-Tensioned Concrete Slab*, PTI Report, Post-Tensioning Institute, Phoenix.

Belhadj, A., Waldron, P. and Blakeborough, A. (1991) Dynamic debonding of grouted prestressing tendons cut during demolition. In *Proceedings of the International Conference on Earthquake, Blast and Impact* (Manchester, 1991) (ed. Society for Earthquake and Civil Engineering Dynamics), Elsevier Applied Science, London, pp. 411–420.

BS 6187: 1982 *Code of practice for demolition*, British Standards Institution, London.

BS 8110: 1985 *Structural use of concrete; Part 1, Code of Practice for design and construction*, and *Part 2, Code of Practice for special circumstances*, British Standards Institution, London.

CAN3-S16.1-M89 *Steel structures for building (limit states design) - Appendix G: Guide for floor vibrations*, Canadian Standards Association, Toronto.

Caverson, R.G., Waldron, P. and Williams, M.S. (1994) Review of vibration guidelines for suspended concrete slabs. *Canadian Journal of Civil Engineering*, **21**, No. 6.

Chana, P.S. (1993) A prefabricated shear reinforcement system for flat slabs. *Proceedings of the Institution of Civil Engineers: (Structures and Buildings)*, **99**, 345–358.

Chandler, J.W.E. (1982) *Design of floors on ground*, Technical Report 550 Cement & Concrete Association

Chandler, J.W.E. and Neal, F.R. (1988) *The design of ground-supported concrete industrial floor slabs*, Interim Technical Note 11, British Cement Association, Wexham Springs.

Choi, E.C.C. (1992) Live load in office buildings. *Proceedings of the Institution of Civil Engineers (Structures and Buildings)*, **94**, 299–322.

Concrete Society (1988), *Concrete industrial ground floors*, Technical Report 34, Concrete Society, Wexham Springs.

Concrete Society (1994), *Post-Tensioned Floors - Design Handbook*, Technical Report 43, Concrete Society, Wexham Springs.

Dowrick, D.J. (1987), *Earthquake resistant design*, 2nd edn, John Wiley & Sons, New York

Fatemi-Ardakani, A., Burley, E., and Wood L.A. (1989) A method for the design of ground slabs loaded by point loads. *The Structural Engineer*, **67**, No.19, 341–345.

Fèdèration Internationale de la Prècontrainte (1982) *Demolition of Reinforced and Prestressed Concrete Structures: Guide to Good Practice*, FIP, Wexham Springs.

Health and Safety Executive (1984) *Health and Safety in Demolition Work, Part 3: Techniques*, Guidance Notes GS29/3, HSE, London.

Institution of Structural Engineers and The Concrete Society (1987), *Guide to the structural use of lightweight aggregate concrete*, London

Key, D. (1988) *Earthquake design practice for buildings*, Thomas Telford Ltd, London.

Libby, J.R. (1990) *Modern Prestressed concrete*, 4th edn, Van Nostrand Reinhold, New York

Lin, T.Y. and Burns, N.H. (1982) *Design of prestressed concrete structures*, 3rd edn, John Wiley & Sons, New York

Mitchell, G.R. and Woodgate, R.W. (1971) *Floor Loadings in Office Buildings–the Results of a Survey*. Current Paper 3/71, Building Research Station, Garston.

Naeim, F. (ed.) (1989) *The seismic design handbook*, Van Nostrand Reinhold, New York.

Neville, A.M. (1981) *Properties of concrete*, 3rd edn, John Wiley & Sons, New York

Nilson, A.H. (1987) *Design of prestressed concrete*, 2nd edn, John Wiley & Sons, New York

Nilson, A.H. and Walters, D.B. (1975) Deflection of two-way floor systems by the equivalent frame method. *Journal of the American Concrete Institute*, **72**, 210–218.

Oh, B.H. (1992) Flexural analysis of reinforced concrete beams containing steel fibers. *Journal of Structural Engineering, American Society of Civil Engineers*, **118**, 2821–2836

Pavic, A., Williams, M.S. and Waldron, P. (1994) Dynamic FE model for post-tensioned concrete floors calibrated against field test results. *Proceedings of 2nd International Conference on Engineering Integrity Assessment*, Glasgow, 357–366.

Pernica, G. and Allen, D.E. (1982) Floor vibration measurements in a shopping centre. *Canadian Journal of Civil Engineering*, **9**, 149–155.

Probst, E.H. (ed.) (1951) *Civil Engineering Reference Book*, Butterworths, London.

Regan, P.E. (1985) The punching resistance of prestressed concrete slabs. *Proceedings of the Institution of Civil Engineers*, Part 2, **79**, 657–680.

Rice, E.K. and Kulka, F. (1960) Design of prestressed lift-slabs for deflection control. *Journal of the American Concrete Institute*, **57**, 681–693.

Ringo, B.C. and Anderson, R.B. (1992) *Designing floor slabs on grade*, PTI Report, Post-Tensioning Institute, Phoenix.

Roark, R.J. (1954) *Formulas for stress and strain*, 3rd edn, McGraw-Hill, New York.

Schupack, M. (1991) Evaluating buildings with unbonded tendons. *Concrete International*, **13**, No. 10, 52–57.

Springfield, J. and Kaminker, A.J. (1990) Discussion of Williams and Waldron (1989a). *Proceedings of the Institution of Civil Engineers*, Part 2, **89**, 123–125.

Suarez, M.G. and Poston, R.W. (1990) *Evaluation of the Condition of a Post-Tensioned Concrete Parking Structure after 15 Years of Service*, PTI Report, Post-Tensioning Institute, Phoenix.

Steel Construction Institute (1992) *Steel Designers' Manual*, 5th edn, Blackwell Scientific Publications, Oxford.

Tang, T., Shah, S.P. and Ouyang, C. (1992) Fracture mechanics and size effect of concrete in tension. *Journal of Structural Engineering, American Society of Civil Engineers*, **118**, 3169–3185.

Tasuji, M.E., Slate, F.O. and Nilson, A.H. (1978) Stress strain response and fracture of concrete in biaxial loading, *Proceedings of the American Concrete Institute*, **75**, 306–312.

Teller, L.W. and Sutherland, E.C. (1943) The Structural Design of Concrete Pavements, *Public Roads*, **23**.

Timoshenko, S.P. and Woinowsky-Krieger, S. (1959) *Theory of Plates and Shells*, 2nd edn, McGraw-Hill, New York.

Timoshenko, S.P. and Goodier, J.N. (1951) *Theory of Elasticity*, 2nd edn, McGraw-Hill, New York.

Westergaard, H.M. (1948) New formulas for stresses in concrete pavements of airfields. *Transactions of the American Society of Civil Engineers*, **113**, 425–439.

Williams, M.S. and Waldron, P. (1989a) Movement of unbonded post-tensioning tendons during demolition. *Proceedings of the Institution of Civil Engineers*, Part 2, **87**, 225–253.

Williams, M.S. and Waldron, P. (1989b) Dynamic response of unbonded prestressing tendons cut during demolition. *Structural Journal, American Concrete Institute*, **86**, 686–696.

Williams, M.S. and Waldron, P. (1990) Longitudinal stress wave propagation in an unbonded prestressing tendon after release of load. *Computers and Structures*, **34**, 151–160.

Williams, M.S. and Waldron, P. (1994) Dynamic characteristics of post-tensioned and reinforced concrete floors. *The Structural Engineer*, **72**, No. 20, 334–340.

Wyatt, T.A. (1989) *Design Guide on the Vibration of Floors*, Steel Construction Institute, Ascot.

INDEX